Cellular Regulation by Protein Phosphorylation

NATO ASI Series

Advanced Science Institutes Series

A series presenting the results of activities sponsored by the NATO Science Committee, which aims at the dissemination of advanced scientific and technological knowledge, with a view to strengthening links between scientific communities.

The Series is published by an international board of publishers in conjunction with the NATO Scientific Affairs Division

A	Life Sciences	Plenum Publishing Corporation
B	Physics	London and New York
C	Mathematical and Physical Sciences	Kluwer Academic Publishers
D	Behavioural and Social Sciences	Dordrecht, Boston and London
E	Applied Sciences	
F	Computer and Systems Sciences	Springer-Verlag
G	Ecological Sciences	Berlin Heidelberg New York
H	Cell Biology	London Paris Tokyo Hong Kong
I	Global Environmental Change	Barcelona Budapest

NATO-PCO DATABASE

The electronic index to the NATO ASI Series provides full bibliographical references (with keywords and/or abstracts) to more than 30 000 contributions from international scientists published in all sections of the NATO ASI Series. Access to the NATO-PCO DATABASE is possible in two ways:

– via online FILE 128 (NATO-PCO DATABASE) hosted by ESRIN, Via Galileo Galilei, I-00044 Frascati, Italy.

– via **CD-ROM** "NATO-PCO DATABASE" with user-friendly retrieval software in English, French and German (© WTV GmbH and DATAWARE Technologies Inc. 1989).

The CD-ROM can be ordered through any member of the Board of Publishers or through NATO-PCO, Overijse, Belgium.

Cellular Regulation by Protein Phosphorylation

Edited by

Ludwig M. G. Heilmeyer, Jr.

Universität Bochum
Medizinische Fakultät
Institut für Physiologische Chemie
Universitätsstr. 150
W-4630 Bochum, FRG

Springer-Verlag
Berlin Heidelberg NewYork London Paris Tokyo
Hong Kong Barcelona Budapest
Published in cooperation with NATO Scientific Affairs Division

Proceedings of the NATO Advanced Study Institute on Cellular Regulation by Protein Phosphorylation held at Château La Londe les Maures (France) September 5–15, 1990

ISBN 3-540-51776-6 Springer-Verlag Berlin Heidelberg New York
ISBN 0-387-51776-6 Springer-Verlag New York Berlin Heidelberg

Library of Congress Cataloging-in-Publication Data
NATO Advanced Study Institute on Cellular Regulation by Protein Phosphorylation (1990 : La Londe les Maures, France) Cellular regulation by protein phosphorylation / edited by Ludwig M. G. Heilmeyer, Jr. (NATO ASI series. Series H, Cell biology ; vol. 56)
"Proceedings of the NATO Advances Study Institute on Cellular Regulation by Protein Phosphorylation held at the château La Londe les Maures (France), September 5–15, 1990"—T.p. verso. Includes bibliographical references and index.
ISBN 3-540-51776-6 (Springer-Verlag Berlin Heidelberg New York).
ISBN 0-387-51776-6 (Springer-Verlag New York Berlin Heidelberg).
1. Phosphoproteins—Synthesis—Congresses. 2. Phosphorylation—Congresses. 3. Cellular signal trans-duction—Congresses. 4. Post—translational modification—Congresses. 5. Protein kinase—Congresses. 6. Phosphoprotein phosphatases—Congresses. I. Heilmeyer, L. M. G. (Ludwig M. G.), 1937. II. Title. III. Series.
QP552.P5N36 1990 574.87'6—dc20 91-13730

© Springer-Verlag Berlin Heidelberg 1991
Printed in Germany

Typesetting: camera-ready by authors

31/3140-543210 – Printed on acid-free-paper

PREFACE

This volume covers presentations that were made during the second NATO International Advanced Study Institute on "Cellular Regulation by Protein Phosphorylation". The course was held under the auspices of the University of Montpellier on the occasion of its 700th anniversary; it took place in La Londe les Maures, France, within the magnificent setting of L'Agelonde overlooking the Mediterranean Sea.

The first Advanced Institute on "Signal Transduction and Protein Phosphorylation" was run on the Island of Spetsai, Greece, in September, 1986. Its Proceedings appeared in the NATO ASI Series "Signal Transduction and Protein Phosphorylation", Plenum Press, New York/London 1987 (L. M. G. Heilmeyer, ed.).

Two factors prompted the organizers to consider a second conference on cellular regulation by protein phosphorylation. First, the unexpected success of the Spetsai Summer Course to which many applicants could not be admitted because of limited number of slots available. Second, the considerable advances that were made during the intervening years because of the appearance on the scene of new methodologies. For instance, the wide application of the PCR has allowed the identification and characterization of a multitude of new protein kinases and phosphatases and other related molecules.

Signal transduction represents one of the main mechanisms by which intracellular processes can be regulated in response to external stimuli. After the usual introductory lectures on the architecture of regulatory enzymes and the properties of cascade systems, the basic chemical, physical and molecular genetic approaches that bear on the field were reviewed.

These were followed by lectures on: a) signal transduction involving second messengers generated in G-protein-regulated reactions or by the induction of the tyrosine kinases intrinsic to the receptors themselves; b) the structure-function relationship of the cyclic nucleotide-dependent or Ca^{2+}/calmodulin-dependent protein kinases, and the so-called "independent" kinases whose regulation in response to mitogenic signals is still ill-understood; and c) the oncogene-related cellular, viral and receptor-linked tyrosine kinases whose overexpression or mutation can lead to cell transformation. As a sign of changing times, and an indication as to how far the field had evolved since the Spetsai Conference, the 1990 course dealt in some detail with several topics that were barely mentioned earlier, including the control of protein translation, neuronal cell function and cytoskeleton assembly. Finally, new lectures were introduced on the enzyme translocation and cellular reorganization that occurs in response to mitogenic hormones and the involvement of the cdc2 gene product on cell cycle progression.

The organizers would like to express their thanks to the coworkers for their help in the preparation of the manuscripts. We are especially grateful to Mrs. Humuza, who organized these contributions in the camera ready form, and to Mrs. Rosenbaum for her part in the practical organization.

                                              The organizers

Bochum, February 1991

CONTENTS

III.   STRUCTURE - FUNCTION RELATIONSHIP

IV.   CA $^{2+}$ AND CYCLIC NUCLEOTIDE-INDEPENDENT PHOSPHORYLATION

VII.   PROTEIN PHOSPHATASES

VIII. CONTROL OF CELLULAR PROCESSES

I. BACKGROUND
AND GENERAL INTRODUCTION

SIGNAL INTEGRATION IN PHOSPHORYLASE KINASE

Ludwig M. G. Heilmeyer, Jr.
Institut für Physiologische Chemie
Abteilung für Biochemie Supramolekularer Systeme
Ruhr-Universität Bochum
Universitätsstraβe 150
D-4630 Bochum 1
F.R.G.

Why is phosphorylase kinase, the first regulatory protein kinase detected by Fischer and Krebs over 30 years ago (Krebs et al., 1959), more complex and exhibits a higher molecular weight than any of those protein kinases of which over one hundred family members are known today? (Hunter et al., 1986) For example, the cyclic AMP-dependent protein kinase, the mediator of the many cellular actions of cyclic AMP, is much less complex and has only 1/6 of the molecular weight of phosphorylase kinase.

Phosphorylase kinase converts glycogen phosphorylase b̲, the inactive form of this enzyme, into the active a̲ form which degrades glycogen to glucose-1-phosphate. It is a vital process for short term energy supply to the cell controlled by several extracellular signals, like hormones involved in regulation of carbohydrate metabolism, or electrical events activating high energy consuming processes like muscle contraction.

Phosphorylase kinase is located at an imaginary interface between signalling and metabolic pathways [Fig. 1]. Extracellular signals which are transduced and modified by signalling systems, located in plasma membranes, lead to release of second messengers into the cytosol. The change in concentration of a given second messenger influences many cellular processes simultaneously by which physiological responses of the cell are coordinated. If the following signal transmitting cascade includes interconversion of an enzyme signal sensitisation/desensitisation can occur which defines the signal amplitude being able to provoke a response to an extracellular signal (Koshland et al., 1982). Further downstream a signalling pathway, the pleiotropic action of extracellular signals or of intracellular second messengers must be converted into a specific physiological response. This function will be carried out by a specialized protein, a specificator. The specificator does not only translate a pleiotropic

4

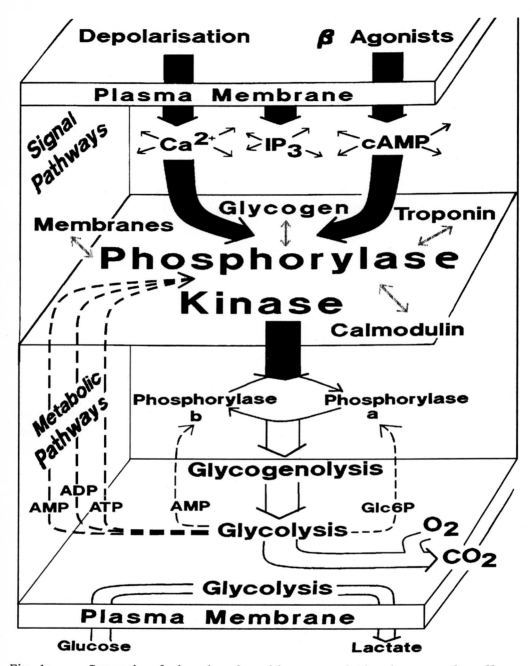

Fig. 1:   Synopsis of phosphorylase kinase regulation in a muscle cell

Incoming signals to phosphorylase kinase from signal pathways are shown in large black arrows, incoming signals from metabolic pathways are indicated by dashed arrows. Metabolic pathways are shown in large colourless arrows. Interactions localizing phosphorylase kinase in specific cellular structures are indicated by hatched arrows.

signal into a specific response but such a molecule is able to select; location of a specificator at the merging point of two or more signalling pathways allows the specificator to trigger a simple additive or synergistic or antagonistic - however specific - cellular answer.

The specificator, phosphorylase kinase, does not only integrate signals from different signal pathways but also receives information from metabolic pathways [Fig. 1]. All these informations must be converted into an appropriate level of phosphorylase a. Therefore, in addition to sites or domains present as targets for signal transmission this enzyme must also provide sites for binding of allosteric effectors.

Finally, phosphorylase kinase is associated with specific cellular structures [Fig. 1]. First, this enzyme binds to glycogen, to which other glycogen metabolizing enzymes are bound, too, thus building up a kind of organelle, the 'glycogen particle' (Meyer, F. et al., 1970). Due to protein/protein interactions, the enzymes associated with this particle behave differently from those enzymes being present solubilized in the cytosol. Secondly, calmodulin interacts with phosphorylase kinase in addition to calmodulin being an integral subunit of phosphorylase kinase. In muscle, these sites might not interact with calmodulin but with the homologous Ca^{2+}-binding protein, troponin C (Cohen et al., 1980). Phosphorylase kinase but not the substrate glycogen phosphorylase is associated with membranes. We have observed that phosphorylase kinase is enriched at the junctional surface of the terminal cysternae (Thieleczek et al., 1987). Thus, phosphorylase kinase must provide sites not only for targeting the action of second messengers and metabolic effectors, but additionally for interactions with macromolecules localizing the enzyme in specific cellular compartments. Therefore, phosphorylase kinase evolved to a more complex and hence bigger enzyme than other protein kinases.

Phosphorylase kinase is a multi-subunit enzyme composed of four kinds of subunits, called α, β, γ and δ [Fig. 2] (Cohen et al., 1978). The subunit δ represents the integral calmodulin which deinhibits the enzyme upon Ca^{2+} saturation leading to expression of activity. If δ does not contain Ca^{2+} the enzyme expresses a very low activity; this activity is called Ca^{2+}-independent activity A_0 [Fig. 2] (Kilimann and Heilmeyer, 1982[1]; 1982[2]; Hessová et al., 1985[1]). Upon increase of the free Ca^{2+} concentration each of the integral subunits δ binds two moles of Ca^{2+} which results in expression of the Ca^{2+} high affinity partial activity A_1 [Fig. 2]. Mg^{2+} is re-quired to saturate calmodulin completely with 4 moles of Ca^{2+}. At neutral pH, Mg^{2+} must be present in the assay at unphysiologically high concentrations of 10-50 mM. It induces two more Ca^{2+}-binding sites on each integral calmodulin i.e. Mg^{2+} allows saturation of calmodulin with four Ca^{2+} ions (Hessova et al., 1985[2], Kohse and Heilmeyer, 1981). Under these conditions the enzyme expresses the Ca^{2+} low affinity partial activity A_2 [Fig. 2].

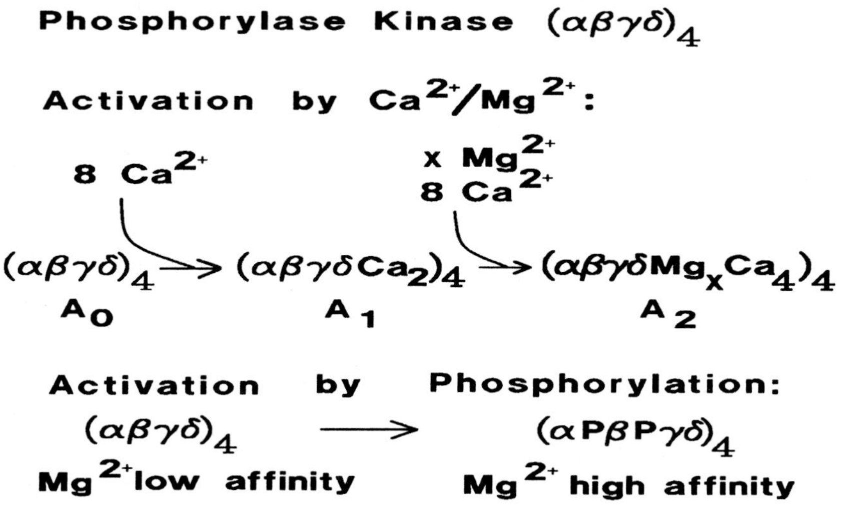

Phosphorylase Kinase $(\alpha\beta\gamma\delta)_4$

Activation by Ca^{2+}/Mg^{2+}:

$$8\ Ca^{2+} \qquad \begin{array}{c} x\ Mg^{2+} \\ 8\ Ca^{2+} \end{array}$$

$$(\alpha\beta\gamma\delta)_4 \longrightarrow (\alpha\beta\gamma\delta Ca_2)_4 \longrightarrow (\alpha\beta\gamma\delta Mg_x Ca_4)_4$$
$$A_0 \qquad\qquad A_1 \qquad\qquad A_2$$

Activation by Phosphorylation:

$$(\alpha\beta\gamma\delta)_4 \longrightarrow (\alpha P\beta P\gamma\delta)_4$$
$$Mg^{2+} \text{low affinity} \qquad Mg^{2+} \text{high affinity}$$

Fig. 2:   Synopsis of activation of phosphorylase kinase by Ca^{2+}/Mg^{2+} and phosphorylation

δ represents the integral calmodulin which can be saturated with Ca^{2+} in two steps depending on the Mg^{2+} concentration. The corresponding complexes express the activities A_0, A_1 and A_2 as indicated below. Phosphorylation of the α- and β-subunit on multiple sites is indicated by αP and βP.

It is well known that the enzyme can also be activated by phosphorylation: increase of the enzymatic activity correlates with phosphorylation of the β-subunit (Yeaman and Cohen, 1975), however, phosphorylation of α seems equally important (Pickett-Gies and Walsh, 1970). It is proposed that upon phosphorylation of the β-subunit the enzyme undergoes a transition from a Mg^{2+} low affinity to a Mg^{2+} high affinity state [Fig. 2]. Therefore, in the phosphorylated enzyme the δ-subunit can be saturated with 4 moles of Ca^{2+} even at low physiological Mg^{2+} concentrations expressing the partial activity A_2.

The catalytic center of phosphorylase kinase is clearly located on the γ-subunit since it is active in the isolated form (Kee and Graves, 1986; Crabb and Heilmeyer, 1984) or after expression of its cDNA in E-coli (Chen et al., 1989). This protein is able to form a complex with calmodulin which may occur during assembly of the holoenzyme in the cell, i.e. calmodulin bound to γ may represent the integral calmodulin of phosphorylase kinase. Ca^{2+}-free calmodulin suppresses the activity of γ partially; Ca^{2+} saturation of calmodulin results in full expression of activity, then the activity is as high as that of free γ. It points towards a quite different role of calmodulin in phosphorylase kinase in comparison to other Ca^{2+} calmodulin dependent enzymes: Ca^{2+}-free calmodulin is an inhibitor of the enzymatic activity and Ca^{2+}-binding to calmodulin releases this inhibition (Hessova et al., 1985[2]; Harris et al., 1990). In many other Ca^{2+}/calmodulin dependent enzymes Ca^{2+}-free calmodulin does not interact with the respective apoenzyme and Ca^{2+} saturated calmodulin is an activator. Ca^{2+} acting as deinhibitor is seen similarly in the regulation of the actomyosin ATPase of striated muscle. Ca^{2+}-free troponin C prevents an actin-myosin interaction; upon Ca^{2+} saturation of troponin C, this inhibition is released (Ebashi and Ogawa, 1988).

A chymotryptic fragment of the γ-subunit can be isolated expressing a higher molar activity than the intact isolated γ-subunit; however, the activity of this fragment is still not as high as that of the native enzyme (Harris et al., 1990). The γ-fragment is not affected by Ca^{2+} saturated or Ca^{2+}-free calmodulin. It shows that chymotryptic cleavage releases the calmodulin binding domain from γ. The primary structure of γ was worked out employing classical protein sequence analysis [Fig. 3]. As it is characteristic for protein kinases, a glycine rich cluster is followed by a lysine residue further downstream which seems to be part of the ATP binding site. This domain together with several other subdomains form the catalytic center. Within this catalytic center a lysine residue (151) following the censensus sequence, Arg-Asp-Leu (147-150), present in all protein kinases specifies the γ-subunit to be a Ser/Thr protein kinase [see Fig. 3]. Exchange of this lysine residue against Arg or Ala changes the specificity to a tyrosine protein kinase (Tagaya et al., 1988). Chymotrypsin releases the C-terminal part of γ (Harris et al., 1990). The residual N-terminal fragment is active and insensitive to Ca^{2+}/calmodulin. Thus, the first 290-300 residues of γ seem to build up the catalytic center. Approximately 90-100 C-terminal residues form a regulatory domain binding calmodulin. Two binding regions are present within this C-terminal part stretching from amino acid 287 to 331 and from 332 to 371 (Dasgupta et al., 1989). Anchoring of calmodulin onto two regions in γ might be necessary to prevent dissociation of the Ca^{2+}-free calmodulin from phosphorylase kinase. Again, this situation is very similar to that of troponin C. Troponin C is also anchored to troponin I at two sites; analogously, Ca^{2+}-free troponin C does not dissociate from holotroponin (Leszyk et al., 1990).

TRDAALPGSH	STHGFYENYE	PKEILGRGVS	30
SVVRRCIHKP	TCKEYAVKII	DVTGGGSFSA	60
EEVQELREAT	LKEVDILRKV	SGHPNIIQLK	90
DTYETNTFFF	LVFDLMKKGE	LFDYLTEKVT	120
LSEKETRKIM	RALLEVICAL	HKLNIVHRDL	150
KPENILLDDD	MNIKLTDFGF	SCQLDPGEKL	180
REVCGTPSYL	APEIIECSMN	DNHPGYGKEV	210
DMWSTGVIMY	TLLAGSPPFW	HRKQMLMLRM	240
IMSGNYQFGS	PEWDDYSDTV	KDLVSRFLVV	270
QPQKRYTAEE	ALAHPFFQQY	VVEEVRHFSP	300
RGKFKVICLT	VLASVRIYYQ	YRRVKPVTRE	330
IVIRDPYALR	PLRRLIDAYA	FRIYGHWVKK	360
GQQQNRAALF	ENTPKAVLFS	LAEDDY	386

Fig. 3: Primary sequence of the γ-subunit of rabbit sceletal muscle phosphorylase kinase.

Subdomains of the catalytic center, i. e. the glycine rich cluster amino acid 26 to 31, lysine 57 in the catalytic center and subdomains present in many other protein kinases, too, like RDL 147-150, DFG 177-179, APE 191-193, MW 212-213 and PFW 227-230. All these subdomains form the catalytic center. They are indicated in hatched boxes. Chymotryptic cleavage sites are indicated by vertical bars. The calmodulin binding domain is underlined, specific calmodulin binding regions comprise amino acid 287-331 and 332-371. The glutamate specific protease peptide binding to calmodulin is indicated by hatched boxes.

By combining protein and cDNA sequence data we determined the structure of the β- and α-subunits of phosphorylase kinase comprising 1.092 and 1.237 amino acid residues, respectively [Fig. 4 and 5]. Comparison of the sequence of α and β indicates clearly that these two proteins are homologous, thus originating from a common ancestor. It is also clearly visible that in each of the two subunits segments are present which are not present in the other subunit and vice versa. For example, the segments C and E each represent a long stretch of amino acids which are present only in the α-subunit. In contrast, the segment A is only present in the β-subunit [Fig. 4 and 5].

The β-subunit was digested with glutamate specific protease to identify calmodulin binding regions. The major peptide which is retained on calmodulin-Sepharose stretches from amino acid residue 896 – 905 [Gerschinski et al., unpublished data], which was not predicted employing the homology argument to other calmodulin binding proteins.

It was proposed that activation of the enzyme is correlated with phosphorylation of the β-subunit by the cyclic AMP-dependent protein kinase (Yeaman and cohen, 1975). The known sequence of this phosphorylation site locates the phosphorylatable serine residue at position 26 in the primary structure [Fig. 4]. A motive of basic amino acid residues precedes this serine residue therefore being an excellent substrate for the cyclic AMP-dependent protein kinase. The same pattern of basic residues precedes serine at position 700 [Fig. 4]. Indeed, serine 700 is a substrate of the cyclic AMP-dependent protein kinase, too, and both these serine residues are phosphorylated equally fast. Therefore, which site is more important in activating the enzyme has to be determined. However, serine β 700 seems to be phosphorylated to a higher degree in isolated phosphorylase kinase than serine β 26. Therefore, β 26 phosphorylation might correlate better with activation than that of β 700. Two additional phosphorylation sites are present in β: Serine 11 is phosphorylated during selfphosphorylation of the enzyme [Fig. 4]. Phosphate was never found to be endogenously present at this particular site. It casts doubt that serine 11 is phosphorylated in the intact cell. In contrast, phosphate is present in the enzyme as isolated in positions which can be phosphorylated by the cyclic AMP-dependent protein kinase, i.e. serine 26 and 700, at different levels.

```
          ← A →        Ps        . . .      ..Pa              I    ← B →
[AGATGLMAEV  SWKVLERRAR  TKRSGSVYEP  LKSINLP RPD  NETLW    45
 DKLDYYYKIV  KSTLLLYQSP  TTGLFPTKTC  GGDQTAKIHD  SLYCA    90
 AGAWALALAY  RRIDDDKGRT  HELEHSAIKC  MRGILYCYMR  QADKV   135
 QQFKQDPRPT  TCLHSLFNVH  TGDELLSYEE  YGHLQINAVS  LYLLY   180
 LVEMISSGLQ  IIYNTDEVSF  IQNLVFCVER  VYRVPDFGVW  ERGSK   225
 YNNGSTELHS  SSVGLAKAAL  EAINGFNLFG  NQGCSWSVIF  VDLDA   270
 HNRNRQTLCS  LLPRESRSHN  TDAALLPCIS  YPAFALDDDV  LYNQT   315
 LDKVIRKLKG  KYGFKRFLRD  GYRTSLEDPK  RRYYKPAEIK  LFDGI   360
 ECEFPIFFLY  MMIDGVFRGN  PKQVKEYQDL  LTPVLHQTTE  GYPVV   405
 PKYYYVPADF  VEYEKRNPGS  QKRFPSNCGR  DGKLFLWGQA  LYIIA   450
 KLLADELISP  KDIDPVQRYV  PLQNQRNVSM  RYSNQGPLEN  DLVVH   495
 VALVAESQRL  QVFLNTYGIQ  TQTPQQVEPI  QIWPQQELVK  AYFHL   540
 GINEKLGLSG  RPDRPIGCLG  TSKIYRILGK  TVVCYPIIFD  LSDFY   585
 MSQDVLLLID  DIKNALQFIK  QYWKMHGRPL  FLVLIREDNI  RGSRF   630

 NPMLDMLAAL  KNGMIGGVKV  HVDRLQTLIS  GAVVEQLDFL  RISDT   675
                           . . .      . .   Pa
 EELPEFKSFE  ELEPPKHSKV  KRQSSTSNAP  ELEQQPEVSV  TEWRN   720
                        ← D →
 KPTHE ILQKL  NDCSCLASQT  ILLGILLKRE  GPNFITQEGT  VSDHI   765
 ERLYRRAGSK  KLWLAVRYGA  AFTQKFSSSI  APHITTFLVH  GKQVT   810
 LGAFGHEEEV  ISNPLSPRVI  KNIIYYKCNT  HDEREAVIQQ  ELVIH   855
 IGWIISNNPE  LFSGMLKIRI  GWIIHAMEYE  LQIRSGDKPA  KDLYQ   900
                                          ← F →
 LSPSE VKQLL  LDILQPQQ NG  RCWLNKRQID  GSLNRTPTGF  YDRVW   945
 QILERTPNGI  IVAGKHLPQQ  PTLSDMTMYE  MNFSLLVEDM  LGNID   990
 QPKYRQIVVE  LLMVVSIVLE  RNPELEFQDK  VDLDKLVKEA  FHEFQ  1035
 KDESRLKEIE  KQDDMTSFYN  TPPLGKRGTC  SYLTKVVMNL  LLEGE  1080
          Pe
 VKPSNEDSCL  VS                                         1092
```

Fig. 4:   Primary structure of the β-subunit

Predicted calmodulin binding domains by sequence homology to other calmodulin binding proteins are underlined by a dashed line. The peptide isolated by chromatography on calmodulin-Sepharose following glutamate specific protease digestion is shown in hatched boxes. Phosphorylation sites are indicated with P, subscript S represent a selfphosphorylation site, subscript A a site phosphorylated by the cyclic AMP-dependent protein kinase, subscript E an endogenous phosphorylation site. Points indicate basic amino acid residues determining the sites 26 and 700 as substrates for the cyclic AMP-dependent protein kinase. The regions A - F are assigned analogously to those in the α subunit (see Fig. 5).

← $\mathscr{B}$ →
```
[MRSRSNSGVR  LDSYARLVQQ  TILCHQNPVT  GLLPASYDQK  DAWVR     45
 DNVYSILAVW  GLGLAYRKNA  DRDEDKAKAY  ELEQSVVKLM  RGLLH     90
 CMIRQVDKVE  SFKYSQSTKD  SLHAKYNTKT  CATVVGDDQW  GHLQL    135
 DATSVYLLFL  AQMTASGLHI  IHSLDEVNFI  QNLVFYIEAA  YKTAD    180
 FGIWERGDKT  NQGISELNAS  SVGMAKAALE  ALDELDLFGV  KGGPQ    225
 SVIHVLADEV  QHCQSILNSL  LPRASTSKEV  DASLLSVISF  PAFAV    270
 EDSKLVEITK  QEIITKLQGR  YGCCRFLRDG  YKTPKEDPNR  LYYEP    315
 AELKLFENIE  CEWPLFWTYF  ILDGVFSGNA  EQVQEYREAL  EAVLI    360
 KGKNGVPLLP  ELYSVPPDKV  DEEYQNPHTV  DRVPMGKLPH  MWGQS    405
 LYILGSLMAE  GFLAPGEIDP  LNRRFSTVPK  PDVVVQVSIL  AETEE    450
 IKAILKDKGI  NVETIAEVYP  IRVQPARILS  HIYSSLGCNN  RMKLS    495
 GRPYRHMGVL  GTSKLYDIRK  TIFTFTPQFI  DQQQFYLALD  NKMIV    540
 EMLRTDLSYL  CSRWRMTGQP  TITFPISQTM  LDEDGTSLNS  SILAA    585
 LRKMQDGYFG  GARIQTGKLS  EFLTTSCCTH  LSFMDPGPEG  KLYSE    630
 DYDDNYDELE  SGDWMDGYNS  TSTARCGDEV  ARYLDHLLAH  TAPHP]   675
 KLAPASQKGG  LNRFRAAVQT  TCDLMSLVTK  AKELHVQNVH  MYLPT    720
 KLFQASRPPL  NLLDSSHPSQ  EDQVPTVRVE  VHLPRDQSGE  VDFQA    765
 LVLQLKETSS  LQEQADILYM  LYTMKGPDWD  TELYEEGSAT  VRELL    810
 TELYGKVGKI  RHWGLIRYIS  GILRKKVEAL  DEACTDLLSH  QKHLT    855
 VGLPPEPREK  TISAPLPYEA  LTRLIEEACE  GDMNISILTQ  EIMVY    900
 LAMYMRTQPG  LFAEMFRLRI  GLIIQVMATE  LAHSLRCSAE  EATEG    945
 LMNLSPSAMK  NLLHHILSGK  EFGVERSVRP  TDSNVSPAIS  IHEIG    990
 AVGATKTERT  GIMQLKSEIK  QVEFRRLSIS  TESQPNGGHS  LGADL   1035
 MSPSFLSPGT  SVTPSSGSFP  GHHTSKDSRQ  GQWQRRRRLD  GALNR   1080
 VPIGFYQKVW  KVLQKCHGLS  VEGFVLPSST  TREMTPGEIK  FSVHV   1125
 ESVLNRVPQP  EYRQLLVEAI  LVLTMLADIE  IHSIGSIIAV  EKIVH   1170
 IANDLFLQEQ  KTLGADDIML  AKDPASGICT  LLYDSAPSGR  FGTMT   1215
 YLSKAAATYV  QEFLPHSICA  MQ                            1237
```

Fig. 5:   Primary structure of the α-subunit.

Subdomains homologous to other protein kinases are expressed in hatched boxes. Lysine 588 is marked with an F to indicate binding of fluorescein isothiocyanate. Calmodulin binding domains homologous to other calmodulin binding proteins are underlined with a dashed line. The isolated calmodulin binding peptide 546-573 is underlined by a solid line. The preferential chymotryptic cleavage site is indicated by a vertical bar. Phosphorylation sites are indicated with P, subscripts are identical to those in Fig. 4. The regions A – F are assigned analogously to those in the β subunit (see Fig. 4)

Apart from these few phosphorylation sites no information is available on functional domains or binding sites in the β-subunit. However, the region B might contain a nucleotide binding site since this part of the molecule is homologous to that of the α-subunit. The region B of α contains an ATP/Mg^{2+}-binding site (see below). There are indications that one or even two moles of ADP bind to the β-subunit (Cheng et al., 1985). Therefore, it might be possible that their binding sites are located in this N-terminal region of β.

Surprisingly, the α-subunit of phosphorylase kinase can be labelled specifically by fluorescein isothiocyanate. The covalent modification is accompanied by partial inhibition of the enzyme's activities. The activity A_0 is inhibited to a much lesser extent than the activities A_1 and A_2 (Zaman et al., 1989). Kinetic studies reveal that ATP and fluorescein iso-thiocyanate bind competitively on one and the same site; the affinity for ATP can be calculated to be approximately 700 micromolar. This value is in the range of Km values determined for ATP by several groups. Further-more, not only ATP competes with fluorescein isothiocyanate binding but also the substrate ATP-Mg^{2+}. It shows that the holoenzyme contains two substrate binding sites: one located on the α- and a second one on the γ -subunit. It is remarkable that only one fluorescent peptide is obtained upon labelling with the very unspecific reagent fluorescein isothiocyanate. The sequence of this peptide shows that lysine residue 588 is modified [Fig. 5] (Harris et al., 1990). Sequence comparison of this region of α or of the whole α- or β-sequence with that of other known members of the protein kinase family show no apparent homology. Subdomains are recog-nizable which are characteristic for catalytic centers of the protein ki-nase family, e.g. a glycine rich cluster, Gly-X-Gly-X-X-Gly, or Asp-Phe-Gly, Leu-Asp-Ala and other subdomains of ATP-Mg^{2+}-binding sites [Fig. 4 and 5]. In the protein kinase family these subdomains are located C-terminally to the glycine rich cluster but not N-terminally as seen in α. It shows that α belongs certainly to a different family of proteins than the known protein kinase family.

What might be the function of such a substrate binding domain on the α-subunit? Somehow, binding of the substrate ATP-Mg^{2+} onto the α-subunit is involved in expression of catalytic activity. It is known that the prod-uct ADP may bind even stronger to the γ -subunit than the substrate ATP

(Cheng and Carlson, 1988). Therefore, in the living cell the γ-subunit might be saturated with ADP. As a prerequisite for activity expression an exchange of ADP against ATP on γ must occur. One might assume that binding of ATP-Mg^{2+} to the a-subunit triggers such an ADP-ATP exchange like in a signal transducing G-protein (Gilman, 1987) [Fig. 6]. Therefore, phosphorylase b to a conversion can occur only when a is saturated with ATP/Mg^{2+}. If this is the case overall expression of activity is dependent on ATP-Mg^{2+}-binding to a which might be controlled by other domains of a.

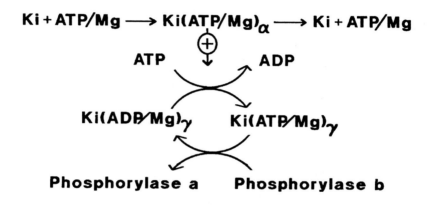

Fig. 6:   Synopsis of ATP/Mg^{2+}-binding to the a-subunit.
An arrow with the ⊕ indicates the effect of ATP/Mg^{2+}-binding to a allowing an ADP/ATP exchange on the γ-subunit and consequently conversion of phosphorylase b to a.

No indication was obtained that sites on a which can be phosphorylated by phosphorylase kinase itself, i. e. during selfphosphorylation, carry phosphate in the freshly isolated enzyme. Three such selfphosphorylation sites have been identified (serine 972, 985, 1007); phosphate is endogenously present only at one serine residue which has been shown previously to be phosphorylated by the cyclic AMP-dependent protein kinase (serine 1018) [Fig. 5] (Yeaman and cohen, 1975). Endogenously, phosphate is present at four serine residues further downstream (serine 1020, 1023, 1030, 1042) (Meyer, H.E. et al., 1990). A pattern emerges: either phosphate is endogenously present in the first (1020) and the last one (1042) of these four serine residues, then there is no phosphate present in the two middle positions. Alternatively, phosphate is present in the two middle positions (1023, 1030), then there is no phosphate in the first and last position [Fig. 5]. Meanwhile, it has been shown that the serine residue following the one phosphorylated by the cyclic AMP-dependent protein kinase (serine 1020) can be phosphorylated by a 5'AMP activated protein kinase (Carling and Hardie, 1989). No information is available which protein kinase(s) incorporates phosphate into the other positions found to be phosphorylated in the intact cell. None of these sites are identical to those phosphorylated by phosphorylase kinase itself. Thus, we have no proof that selfphosphorylation occurs in the intact cell.

All these newly detected phosphorylation sites are located in region E which we call the multiphosphorylation loop. Similarly, the region A, being present in the β-subunit exclusively, contains two phosphorylation sites, and β 700 is just shortly before the beginning of region C. The region B in the α-subunit contains the ATP-Mg^{2+}-binding domain as well as a calmodulin binding domain whereas calmodulin binding on β may occur in region D [Fig. 4 and 5].

The available informations on the structure of the phosphorylase kinase subunits allow to formulate the following hypothesis:
The regulatory potential of the γ-calmodulin complex which might represent a type of archaic Ca^{2+}-dependent protein kinase is too low to handle all the signals regulating phosphorylase b to a conversion [compare Fig. 1]. Clearly, the two subunits α and β augment the regulatory potential of this enzyme. How then the α and β-subunits could influence the enzymatic activity? There are two possible modes: either by modulating Ca^{2+}-binding to the integral calmodulin or by modulating substrate binding to the catalytic site on the γ-subunit [Fig. 7].

Modulation of Ca^{2+}-binding to the integral calmodulin is exerted by Mg^{2+}. It is postulated that phosphorylation of serine β 26 or β 700 increases the affinity for Mg^{2+}. Since Mg^{2+}-binding to phosphorylase kinase is a prerequisite for full saturation of the integral calmodulin with Ca^{2+} phosphorylation of β on these two sites modulates activity expression via modulation of Mg^{2+} and hence Ca^{2+}-binding. Assuming that the N-terminal serine β 26 residue is located in a calmodulin binding domain one could visualize that binding of calmodulin to this site influences phosphorylation of serine 26 but not phosphorylation of serine 700. ADP probably

binds onto the β-subunit, too. Since the β-subunit might have been designed to influence the integral calmodulin ADP might modulate Mg^{2+}-
binding, i.e. Ca^{2+}-binding as well.

The α-subunit provides an ATP-Mg^{2+}-binding site. At present it cannot be
decided if this binding site forms a part of a protein kinase itself or if
it constitutes an allosteric ATP-Mg^{2+}-binding site. Assuming an allosteric
function, the α-subunit has been designed to modulate substrate binding
onto the γ-subunit and thus the expression of activity. Binding of Ca^{2+}
calmodulin to the α-subunit may also modulate ATP-Mg^{2+}-binding on α.
Similarly, multiple phosphorylation of the α-subunit could also influence
ATP-Mg^{2+}-binding.

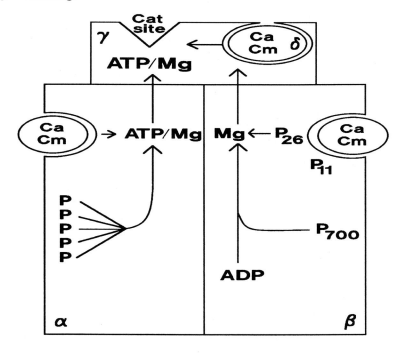

Fig. 7:   Synopsis on overall activity regulation of the γ-subunit.
Calmodulin binding to the γ-, α- and β-subunit are indicated with Ca^{2+}
Cm. The catalytic site on γ is indicated by the substrate binding site. It
binds ADP as well. The ATP/Mg^{2+}-binding site in α represents the site labelled by fluorescein isothiocyanate. Multiple P indicate the multiphosphorylation loop on α. Binding of Mg^{2+} to the β-subunit it hypothetical. P26
and P11 indicate the phosphorylation sites in the A region of the β-subunit, P700 the phosphorylation site of serine in position 700. ADP binding
on the β-subunit is proposed from the homologous regions in the N-terminal region of α.

In general, association of the γ-δ complex with these two high molecular weight proteins, the α- and β-subunits, enhances the regulatory potential of phosphorylase kinase by modulation of substrate binding on γ via the α-subunit or binding of Ca^{2+} on δ via the β-subunit [Fig. 7]. A multiplicity of species can be produced by phosphorylating the inactive $\alpha\beta\gamma\delta$ complex to different degrees. The enzymatic activity of each of these species is modulated furthermore by allosteric effectors such as ATP/Mg^{2+}, Mg^{2+} and Ca^{2+}. Each species then contributes to the phosphorylase kinase activity level assuring a phosphorylase a steady state according to the sum of all incoming signals regulating glycogen degradation.

PHYSICAL APPROACHES TO CONFORMATION AND ASSEMBLY OF BIOLOGICAL MACROMOLECULES

M. H. J. Koch
EMBL c/o DESY
Notkestrasse 85
D-2000 Hamburg 52
Federal Republic of Germany

In this lecture I would like to make a few remarks concerning the nature and the use of the results of structural investigations on biological macromolecules. Hopefully, they will encourage the reader to look up some of the references and perhaps even to take a slightly different look at some of the problems of the structure and function of biological macromolecules and their assemblies.

A. Structures and forces

For engineers a structure is the response of a system to the forces that act upon it (Gordon, 1986). At the macroscopic level this is illustrated by the fact that most objects around us have some form of bilateral symmetry mainly because we live in the field of gravity, whereas small marine animals, for instance, that are not so much subject to this force tend to have a spherical shape.

A rigid structure is one that can only provide support (i.e. delay an energetically more favourable event) but not perform any active (e.g. control) function. Hence, what matters for function is not so much structure as structural change.

At the molecular level the understanding of the relationship between structure and function should thus be concerned with the polymorphism of charge distributions and of the associated fields. In this context, a detailed geometrical description of a structure may provide the basis for understanding the constraints that exist on the changes, but this requires to switch from the world of form to the world of forces. There is much to be gained from this transition. Consider, for instance, the architecture of many oligomeric enzymes. In the world of form the requirement for symmetry is a postulate or, quoting Max Perutz

NATO ASI Series, Vol. H 56
Cellular Regulation by Protein Phosphorylation
Edited by L. M. G. Heilmeyer, Jr.
© Springer-Verlag Berlin Heidelberg 1991

(1989), "The postulate of symmetry also seemed to have some aesthetic fascination for Monod, like Ptolemy's spheres, quite independent of its formal advantages."

In the world of forces things are simpler to understand and the origin of the postulate becomes clearer. Consider the forces in an assembly of the simplest asymmetric objects: dipoles. Dipoles tend to orient in an antiparallel manner resulting in objects with twofold symmetry. (The alternative head to tail arrangement leads to filamentous structures which we shall not discuss here.) As illustrated in Fig. 1, assembly requires a complementarity of geometry (i.e. the humps fitting into the bumps) but even more a complementarity of polarity (positive corresponding to negative) at the interface. If this is not the case, formation of an $\alpha\beta$ complex with a more complementary surface is more likely than that of an α_2 complex. Such assemblies can undergo concerted shape changes provided the fundamental asymmetry of the subunits is conserved.

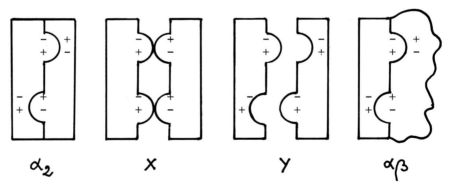

α_2 X   Y $\alpha\beta$

Fig. 1: Examples of interfaces with complementary geometry and polarity (α_2 and $\alpha\beta$) or only one type of complementarity (X,Y), which does not suffice for assembly.

The postulate of symmetry in the world of form is thus a direct consequence of observable asymmetries in the world of forces. The symmetry arises and is maintained in oligomeric enzymes because it corresponds to an energy minimum in the packing of in-

trinsically asymmetric subunits. Transitions between states
(T→R) in these systems are triggered by an asymmetry resulting,
for instance, from the binding of a substrate at one site.
Curie's famous sentence "C'est la dissymétrie qui crée le
phénomène" (Curie, 1884) also applies to allosteric enzymes!

Not every biological system is that simple and ribosomes or membrane proteins are neither symmetric nor do they consist of
identical subunits, although the simultaneous complementarity
of geometry and polarity at the interfaces is still required.
Another example of asymmetry is that of bacteriorhodopsin and
several other membrane proteins which have an odd number of
antiparallel helices (7) perpendicular to the membrane surface
and should thus have a net dipole moment in this direction.
Since membranes tend to be polarized, large forces can act on
these objects. It is not clear whether this has any physiological significance but it is known that enzymatic activity can be
coupled to electric fields (Westerhoff et al. 1986) and such
features may play a role in this phenomenon.
Unfortunately, with few exceptions, textbooks do not explicitly
consider the structure of biological macromolecules in the world
of forces, whereas the aesthetic satisfaction associated with
symmetry is extensively exploited.

The difference between small molecules and macromolecules

In a molecule with a small number of atoms (N) the number of degrees of freedom (3N-6) and of intermolecular interactions
(N·(N-1)/2) is also small and conformational changes require a
large relative change in energy ($\Delta E/E$). The notion of "the"
structure or of a few distinct conformations is valid. The small
number of parameters and large constraints make computational
predictions of conformation very powerful. Note that, although
several thousand structures of organic compounds are known to
very high resolution, relatively little use of this information
is made in organic chemistry textbooks. The reason is partly
that substitutions, for instance, which have large effects on

rate constants lead to geometric effects that are often at the limit of the precision that can be achieved. Careful analysis indicates, however, that these effects are significant (Bürgi and Dunitz, 1988). Entropic effects usually only play a secondary role in the behaviour of these molecules.

In the case of macromolecules the number of interactions becomes very large, the relative change in energy ($\Delta E/E$) accompanying conformational changes is small and entropy becomes dominant. Structures become more "flexible" and there is ample experimental evidence that the notion of "the" structure has to be replaced by that of conformational space (for a review see Frauenfelder et al., 1988).
In these systems, many reactions are entropy-controlled and water molecules and osmotically free counterions which remain invisible even in high resolution structure determinations, play a major role. Concomitantly, experimental and even more computational methods become less powerful and structure prediction has been a continuously receding goal. At present, using the most powerful methods the conformation of about 60 % of the residues is predicted correctly from the sequence. (Given that a half truth is a full lie, the fascetious experimentalist is of course tempted to say that 100 % of the calculated protein conformations are wrong!)

Processes like protein folding result from the balance between hydrophobic effects and the opposing reduction in conformational entropy (for a review see Dill, 1990). Neither of these effects are directly accessible through X-ray structural observations, nor can they effectively be dealt with in calculations. It is thus a real challenge to find out what the missing forces are and how to treat them in the calculations. New approaches using, for instance, NMR and protein engineering (Matouschek et al., 1990) are certainly required.

B. Function and coherent change

The basis of function is change since to perform work (W=Fdl) or to transmit one bit of information (0→ 1) a change is required, but not every change is useful. What matters is the coherent change (displacement) as opposed to the incoherent change (agitation or uncorrelated motion). Obviously, coherent geometrical changes occur because the forces acting on the system have changed but, as already mentioned, the causes for these changes can be much more subtle than usual geometrical structural methods allow to detect. The thermodynamics of these effects is, however, well understood (for a review see Eisenberg, 1990).

Some devices can select a component of the incoherent motion so that it results in a displacement. In a piston, for instance, large forces prevent the deformation of the wall of the cylinder. If the walls were made out of a thin rubber sheet the piston would certainly not work!

Proteins or macromolecular assemblies like muscle must do something similar. Unfortunately, very little hard experimental evidence exists on motions in biological macromolecules, especially on coherent or functionally important motions, as they are sometimes called (Frauenfelder et al., 1988). In absence of strong experimental constraints the scope of molecular dynamics calculations (McCammon and Harvey, 1987) remains limited.

The continued use of descriptions of coherent motions in enzymes (i.e. their mechanism), inferred from a comparison of structures corresponding to different states, in terms of mechanical devices (levers and springs) is indicative of the fact that no adequate language has yet evolved to describe these phenomena. The picture which is emerging is, however, very complex.

Coherent changes and kinetics

In recent years, developments in time resolved X-ray scattering and diffraction methods have raised high expectations for kinetic crystallography (for reviews see Moffat, 1990 or Hajdu and Johnson, 1990). How realistic is it to "see" a reaction? To answer this question first consider the Maxwell-Boltzmann energy distribution. The molecules that react are those with an energy larger than the activation energy, which is quite different from the average energy, especially if the reaction is slow.

Next consider, to simplify matters, the equation giving the rate for a unimolecular reaction (see e.g. Fersht (1985)):

$$- \frac{d[X]}{dt} = \left(\frac{kT}{h} \right) [X] \, exp \left(- \Delta G^{\ddagger} / RT \right)$$

Clearly, the rate of a reaction measures the rate at which the equilibrium between the initial and final state is established, not the time it takes for a molecule to cross the energy barrier, which is given by h/kT and is of the order of 10^{-11} s. So the actual reaction corresponds to a few molecules ($[X] \, exp(-\Delta G^{\ddagger}/RT)$) crossing the barrier very fast. These are obviously very unfavourable conditions to apply X-ray methods which yield a space and time averaged structure.

Hoping to make a movie of a reaction with these methods is thus not realistic. One can, however, in favourable cases, find out which part of a molecule is moving as illustrated in the first example below.

If there are no intermediates that can be trapped and studied in a static manner, it suffices to measure the initial and final states accurately and, in a separate experiment, the time course of the intensity of a few reflections. If one suspects where the changes occur, it is possible to predict which reflections will be most sensitive to the changes. It is thus not indispensable to measure all reflections and as illustrated by the example of bacteriorhodopsin it should even be possible to use microcrystalline (powder) samples.

Spectroscopic methods thus keep their advantage in following changes, especially because they allow to monitor processes through selective windows. Interpretation of the results in absence of a geometrical structure is, however, usually very risky especially as the rate of spectroscopic changes may not be that of the shape changes.

Biological macromolecules as devices

Understanding the behaviour of a cascade (i.e. an energy flow) like, for instance, in phosphorylation, does not require know- ledge of the detailed geometrical structure of the components. This knowledge is only required if one wants to modify the de- vices (e.g. by protein engineering) in this cascade. Remember though that it is a charge distribution one is trying to en- gineer, not only a geometry.

Devices can be passive, in which case the output solely depends on the input and on the characteristics of the device, or they can be active. The main characteristic of an active device is its gain (energy at the outputs/energy at the inputs) and the manner in which its output is controlled by the potential <u>at some other point</u> in the system. In the case of an enzyme this may occur, for instance, through the binding of molecules syn- thesized elsewhere, through covalent modification or by direct feedback of the reaction products or through variations of the potential across membranes.

To take an analogy, the properties of a complex electronic cir- cuit can be described without reference to the detailed geome- trical structure of its components. Moreover, with present structural methods one would not detect geometrical changes in the structure of silicon when the bias of a transistor changes. Circuits initially implemented with one type of structure (valves) can equally well be made with geometrically totally different structures (transistors).

Biological systems which are intermediates in an evolutionary selection process cannot so easily switch to the use of funda- mentally new devices. Still, new devices may evolve that solve an old problem (analogy) or old structures may, with little mo- dification, provide the answer to a new challenge (homology). This corresponds to time's arrow (homology) and time's cycle (analogy) which one also finds at the macroscopic level (Gould, 1988). Note that whereas it is very fashionable to look for sequence homology much less effort is made looking for analo- gies although this may be more productive if one wants to learn to design molecular devices. Analogous organs, like the eyes

of very distant species, like flies and humans, tell more about
the fundamental physical constraints of vision than a detailed
comparison of eyes in closely related species. The same is probably true at the molecular level.

There are thus intrinsic limitations to a geometrical structural approach which can only be circumvented by integrating the
experimental and theoretical approaches that give information
about the changes in the forces that act upon the system.
There is also another and more technical limitation to the geometrical structural approach which appears to have been hitherto insufficiently recognized by many biochemists.

C. Structural methods and inverse problems: The need for consistency

Finding the structure corresponding to a given diffraction pattern is an inverse problem, i.e. a problem which has no unique
solution. Given the nature of the problem, there are thus many
possibilities for errors and in the recent past a number of er-
roneous structures have appeared. This has even caused some
alarm among practitioners of this art (Brandén and Jones, 1990).
The causes for error are multiple. The quality of the data may
be poor because the crystals are not good or because the measurements were not accurate, software packages are used as black
boxes although they are not yet foolproof, databanks are not
yet equipped with tools to check the quality of their input,
etc. In these circumstances it is thus very important that the
coordinates of the atoms and, if need be, the experimental data
be made available before anybody is credited with having solved -
or rather plausibly interpreted - a structure. The problem is
not that regretable mistakes are made, but what is wrong is for
a subset of the scientific community to impede or delay the independent consistency checks to which scientific knowledge has
to be subjected. Fortunately, a number of journals have recently taken steps to make sure that this would be the case. With
the coordinates one can attempt to switch to the world of for-

ces whereas, despite all its aesthetic appeal, a beautiful colour picture is of little use in the world of forces.

D. Examples

The examples described during the lecture were taken from the literature and from recent work. They will not be treated in detail here but recent references are given below.

1) Detecting coherent changes during a reaction: the photocycle of bacteriorhodopsin (BR) (Henderson et al., 1990, Ames and Mathies, 1990, Dencher et al., 1990, Koch et al., 1991). The BR membrane is a particularly interesting device which converts light into a proton gradient. For the phenomenon to be irreversible the system has to go through a cycle of reactions, which consists in this case of at least seven steps. Using a mutant where some of the reactions are slower than in the wild type, it is possible to observe coherent conformational changes occuring during the photocycle.
Under certain conditions repeated light flashes can desensitize the membrane by pumping the BR molecules into very long lived states. This is a sign that one has homogeneous proteins with different reaction pathways within each individual molecule. This is different from the flash experiments on CO myoglobin and corresponds to a complex hierarchy of reaction pathways (Frauenfelder et al., 1988). The system is a good candidate for higher resolution observations using electron microscopy or crystals if they can be obtained with sufficiently high order.

2) Building a consistent model: Chromatin condensation (Koch, 1989): Reviews an example of how the combination of various static and dynamic structural and other observations leads to a more consistent model for the structure of chromatin. The structural equilibria in this system are mainly controlled by counterion condensation. Because of the large structural changes involved it would not be possible to study such systems at high resolution even assuming that they could be crystallized. In

such a case, as in the following example, scattering methods are
more useful.

3) Complex reaction pathways and structural oscillations in microtubules (Mandelkow et al., 1989, Tabony and Job, 1990). Depending on the conditions proteins like tubulin can form dissi-
pative structures which require a net input of energy for their
formation and maintenance. In the case of microtubules, hydrolysis of GTP provides the necessary energy and both time dependent
and steady-state spatial structures have been observed. In the
latter case (Tabony and Job, 1990) it is not entirely clear
whether GTP is required to maintain the spatial structure as
such, or whether it only influences the stability of the microtubules. This behaviour observed in-vitro is reminiscent of the
phenomena accompanying spindle formation during mitosis. Dissi-
pative structures are the key to many biological functions -
they do not occur in crystals.

4) The power of images or beautiful wrong structures. The paper
by George and George (1988) illustrates how the discrepancy between X-ray spectroscopy (EXAFS) and X-ray crystallography results led to a revision of the crystal structure of Ferredoxin.
For other revised structures see for instance Human Serum Albumin (Carter and He, 1990) and Eco RI (Kim et al., 1990). Note
that some of these revisions are quite drastic and in the case
of Ferredoxin involve regions of high electron density in the
molecule (Fe-S cluster) that are expected to be defined with
higher precision. Accuracy is, of course, something else!

E. Conclusion

The emphasis on the world of form in the early days of structural research at the molecular level was certainly justified given that the main problem was to develop methods to solve structures. The aesthetic satisfaction that drives many of those who
enjoy solving structures certainly also played a role. As structural information accumulates its operational value should be
assessed more carefully. If one aims at using structural know-

ledge to make new devices, an approach in terms of forces (i.e. asymmetries) is probably more useful.

In this endeavour, no physical technique can claim to provide complete answers. A lot is to be gained in the understanding of biochemical systems by constructing more consistent models allowing to translate the information obtained in the world of form to the world of forces. At present, new structures certainly accumulate much faster than new ideas. This situation provides a great challenge and even greater opportunities for newcomers in the field, be they experimentalists or theorists.

References

Ames JB, Mathies RA (1990) The role of back-reactions and proton uptake during the N→O transition in bacteriorhodopsin's photocycle: A kinetic resonance Raman study. Biochemistry 29:7181-7190

Brandén CI, Jones TA (1990) Between objectivity and subjectivity. Nature 343:687-689

Bürgi HB, Dunitz JD (1988) Can statistical analysis of structural parameters from different crystal environments lead to quantitative energy relationships? Acta Cryst B44:445-457

Carter DC, X-M He (1990) Structure of Human Serum Albumin. Science 249:302-303

Curie P (1884) Sur la symétrie. Soc Mineralog France Bull Paris 7:418-457

Dencher NA, Dresselhaus D, Zaccai G, Büldt G (1989) Structural changes in bacteriorhodopsin during proton translocation revealed by neutron diffraction. Proc Natl Acad Sci USA 86:7876-7879

Dill KA (1990) Dominant forces in protein folding. Biochemistry 29:7133-7155

Eisenberg H (1990) Thermodynamics and the structure of biological macromolecules. Eur J Biochem 187:7-22

Fersht A (1985) Enzyme structure and mechanism. 2nd ed, WH Freeman and Co New York

Frauenfelder H, Parak F, Young RD (1988) Conformational substates in proteins. Ann Rev Biophys Chem 17:451-479

George GN, George SJ (1988) X-ray crystallography and the spectroscopic imperative: The story of the [3Fe-4S] clusters. TIBS 13:369-370

Gordon JE (1986) Structures or Why things don't fall down. Penguin Books Ltd London

Gould SJ (1988) Time's arrow Time's cycle. Penguin Books Ltd London

Hajdu J, Johnson LN (1990) Progress with Laue diffraction studies on protein and virus crystals. Biochemistry 29:1669-1678

Henderson R, Baldwin JM, Ceska TA, Zemlin F, Beckmann E, Downing KH (1990) Model for the structure of bacteriorhodopsin based on high-resolution electron cryo-microscopy. J Mol Biol 213:899-929

Kim Y, Grable JC, Love R, Greene PJ, Rosenberg JM (1990) Refinement of Eco RI endonuclease crystal structure: A revised protein chain tracing. Science 249:1307-1309

Koch MHJ (1989) Structure and condensation of chromatin. In: Saenger W, Heinemann U (eds) Protein-Nucleic Acid Interactions. McMillan London p163

Koch MHJ, Dencher NA, Oesterhelt D, Plöhn H-J, Rapp G, Büldt G (1991) Time-resolved X-ray diffraction study of structural changes associated with the photocycle of bacteriorhodopsin. EMBO J (in press)

Mandelkow E, Lange G, Mandelkow EM (1989) Biopolymers in solution. In: Mandelkow E (ed) Topics in Current Chemistry Vol 151. Springer Verlag Berlin p9

McCammon JA, Harvey SC (1987) Dynamics of proteins and nucleic acids. Cambridge University Press Cambridge

Moffat K (1989) Time-resolved macromolecular crystallography. Ann Rev Biophys Biophys Chem 18:309-332

Perutz MF (1989) Mechanisms of cooperativity and allosteric regulation in proteins. Quart Revs Biophys 22:139-236

Tabony J, Job D (1990) Spatial structures in microtubular solutions requiring a sustained energy source. Nature 346: 448-451

Westerhoff HV, Tsong TY, Chock PB, Chen Y, Astumian RD (1986) How enzymes can capture and transmit free energy from an oscillating electric field. Proc Natl Acad Sci USA 83:4734-4738

A Vibrational Raman Spectroscopic Study of Myosin and Myosin - Vanadate Interactions

J.J.C. Teixeira-Dias[1], E.M.V. Pires[2], P.J.A. Ribeiro-Claro[1],
L.A.E. Batista de Carvalho[1], M. Aureliano[2] and Ana Margarida Amado[1]
Department of Chemistry[1], Department of Zoology[2]
University of Coimbra
P-3049 Coimbra
Portugal

1. Introduction

As it is well known, a Raman spectrum is a distribution of frequencies of photons inelastically scattered by molecules. Each Raman frequency shift corresponds to the energy transferred between the radiation and the molecule and to the frequency of an infrared absorption maximum. Raman spectroscopy gives detailed information on the vibrational motions of atoms in molecules. As these vibrations are sensitive to chemical changes, the vibrational spectrum can be used to monitor molecular chemistry. Compared with the infrared absorption technique, Raman spectroscopy presents several advantages in the studies of biological systems, among which the most relevant are the low intensity of water bands and the greater sensitivity to important vibrational modes.

The Raman spectrum of a protein consists of bands originated in vibrational modes from the peptide backbone and amino acids side chains (Carey, 1982; Tu, 1982). Some of these bands are sensitive to the average conformation of the protein, as is illustrated in Table 1, for Amide I (essentially a C=O stretching vibration of the peptide) and Amide III (C-N-H in plane bending, mixed with C-N stretching vibration) bands.

Table1- Frequencies (cm^{-1}) of Amide I and Amide III modes (Carey, 1982)

Secundary Structure	Amide I	Amide III
α-helix	1645-1660	1265-1300
β-sheet	1665-1680	1230-1240
unordered[(1)]	1660-1670	1240-1260

(1) broader than those of α-helix and β-sheet

NATO ASI Series, Vol. H 56
Cellular Regulation by Protein Phosphorylation
Edited by L. M. G. Heilmeyer, Jr.
© Springer-Verlag Berlin Heidelberg 1991

Vanadate (V(V)) and phosphate ions are known to compete for the binding to proteins, as they exhibit important size and structural similarities (Chasteen, 1983). In particular, it has been reported that vanadate ions, namely, monovanadate ones, strongly inhibit the myosin ATPase activity (Goodno, 1979). In order to understand the conformational effects of the vanadate ion on myosin, a Raman spectroscopic study of myosin and myosin-monovanadate is herein reported.

2. Experimental

Myosin was isolated from the rabbit white skeletal muscle (Pires, 1977). Vanadate ion solutions (ca. 50 mM) were prepared as described elsewere (Aureliano, in preparation).

Raman spectra were recorded on a Ramalog double spectrometer, 0.85 m, f/7.8, Spex model 1403. The light source was a Spectra-Physics Ar^+ laser, whose output at 514.5 nm was adjusted to provide 200 mW at the sample position. Samples were sealed in Kimax glass capilary tubes with inner diameter 0.8 mm. The Raman Spectrometer was fully controlled by a DM1B microcomputer and the Raman data were stored on disk for futher processing, namely, for a 13-point Savitzky-Golay smoothing.

3. Results and Discussion

The Raman spectra of myosin (a), partially denaturated myosin (b) and myosin--monovanadate (c) are presented in the Figure. The Raman shifts of the bands and their approximate descriptions are listed in Table 2.

It is well known that the band at 1003 cm^{-1}, ascribed to the C-C stretching vibrations in the phenylalanine side chains (Tu, 1982), is very insensitive to the chemical environment of these chains. Thus, this band can be taken as a reference for intensity comparisons. In particular, the I_{937}/I_{1003} ratio decreases when the myosin is partially denaturated (compare spectra (a) and (b)), indicating a decrease of the α-helix content (Carey, 1982).

Other bands which are signatures of a secundary structure of myosin occur at 1240 (β-sheet, random coil), 1281 (β-sheet), 1303 and 1320 (α-helix) cm^{-1}. The occurrence of a feature at 1240 cm^{-1} in the partially denaturated myosin (b) points to a random coil or a β-sheet structure. In addition, in the myosin-monovanadate spectrum (c) a feature at 1303 cm^{-1} suggests a prevalent α-helix structure for this system. This conclusion is also indicated by the increase of the I_{937}/I_{1003} intensity ratio (spectrum (c) compared with spectrum (a)).

Table 2 - Observed frequencies and approximate description of bands in the myosin Raman spectra

$\Delta v/cm^{-1}$	Residue	Approximate description	Secundary Structure
		Mode	
826	Tyr	ring overtone	
855	Tyr	sym. ring stretch	
877			
904		$C_\alpha C$ stretch	
937	Lys, Asp	$C_\alpha C$ stretch	α-helix
957		$C_\alpha C$ stretch	random coil
990			
1003	Phe	CC ring stretch	
1032	Phe	in-plane ring bend	
1051	Glu, Arg, Pro	CH_2 twist (?)	
1059	Lys, Arg	CH_2 twist; $C_\alpha N$ stretch	
1081		$C_\alpha N$ stretch	
1101		$C_\alpha N$ stretch	
1127		$C_\alpha N$ stretch	
1157	Leu, Val; Phe, Tyr	CH_3 antisym. rock; CH rock	
1173	Leu, Val; Phe, Tyr	CH_3 antisym. rock; CH rock	
1208	Tyr, Phe		
1240		Amide III	β-sheet, random coil
1264		Amide III	α-helix
1281		Amide III	β-sheet
1303		Amide III	α-helix
1320		Amide III	α-helix
1340		CH bend, CH_2 twist	
1404	Asp, Glu	COO^- sym. stretch	
1425		NH bend	
1450		CH_3, CH_2, CH bend	

The pair of bands at 826 and 855 cm^{-1} are ascribed to a Fermi resonance doublet caused by the phenol side chain of tyrosine amino acids in the protein. Its intensity ratio, I_{855}/I_{826}, is sensitive to the environment (Carey, 1982; Tu, 1982). In particular, when the residue is buried within the protein, in

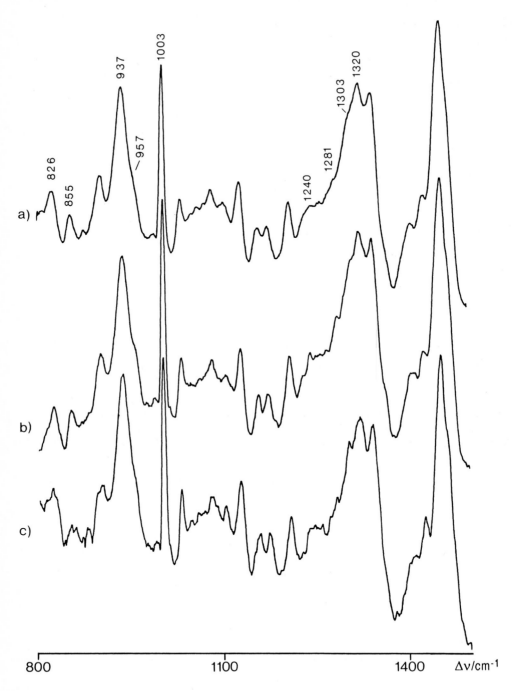

Figure - Raman spectra of myosin in three sequential stages: a) native myosin, b) partially denaturated myosin and c) partially denaturated myosin with monovanadate.

a hydrophobic region, the phenolic group is in the unionized form and the intensity ratio is high. On the other hand, when a tyrosine residue is on the surface of the protein, in aqueous solution, at basic pH, the phenolic hydroxyl group is ionized, causing a comparatively low intensity ratio (Siamwiza, 1975), as in the Raman spectrum of myosin-monovanadate (c).

In conclusion, the Raman spectrum of partially denatured myosin shows clear evidence of conformational changes upon denaturation, namely, decrease of α-helix structure and increase of random coil content. In addition, comparison of Raman spectra of partially denatured myosin without and with monovanadate points to a structure organizing effect of vanadate ions, as revealed by the increase of α-helix content.

4. References

Aureliano M Effect of LC2 Phosphorylation and Vanadium (V) Oligomers on Myosin-Actin Interaction Studied by Flow Microcalorimetry and [51]V-NMR (in Port.). MSc. Thesis, Coimbra (in preparation).

Carey P R (1982) Biochemical Applications of Raman and Resonance Raman Spectroscopies, Academic Press, New York London Paris San Diego San Francisco São Paulo Sydney Tokyo Toronto.

Chasteen N D (1983) The Biochemistry of Vanadium. In: Structure and Bonding, 53. Springer-Verlag, Berlin Heidelberg, p 105.

Goodno C C (1979) Inhibition of Myosin ATPase by Vanadate. Proc. Natl. Acad.Sci. 76: 2620-2624.

Pires E M V, Perry S V (1977) Purification and Properties of Myosin Light-Chain Kinase from Fast Skeletal Muscle. Biochem J 167: 137-146.

Siamwiza M N, Lord R C, Chen M C (1975) Interpretation of the Doublet at 850 and 830 cm[-1] in the Raman Spectra of Tyrosyl Residues in Proteins andCertain Model Compounds. Biochem 14: 4870-4876.

Tu A (1982) Raman Spectroscopy in Biology, John Wiley and Sons, New York Chichester Brisbane Toronto Singapore.

PHOSPHORYLATION AND THE FREQUENCY ENCODING OF SIGNAL-INDUCED CALCIUM OSCILLATIONS

Albert Goldbeter and Geneviève Dupont
Faculté des Sciences
Université Libre de Bruxelles
Campus Plaine, C.P. 231,
B-1050 Brussels, Belgium

1. Introduction

Most theoretical studies of phosphorylation-dephosphorylation cycles have been devoted to the case where the rates of the kinase and phosphatase evolve from one steady value to another as a result of the action of some effector such as cAMP. Thus, in the case of a kinase activated by cAMP, at a given value of the maximum phosphatase rate, a step increase in cAMP will result in the evolution of the level of phosphorylated protein to a new, higher stationary value. The mathematical analysis of monocyclic and muticyclic cascades of covalent modification has revealed their high potential for amplification and sensitivity (Stadtman and Chock, 1977, 1978; Goldbeter and Koshland, 1981, 1982, 1984, 1990; Cardenas and Cornish-Bowden, 1989). In particular, sigmoidal response curves whose steepness exceeds that of allosteric enzymes characterized by Hill coefficients larger than 10 can be generated when the converter enzymes become progressively saturated by their substrate — hence the name of "zero-order ultrasensitivity" given to this phenomenon (Goldbeter and Koshland, 1981) which has been demonstrated in several experimental systems (La Porte and Koshland, 1983; Meinke et al., 1986). Given that sharp thresholds are associated with all-or-none responses and thereby provide highly sensitive regulatory mechanisms, zero-order ultrasensitivity could play an important role in the control of cellular processes in both normal and pathological conditions.

While the monotonous switching of the kinase and phosphatase rates from one steady state to another represents the most common response to some external stimulation, transient or non-steady state phenomena might also be important. Thus a brief, transient rise in hormone might lead to the synthesis of a pulse of cAMP that will in turn elicit a burst in protein phosphorylation through the control of a cAMP-dependent kinase. On the other hand, oscillations in some intracellular effector could also affect phosphorylation. Such a situation might in fact be relatively common, given the wide occurrence of oscillations in intracellular Ca^{2+} which are triggered in a variety of cells by hormones or neurotransmitters (Woods et al., 1987; Berridge et al., 1988; Berridge and Galione, 1988). Through control of a protein kinase by Ca^{2+}, such oscillations would lead to periodic variations in the level of phosphorylation of one or more target proteins. Such a mechanism could provide a molecular basis for the frequency-encoding of signal-induced Ca^{2+} oscillations.

The purpose of this paper is to briefly discuss the possible relationship between phosphorylation and signal transduction based on intracellular Ca^{2+} oscillations. Following a previous study carried out in a model recently proposed for the oscillatory phenomenon (Goldbeter et al., 1990), we shall investigate the effect of oscillating levels of cytosolic Ca^{2+} on the activity of a Ca^{2+}-dependent kinase, as well as the dependence of the mean level of phosphorylated protein on the frequency of Ca^{2+} oscillations.

NATO ASI Series, Vol. H 56
Cellular Regulation by Protein Phosphorylation
Edited by L. M. G. Heilmeyer, Jr.
© Springer-Verlag Berlin Heidelberg 1991

2. Signal transduction based on calcium oscillations

In an increasing number of instances, it is found that hormones or neurotransmitters act on target cells by triggering a train of cytosolic Ca^{2+} spikes. The oscillations occur in a variety of cells, with periods ranging from less than 1 second to minutes (see Berridge et al., 1988, and Berridge and Galione, 1988, for recent reviews). The mechanism of Ca^{2+} oscillations appears to involve the synthesis of inositol 1,4,5-trisphosphate (IP3) that follows binding of the stimulatory ligand to the membrane receptor. According to Berridge, the rise in IP3 in turn triggers the release of a certain amount of Ca^{2+} from an IP3-sensitive intracellular store (Berridge and Galione, 1988; Berridge et al., 1988). The analysis of a two-variable model taking into account the Ca^{2+} input from the external medium, the extrusion of Ca^{2+} from the cell, and the exchange of Ca^{2+} between the cytosol and an IP3-insensitive store, shows that sustained oscillations may arise through a mechanism of Ca^{2+}-induced Ca^{2+} release from the IP3-insensitive store once the signal-induced rise in cytosolic Ca^{2+} triggered by IP3 has reached a sufficient level (Dupont and Goldbeter, 1989; Goldbeter et al., 1990).

The model accounts for the observation that the frequency of the Ca^{2+} spikes increases with the extent of stimulation; the latter is measured by the saturation function (β) of the IP3 receptor, given that the level of IP3 rises with the external signal. Also accounted for by this model is the observed linear correlation between the period of Ca^{2+} oscillations and their latency, i.e., the time required for the appearance of the first Ca^{2+} peak after stimulation (Dupont et al., 1990). In contrast with an alternative model based on a feedback of Ca^{2+} on IP3 production (Meyer and Stryer, 1988), a specific prediction of the analysis is that sustained oscillations in cytosolic Ca^{2+} may occur in the absence of a concomitant, periodic variation in IP3.

3. Calcium oscillations and phosphorylation

The simplest situation in which oscillations in cytosolic Ca^{2+} might affect phosphorylation is that where a Ca^{2+}-dependent protein kinase would be turned on and off, periodically, in the course of Ca^{2+} transients. Such a situation would likely give rise to periodic variations in the level of phosphorylated substrate, as schematized in Fig. 1. Considered in this scheme is a protein W phosphorylated by a kinase K into the form W* which can be dephosphorylated by a phosphatase P; kinase K is activated by cytosolic Ca^{2+}. To investigate in a qualitative as well as quantitative manner how oscillations in Ca^{2+} affect the phosphorylation of W, we resort to the model proposed for signal-induced Ca^{2+} oscillations, and couple its dynamics to that of a monocyclic phosphorylation system in which the V_{Max} of the phosphatase remains constant while the kinase is activated by cytosolic Ca^{2+}. The kinetic equations governing such a phosphorylation system controlled by Ca^{2+} oscillations are given as eqns (1)-(3) in (Goldbeter et al., 1990).

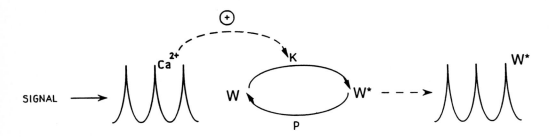

Fig. 1: Scheme for the control of phosphorylation by Ca^{2+} oscillations (see text).

When the kinetic equations of the model are integrated for parameter values yielding sustained Ca^{2+} oscillations, the amount of phosphorylated protein undergoes a periodic variation in the course of time (Fig. 2). The peak in the fraction of phosphorylated protein, W*, shortly follows the maximum of the Ca^{2+} spike, as expected from the fact that the kinase is activated by Ca^{2+}.

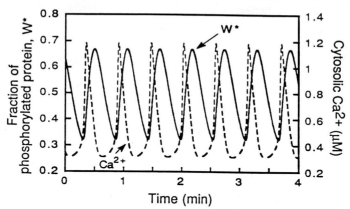

Fig. 2: Typical oscillations in phosphorylated protein (W*) accompanying Ca^{2+} oscillations in the phosphorylation system schematized in Fig. 1 (see Goldbeter et al. (1990) for details on the system of equations used for simulations).

4. Frequency encoding based on phosphorylation

The question arises as to how the mean level of phosphorylated protein, <W*>, varies with the characteristics of Ca^{2+} oscillations. In particular, how does <W*> depend on the frequency of Ca^{2+} transients, given that this frequency changes (while the amplitude of Ca^{2+} spikes remains largely constant) at different levels of stimulation?

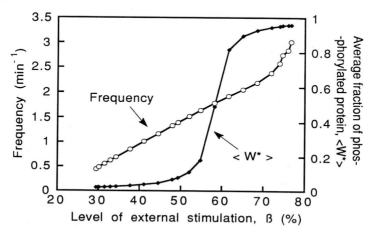

Fig. 3: Variation of the mean fraction of a putative protein phosphorylated by a Ca^{2+}-dependent kinase as a function of the stimulus level in the model for signal-induced Ca^{2+} oscillations based on Ca^{2+}-induced Ca^{2+} release. Also indicated is the variation of the frequency of Ca^{2+} oscillations (redrawn from Goldbeter et al., 1990).

The results of simulations shown in Fig. 3 indicate that the mean level of phosphorylated protein increases with the frequency of Ca^{2+} transients which rises in turn with the level of the stimulus. An efficient encoding depends, however, on the parameters of the kinase and phosphatase. In particular, the dependence of the mean fraction of phosphorylated protein on the frequency of Ca^{2+} spikes (and, hence, on the level of external stimulation) becomes steeper when the converter enzymes involved in covalent modification become saturated by the target protein. Therefore, the phenomenon of zero-order ultrasensitivity which provides increased responsiveness to a step change in the ratio of the kinase to phosphatase maximum rates also favors the efficient encoding of signal-induced Ca^{2+} oscillations in terms of their frequency.

5. Discussion

In many cell types, stimulation by a hormone or neurotransmitter triggers a train of Ca^{2+} spikes. The possibility of encoding these spikes in terms of their frequency has been raised by several authors (see, e.g., Rapp and Berridge, 1981; Woods et al., 1987; Berridge et al., 1988; Meyer and Stryer, 1988). The analysis of a minimal model shows that protein phosphorylation by a Ca^{2+}-dependent kinase provides a molecular mechanism for the frequency encoding of agonist-induced Ca^{2+} oscillations (Goldbeter et al., 1990).

The results of Fig. 3 indicate that the mean level of phosphorylated protein increases in a sigmoidal manner with the frequency of oscillations, i.e., with the intensity of the stimulus. These results were obtained in conditions of zero-order ultrasensitivity when changing the frequency of Ca^{2+} transients by increasing the level of stimulation measured by the saturation function β of the IP_3 receptor. Such a parameter change has in fact two effects: (i) an increase in the frequency of Ca^{2+} spiking, and (ii) an increase in the (unstable) steady-state level of cytosolic Ca^{2+}. These two effects combine to raise the mean level of protein phosphorylated over a period. To test the effect of an increase in frequency in the absence of any change in the steady-state level of cytosolic Ca^{2+} one can alter another parameter such as the threshold constant K_R that characterizes the release of Ca^{2+} from the IP_3-insensitive store (see Goldbeter et al., 1990). Results similar to those of Fig. 3 are obtained in such conditions, provided that Ca^{2+} ions control the kinase in a cooperative rather than Michaelian manner.

Encoding intracellular or extracellular signals by frequency rather than by the sole amplitude presents a number of advantages, e.g., with regard to accuracy (Rapp et al., 1981). Rhythmic, pulsatile signals abound in cell physiology, as exemplified by the large number of hormones which are delivered in a pulsatile rather than continuous manner (Knobil, 1981; Crowley and Hofler, 1987; Weigle, 1987). Desensitization in target cells has been proposed as a mechanism that may allow for the frequency encoding of both pulsatile hormonal signals and pulsatile signals of cAMP in *Dictyostelium* (Goldbeter and Li, 1989; Li and Goldbeter, 1989, 1990). In the latter organism, desensitization is mainly achieved through receptor phosphorylation (Vaughan and Devreotes, 1988). Thus, besides Ca^{2+} oscillations, the latter example provides a different illustration of the role of protein phosphorylation in the frequency encoding of signals in intercellular communication.

Acknowledgements: This work was supported by the Belgian National Incentive Program for Fundamental Research in the Life Sciences (Convention BIO/08), launched by the Science Policy Programming Services of the Prime Minister's Office (SPPS).

References

Berridge MJ., Galione A (1988) Cytosolic calcium oscillators. FASEB J 2:3074-3082

Berridge MJ, Cobbold PH, Cuthbertson KSR (1988) Spatial and temporal aspects of cell signalling. Phil Trans R Soc Lond B 320:325-343

Cardenas ML, Cornish-Bowden A (1989) Characteristics necessary for an interconvertible enzyme cascade to generate a highly sensitive response to an effector. Biochem J 257:339-345

Crowley WF, Hofler JG (eds) (1987) The episodic secretion of hormones. Wiley, New York

Dupont G, Goldbeter A (1989) Theoretical insights into the origin of signal-induced calcium oscillations. In: Goldbeter A (ed) Cell to cell signalling: from experiments to theoretical models. Academic Press, London, p 461

Dupont G, Berridge MJ, Goldbeter A (1990) Latency correlates with period in a model for signal-induced Ca^{2+} oscillations based on Ca^{2+}-induced Ca^{2+} release. Cell Regul 1: 853-861

Goldbeter A, Dupont G, Berridge MJ (1990) Minimal model for signal-induced Ca^{2+} oscillations and for their frequency encoding through protein phosphorylation. Proc Natl Acad Sci USA 87:1461-1465

Goldbeter A, Koshland DE Jr (1981) An amplified sensitivity arising from covalent modification in biological systems. Proc Natl Acad Sci USA 78:6840-6844

Goldbeter A, Koshland DE Jr (1982) Sensitivity amplification in biochemical systems. Quart Rev Biophys 15:555-591

Goldbeter A, Koshland DE Jr (1984) Ultrasensitivity in biochemical systems controlled by covalent modification: interplay between zero-order and multistep effects. J Biol. Chem. 262:4460-4471

Goldbeter A, Koshland DE Jr (1990) Zero-order ultrasensitivity in interconvertible enzyme systems. In: Cardenas ML, Cornish-Bowden A (eds) Control of metabolic processes. Plenum, New York, p 173

Goldbeter A, Li YX (1989) Frequency coding in intercellular communication.In: Goldbeter A (ed) Cell to cell signalling: from experiments to theoretical models. Academic Press, London, p 415

Knobil E (1981) Patterns of hormone signals and hormone action. New Engl J Med 305:1582-1583

LaPorte DC, Koshland DE Jr (1983) Phosphorylation of isocitrate dehydrogenase as a demonstration of enhanced sensitivity in covalent modification. Nature 305:286-290

Li YX, Goldbeter A (1989) Frequency specificity in intercellular communication: The influence of patterns of periodic signaling on target cell responsiveness. Biophys J 55:125-145

Li YX, Goldbeter A (1990) Frequency encoding of pulsatile signals of cAMP based on receptor desensitization in *Dictyostelium* cells. J Theor Biol 146:355-367

Meinke MH, Bishop JS, Edstrom RD (1986) Zero-order ultrasensitivity in the regulation of glycogen phosphorylase. Proc Natl Acad Sci USA 83:2865-2868

Meyer T, Stryer L (1988) Molecular model for receptor-stimulated calcium spiking. Proc Natl Acad Sci USA 85:5051-5055

Rapp PE, Berridge MJ (1981) The control of transepithelial potential oscillators in the salivary gland of *Calliphora erythrocephala*. J Exp Biol 93:119-132

Rapp PE, Mees AI, Sparrow CT (1981) Frequency encoded biochemical regulation is more accurate than amplitude dependent control. J Theor Biol 90:531-544

Stadtman ER, Chock PB (1977) Superiority of interconvertible enzyme cascades in metabolic regulation: analysis of monocyclic systems. Proc Natl Acad Sci USA 74:2761-2765

Stadtman ER, Chock PB (1978) Interconvertible enzyme cascades in metabolic regulation. Curr Top Cell Regul 13:53-95

Vaughan R, Devreotes PN (1988) Ligand-induced phosphorylation of the cAMP receptor from *Dictyostelium discoideum*. J Biol Chem 263:14538-14543

Weigle DS (1987) Pulsatile secretion of fuel-regulatory hormones. Diabetes 36:764-775

Woods NM, Cuthbertson KSR, Cobbold PH (1987) Agonist-induced oscillations in cytoplasmic free calcium concentration in single rat hepatocytes. Cell Calcium 8:79-100

II. METHODOLOGY

DETERMINATION OF PHOSPHORYLATED AMINO ACIDS IN PROTEIN SEQUENCES

Helmut E. Meyer, Edeltraut Hoffmann-Posorske, Horst Korte, Thomas R. Covey[1] and Arianna Donella-Deana[2]

Institut für Physiologische Chemie
Ruhr-Universität Bochum
D-4630 Bochum
Germany

INTRODUCTION

Identification of O-phosphorylated amino acids within the primary structure of a distinct protein is very important to understand the mechanism(s) by which the functions of proteins are regulated by this posttranslational modification. However, in many cases the only known way to do this is to label the protein with ^{32}P-phosphat. This procedure is often very tedious or sometimes impossible to achieve.

Therefore, we have established a series of new methods which permit the localization of phosphoserine, phosphothreonine and phosphotyrosine in the lower picomolar range *without* the need of radioactive labelling.

1. PHOSPHOSERINE

A highly specific method for determination of *phosphoserine* is the chemical modification to *S-ethyl-cysteine*. This amino acid derivative can be determined by amino acid or sequence analysis (Meyer et al. 1990, 1991a). Fig. 1 shows the amino acid analysis of a Lys-C-peptide from the beta subunit of phosphorylase kinase after modification of the endogenously present phosphoserine to S-ethyl-cysteine. The same peptide was analyzed by gas-phase sequence analysis (Fig. 2) after the same modification procedure. The chromatogram of degradation step 4 demonstrate the presence of S-ethyl-cysteine and identifies phosphoserine in this position of the sequence. The high specificity of this method is evident; false positive results from distinct cysteine residues which might react in the same manner can be prevented by prior oxidation of the protein or peptide.

---

[1]SCIEX, Toronto, Ontario, Canada; [2]Universita di Padova, Italy

NATO ASI Series, Vol. H 56
Cellular Regulation by Protein Phosphorylation
Edited by L. M. G. Heilmeyer, Jr.
© Springer-Verlag Berlin Heidelberg 1991

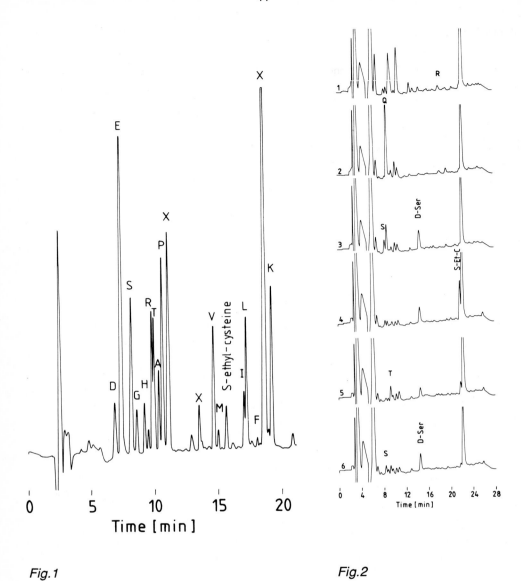

Fig.1

Fig.2

Fig. 1: **Amino acid analysis of a phosphoserine-containing peptide.** 50 pmol of the purified peptide was modified, hydrolyzed and coupled with phenylisothiocyanate. 75% of the resulting phenylthiocarbamyl-amino acids were analyzed. (SEC = S-ethyl-cysteine, X = unidentified impurity).

Fig. 2: **Sequence analysis of a phosphoserine-containing peptide following S-ethyl-cysteine modification.** 400 picomol of the oxidized and S-ethyl-cysteine-modified peptide is analyzed using the gas-phase sequencer. Phenylthiohydantoin-amino acid analyses of degradation steps 1 to 6 are shown.

2. PHOSPHOTHREONINE

Phosphothreonine can be determined after partial hydrolysis, as commonly used for phosphoamino acid analyses, and modification to its phenylthiohydantoin(PTH)-derivative. Capillary electrophoresis of this derivative allows detection at 261 nm in the low picimolar range. Fig. 3 shows the capillary electrophoresis chromatogram using 1 nmol of the phosphothreonine-containing peptide RRREEET(P)EEEAA for partial hydrolysis.

Non-radioactive sequence analysis of phosphothreonine-containing peptides is possible under distinct cirumstances as shown earlier (Dedner et al. 1988), however, a general method is not yet developed.

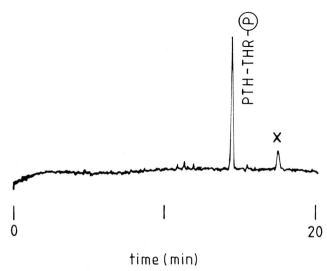

time (min)

Fig. 3: **Phosphothreonine determination by capillary electrophoresis.** 1 nmol of the phosphopeptide RRREEET(P)EEEAA was partially hydrolysed with 6 M HCl at 110°C for 2 h. The liberated amino acids were coupled with phenylisothiocyanate(PITC) and converted to their phenylthiohydantoin-derivatives. After drying, the samples are dissolved in 20 μl 20% aqueous acetonitrile containing 0.35% trifluoroacetic acid and 50 mM sodium chloride. 30 nl are analysed by capillary electrophoresis. Conditions for capillary electrophoresis: Temp. 30°C; buffer 20 mM sodium citrate, pH 2.5; capillary length 72 cm (50 cm to detector), 50 μm ID; voltage -25 kV negative polarity; detection at 261 nm.

3. PHOSPHOTYROSINE

Phosphotyrosine, like phosphothreonine, can be determined by capillary electrophoresis (Meyer et al. 1990b, 1991b). Fig. 4 shows (left site) the results from capillary electrophoresis of two different phosphotyrosine-containing peptides after partial hydrolysis and modification to their PTH-derivatives. Using the same conditions like before, non-radioactive determination of phosphotyrosine is possible. The detection limit of this method is 0.5 pmol/μl PTH-phosphotyrosine.

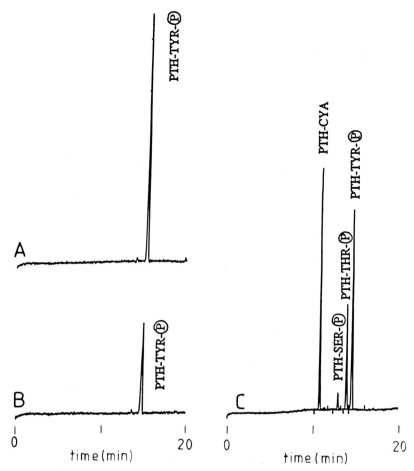

Fig. 4: **Determination of PTH-phosphoamino acids by capillary electrophoresis**. Left: 1 nmol of the peptides **L R R A Y-P L G** (A) or **D R V Y-P I H P F** (B) are partially hydrolyzed with 6 N HCl at 110°C for 2 h. The liberated amino acids were coupled with PITC and transformed to the more stable PTH-derivatives. After drying, the samples are dissolved in 50 μl 20% aqueous acetonitrile containing 0.35% trifluoroacetic acid and 50 mM sodium chloride. 30 nl are analyzed by capillary electrophoresis. Detection at 261 nm. Run time 20 min at -25 kV and 30°C, buffer 20 mM sodium citrate pH 2.5.
Right: 1 nmol of each, cysteic acid, phosphoserine, phosphothreonine and phosphotyrosine, are transferred to their PTH-derivatives. Sample is dissolved in 50 μl 20% aqueous acetonitrile. 30 nl are analyzed by capillary electrophoresis. Detection at 261 nm. Run time 20 min at -20 kV and 30°C, buffer 20 mM sodium citrate pH 2.5.

On the right hand of Fig. 4, the separation of three phosphoamino acids and of cysteic acid as their PTH-derivatives is shown. It demonstrate that capillary electophoresis is a powerfull technique for the determination of these negatively charged amino acids. In addition, under the applied conditions all other PTH-amino acids will not be detected because they a washed out from the capillary by the electroendosmotic flow. This is a real advantage, allowing the

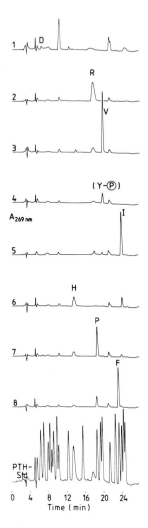

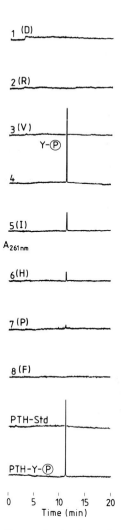

Fig. 5 **Sequence analysis of the peptide D R V Y(P) I H P F employing on-line reversed phase HPLC.** 1 nanomol of the phosphopeptide is taken for the attachment procedure. 47% of the peptide is covalently bound to the solid phase support. Reversed phase HPLC analyses are performed with 50% of the samples using the model 120 on-line PTH-analyser from Applied Biosystems. No signal is visible in degradation step 4 where PTH-phosphotyrosine is expected. Additionally, the chromatogram of the PTH-standard amino acids (62.5 pmol)is shown at the bottom. The other 50% of the sample is analysed by capillary electrophoresis (Fig. 6).

Fig. 6 **Capillary electrophoresis of PTH-amino acids of the peptide D R V Y(P) I H P F.** 50% of each degradation sample were transferred into 500 μl Eppendorf vials, dried under vacuum in a SpeedVac and dissolved in 20 μl of 20% aqueous acetonitrile. Conditions: injection time (vacuum), 10 sec.; temp. 30°C; capillary length 72 cm (50 cm to detector); ID 50 μm; voltage -25 kV negative polarity; buffer, 20 mM sodium citrate, pH 2.5; detector rise time, 1 sec.; detection at 261 nm, 0.008 A.U.F.S.

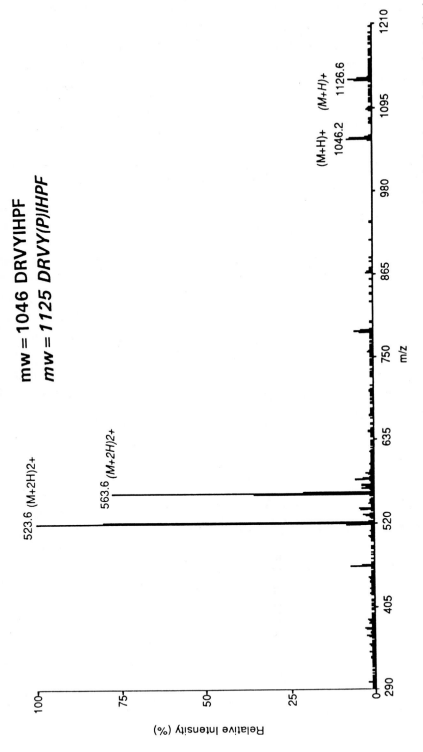

Fig. 7 **Ion spray mass spectrum of DRVYIHPF and DRVY(P)IHPF.** A 1:0.85 stoichiometric mixture of the dephospho- and phosphopeptide is dissolved in 50% aqueous methanol containing 0.1% formic acid. The concentration of the dephosphopeptide is 50 pmol/μl. 3 μl are flow injected into the ion spray mass spectrometer (SCIEX API III triple quadrupole equipped with a standard atmospheric pressure ion source). Almost identical signals are observed for the dephospho- or phosphopeptide.

determination of phosphotyrosine in high molecular weight proteins where phosphotyrosine is present in low stoiciometric amounts.

Non-radioactive sequence analysis of phosphotyrosine-containing peptides can be performed using solid-phase techniques (Meyer et al. 1991b). Fig. 5 shows the reversed-phase HPLC analyses of solid-phase sequence analysis of DRVY(P)IHPF. Butylchloride, normally used as transfer solvent, is replaced by trifluoroacetic acid; this allows the quantitative transfer of anilinothiazolinone-phosphotyrosine to the conversion flask. However, as seen in step 4 of Fig. 5, PTH-phosphotyrosine gives no sharp signal during reversed-phase HPLC. Here again, capillary electophoresis allows the positive identification of PTH-phosphotyrosine as demonstrated in Fig. 6 *without* radioactivity.

4. ION SPRAY MASS SPECTROMETRY OF PHOSPHOPEPTIDES

Complementary to the presented proteinchemical techniques, physicochemical techniques will help to eludicate the phosphorylation pattern of peptides and proteins. As an example, ion-spray mass spectrometry of a 1:0.85 stoichiometric mixture of DRVYIHPF/DRVY(P)IHPF is shown (Fig. 7). Contrary to fast atom bombardment mass spectrometry(MS) or plasma-desorption MS (Craig et al. 1991), ion-spray MS allows the determination of phosphopeptides quantitatively, since there is no selectivity against of phosphorylated species of the same unphosphorylated peptide. Over that, this technique will help to identify other phosphoamino acids, like phosphohistidine or phosphocysteine, which are very unstable, precluding application of conventional proteinchemical techniques.

CONCLUSION

Non-radioactive methods for phosphoamino acid analysis will help to study events of phosphorylation as they happens in the living cell. The presence of even small amounts of radioactivity may alter metabolic reactions in such a way that repair reactions prevail, disturbing the observed reactions and leading to wrong conclusions.

ACKNOWLEDGEMENTS

This work was financially supported by the Deutsche Forschungsgemeinschaft, the Ministerium für Wissenschaft und Forschung des Landes Nordrhein-Westfalen and the Fonds der Chemischen Industrie.

REFERENCES

Anthony G. Craig, Å. Engström, G. Lindeberg, H. Bennich, M. Serwe, E. Hoffmann-Posorske, H. Korte and H. E. Meyer, Plasma Desorption Mass Spectrometry in Monitoring Peptide Synthesis And Phosphorylation Reactions, Proceedings of the 8. MPSA Conference, Kiruna 1990, in press (1991)

Norbert Dedner, Helmut E. Meyer, Chris Ashton and Günter F. Wildner, N-terminal sequence analysis of the 8 kDa protein in Chlamydomonas reinhardii. Localization of the phosphothreonine, FEBS Lett., 236, 77-82 (1988)

Helmut E. Meyer, Gerhard F. Meyer, Herbert Dirks and Ludwig M. G. Heilmeyer, Jr., Localisation of Phosphoserine Residues in the α Subunit of Rabbit Skeletal Muscle Phosphorylase Kinase, Eur. J. Biochem., 188, 367-376 (1990a)

Helmut E. Meyer, Edeltraut Hoffmann-Posorske, Horst Korte, Arianna Donella-Deana, Anna-Maria Brunati, Lorenzo A. Pinna, Jim Coull, John Perich, Robert M. Valerio and R. Basil Johns, Sequence Analysis of Phosphotyrosine-Containing Peptides. Determination of PTH-Phosphotyrosine by Capillary Electrophoresis, Chromatographia, 30, 691-695 (1990b)

Helmut E. Meyer, Edeltraut Hoffmann-Posorske and Ludwig M. G. Heilmeyer, Jr., Determination and Location of Phosphoserine in Proteins and Peptides by Conversion to S-Ethyl-Cysteine. In: Methods in Enzymology, Volume 201 Protein Phosphorylation (T. Hunter and B. M. Sefton eds.) (1991a) in press

Helmut E. Meyer, Edeltraut Hoffmann-Posorske, Arianna Donella-Deana and Horst Korte, Nonradioactive Analysis of Phosphotyrosine-Containing Peptides. In: Methods in Enzymology, Volume 201 Protein Phosphorylation (T. Hunter and B. M. Sefton eds.) (1991b), in press

BACK-PHOSPHORYLATION – A SENSITIVE TECHNIQUE TO STUDY PROTEIN PHOSPHORYLATION IN THE INTACT HEART

Peter Karczewski
Division of Cellular and
Molecular Cardiology
Institute for Cardiovascular Research
Robert-Rössle-Straße 10
O-1115 Berlin
FRG

INTRODUCTION

Reversible protein phosphorylation is well established to be causally involved in numerous processes regulating cellular function. In cardiac muscle phosphoproteins as phospholamban and troponin I are thought to be of importance for regulation of contractility. The in vitro phosphorylation of phospholamban, a protein intrinsic to membranes of sarcoplasmic reticulum, increases the uptake of calcium ions by the sarcoplasmic reticulum (Tada et al. 1975). Phospholamban is a substrate for cAMP-dependent protein kinase (cAMP-PrK) (Tada et al. 1974), a calcium/calmodulin-dependent protein kinase (Le Peuch et al. 1979) and protein kinase C (Movsesian et al. 1984). Troponin I, the inhibitory subunit of the troponin complex, is localized in the myofilaments. Its phosphorylation in vitro by cAMP-PrK has been shown to reduce the calcium sensitivity of contractile proteins (Ray and England 1976).

To assess the physiological significance of protein phosphorylation/dephosphorylation one essential step is to demonstrate that the protein can be phosphorylated and dephosphorylated *in vivo* or in intact cell systems with accompanying functional changes (Krebs and Beavo 1979). To study protein phosphorylation in intact systems the mostly used method is to label the intracellular ATP pool with ^{32}P (for review see Manning et al. 1980). Working with isolated hearts this is achieved by perfusion of the whole organ with buffer solutions containing ^{32}P$_i$. Studies on beating intact cardiac preparations using the ^{32}P$_i$-prelabelling technique have shown that phospholamban (Lindemann et al. 1983) and troponin I (England 1976) undergo phosphorylation in response to ß-adrenergic stimulation. This procedure, however, allows only a limited differentiation between phosphorylation reactions catalysed by different types of protein kinases.

Firstly Forn and Greengard (1978) employed the principle of back-phosphorylation to describe qualitatively protein phosphorylation in brain tissue. In the intact cell the capacity to incorporate phosphate of a given protein is utilized to a degree depending on the activity of protein kinases and phosphoproteinphosphatases which are linked to phosphorylation/dephosphorylation of this protein. The indirect assay of the phosphorylation of a protein *in vivo* consists of two principal steps. The first comprises the fixation of the phosphorylation state of the protein present in the intact cell and the extraction of this protein under specific conditions. The second step is the back-phosphorylation *in vitro* of the remaining phosphorylation capacity of the protein using exogenous, specific protein kinase and [γ-^{32}P]ATP. In that way the phosphorylation state of the protein *in vivo* can be indirectly assessed.

In order to obtain quantitative data on cAMP-PrK-catalysed protein phosphorylation in the intact heart we standardized the back-phosphorylation technique. In the following methodical aspects are discussed and examples are given to illustrate sensitivity and potency of back-phosphorylation.

BASIC METHODICAL ASPECTS

In the myocardium changes in second messenger systems coupled to protein phosphorylation are very rapid. Increases in
the intracellular level of cAMP, activation of cAMP-PrK and
phosphorylation of target proteins are realized in a few
seconds in response to stimulation of ß-adrenergic receptors. To detect the phosphorylation state of a protein acu-
rately reflecting the situation in the intact cell it is
essential to block protein kinase and phosphoproteinphosphatase activities rapidly and completely during tissue fixation, tissue homogenization and tissue fractionation. Typically, heart samples are quick-frozen using Wollenberger
clamps precooled in liquid nitrogen. Clamping the tissue to
a thickness of 0.7 mm decreases sample temperature to -20 °C
in 120 milliseconds (Wollenberger et al. 1960).

Frozen heart samples of 300 - 500 mg are used to prepare
subcellular fractions. Homogenization buffer and all isolation media for further fractionation contained 10 mM EDTA,
25 mM NaF and 50 mM inorganic phosphate. These additions
have been shown to preserve sufficiently the phosphorylation state of proteins during membrane vesicle preparations
from guinea pig heart (Lindemann et al. 1983). In preliminary experiments under these conditions there was no significant dephosphorylation of phospholamban in preparations
from canine heart (Karczewski et al., unpublished results).

To study the cAMP-controlled phosphorylation state of
proteins in the intact heart muscle, the back-phosphorylation technique uses an excess of C-subunit of cAMP-PrK and
high specific radioactivity $[\gamma-^{32}P]$ATP. Assay conditions
were chosen which ensure a complete in vitro topping-up
phosphorylation of the protein studied. It was checked that
the concentration of ATP was kept at a level allowing the
added protein kinase to be fully active and that there was
no significant dephosphorylation of phosphoproteins for the
time of incubation. Under the standard conditions of back-

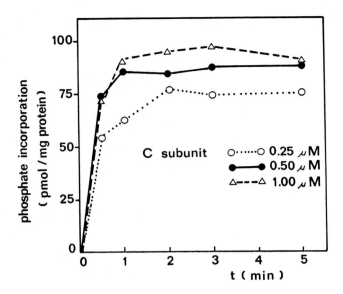

Fig. 1. Dependence of phospholamban phosphorylation *in vitro* on time of incubation and the amount of added C-subunit of cAMP-PrK. Phospholamban-containing crude membranes were prepared from isolated rat hearts perfused under control conditions (Karczewski et al. 1990)

phosphorylation more than 75% of the initial ATP remained after incubation for 20 min with cardiac fractions (Karczewski et al. 1990). In the presence of 1 mM EGTA and 15 mM NaF no significant dephosphorylation of phospholamban or troponin I was detectable up to 20 min of incubation. Fig. 1 shows an example for the dependence of back-phosphorylation of phospholamban in cardiac membranes on time and concentration of added protein kinase. At 0.5 µM C-subunit the phosphorylation of phospholamban reaches completion within 1 min. The amount of phosphate incorporated remained stable at least for 5 min. Thus in this case incubation for 2 min and 0.5 µM C-subunit was routinely used to back-phosphorylate cardiac particulate fractions containing phospholamban.

Electrophoretic separation of phosphorylated cardiac proteins and quantification of phosphate incorporation was performed using conventional methods (Karczewski et al. 1990).

PHOSPHORYLATION OF PHOSPHOLAMBAN IN CANINE HEART IN VIVO

It is well known that ß-adrenergic modulation of cardiac function is mediated by activation of the cAMP system involving cAMP-PrK-dependent phosphorylation of several regulatory proteins (Tada and Katz 1982). In the sarcoplasmic reticulum phospholamban is the predominant substrate for cAMP-PrK. The back-phosphorylation allows to detect cAMP-specific phosphorylation and was employed to study these events in canine heart *in vivo*. It was of interest to look for the magnitude of changes in phospholamban phosphorylation produced by maximal stimulation of the cAMP system.

An extremely low level of ß-adrenergic activity was achieved by treating whole animals with reserpine to deplete hearts from catecholamines. To increase intracellular cAMP through activation of adenylate cyclase high doses of the ß-adrenergic agonist isoprenaline were injected directly into the left ventricle of the myocardium (Karczewski et al. 1986). Comparing hearts from both experimental groups an about 6-fold increase of cAMP and a two-fold rise in the activity of cAMP-PrK indicated a strong reaction in hearts exposed to isoprenaline (Karczewski et al. 1987).

In membrane fractions prepared from the same hearts significant differences were found for $^{32}P_i$ incorporation into phospholamban back-phosphorylated with the C-subunit of cAMP-Prk *in vitro*. Fig. 2 shows a typical autoradiograph of back-phosphorylated cardiac membranes which obviously indicates a decreased incorporation of $^{32}P_i$ into phospholamban as result of isoprenaline exposure. Comparing quantitative data there was a four-fold higher $^{32}P_i$ incorporation *in vitro* into phospholamban in preparations from hearts treated with reserpine compared with hearts exposed to isoprenaline (Tab. 1).

Now the question arose whether the differences obtained by back-phosphorylation *in vitro* were based on different amounts of phosphate bound to phospholamban in the analyzed

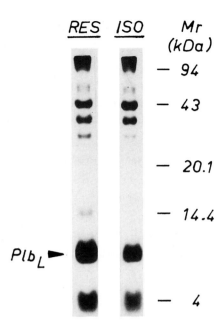

Fig. 2. Autoradiograph of cardiac membrane fractions back-phosphorylated with exogenous C-subunit of cAMP-Prk. Membrane fractions were prepared from hearts of reserpine-treated dogs (RES) and from dog hearts exposed to isoprenaline (ISO). Under the conditions used for electrophoretic separation of membrane proteins phospholamban migrates as low molecular monomer (PlbL) at about 6 kDa indicated by the arrow (Karczewski et al. 1987)

preparations. Therefore membrane fractions from hearts treated with reserpine and isoprenaline were incubated with a phosphoproteinphosphatase. The phosphatase used was shown to remove more than 90 per cent of phosphate bound to phospholamban (Karczewski et al., unpublished results). When phosphatase-treated membranes were back-phosphorylated the differences in phosphate incorporation between phospholamban from reserpine-treated hearts and from hearts exposed to isoprenaline were completely abolished (Tab. 1). Thus back-phosphorylation indeed reflects differences in the phosphorylation state of the protein present in the intact heart.

Apparently there was a loss of phosphorylatable phospholamban in cardiac membranes after treatment with the phosphoproteinphosphatase. Using an EIA based on polyclonal antibodies against phospholamban to assess semi-quantitatively the membrane content of this protein (Holtzhauer 1987) the data were corrected for losses of immuno-reactive phospholamban (Tab. 1). The corrected values are in the range of $^{32}P_i$ incorporation *in vitro* of hearts treated with reserpine. This indicates a very low level of cAMP-controlled phosphorylation of phospholamban in these hearts which was found to be significantly below the level in hearts of untreated control animals (Karczewski et al., unpublished results).

These data clearly show that back-phoshorylation detects specific changes in cAMP-PrK-catalysed phosphorylation of phospholamban in canine heart *in vivo* with high sensitivity.

Table 1. Capacity for cAMP-dependent phosphorylation of phospholamban in membrane fractions prepared from dog hearts treated with reserpine (RES) and exposed to isoprenaline (ISO)

	$^{32}P_i$ incorporation (pmol/mg of protein)		
		P-ase-S-treated	corr. value
RES	221.8 ± 15.5[a]	114.2 ± 16.4[a]	175.0 ± 18.3[b]
ISO	50.1 ± 8.9[a]	93.9 ± 8.8[a]	172.8 ± 14.5[b]
RES/ISO	4.4	1.2	1.0

Values are means for three hearts ± S.E.M.
[a] Karczewski et al. 1986
[b] Karczewski et al., unpublished results

DOSE–RESPONSE RELATIONSHIPS OF PHOSPHOLAMBAN AND TROPONIN I PHOSPHORYLATION IN PERFUSED RAT HEARTS

Studies on intact myocardium with the $^{32}P_i$-prelabelling technique gave strong evidence that cAMP-controlled phosphorylation of regulatory proteins, such as phospholamban in the sarcoplasmic reticulum (Lindemann et al. 1983), troponin I in the myofilaments (England 1976) and the 15 kDa protein in the sarcolemma (Presti et al. 1985), may explain at least a part of catecholamine effects on the heart.

In order to obtain more detailed information on cAMP-dependent protein phosphorylation in intact beating cardiac preparations, a study of dose-response relationships of

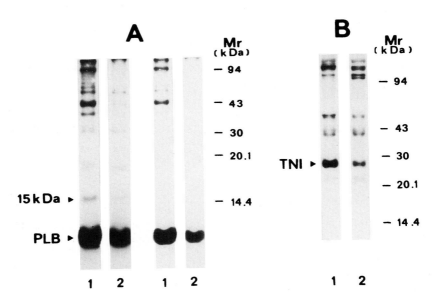

Fig. 3. Autoradiographs of cardiac membrane fractions (A) and troponin I (TNI)-containing soluble fractions (B) back-phosphorylated with exogenous C-subunit of cAMP-PrK. Fractions were prepared from control (1) and isoprenaline perfused rat hearts (2). The two sets of lanes with cardiac-membranes (A) represent the same gel exposed for 16 h (right) and 36 h (left) to show differences in phosphorylation of both phospholamban (PLB) and the 15 kDa protein (Karczewski et al. 1990)

Table 2. Capacity for cAMP-dependent phosphorylation of phospholamban (PLB), a 15 kDa protein and troponin I (TNI) in cardiac fractions before and after exposure to isoprenaline (Karczewski et al. 1990)

Condition	$^{32}P_i$ incorporation (pmol/mg of protein of the tissue fraction)		
	PLB	15 kDa protein	TNI
Control	102.1 ± 3.7 (17)	6.6 ± 0.3 (15)	74.2 ± 3.6 (11)
Isoprenaline (0.1 µM)			
after 0.5 min	38.9 ± 2.0 (5)	3.9 ± 0.3 (4)	38.8 ± 3.0 (4)
after 2 min	39.6 ± 5.4 (4)	3.3 ± 0.1 (3)	42.4 ± 2.9 (4)

Membrane fractions and contractile-proteins-containing soluble fractions were preparaed from control and from isoprenaline-perfused hearts (exposure time 0.5 and 2 min). Means ± S.E.M. are given for the numbers of experiments in paratheses.

phospholamban and troponin I was undertaken. For this purpose isolated rat hearts were exposed to increasing concentrations of isoprenaline and quick-frozen in the maximum of contractile response. In these hearts there was a dose-related increase in the tissue cAMP content from 3.5 pmol/mg of protein in controls to 14.8 pmol/mg of protein in hearts exposed to 10^{-6} M isoprenaline (Karczewski et al. 1990).

The phosphorylation of phospholamban and troponin I was assayed in subcellular fractions isolated from the same hearts. Fig. 3 presents typical autoradiographs of back-phosphorylated membranes and contractile-proteins-containing fractions of hearts under control perfusion and with exposure to isoprenaline clearly showing a decreased incorporation of $^{32}P_i$ into both phospholamban and troponin I in response to the drug. The phosphorylation of a membrane-bound 15 kDa protein, which may be identical to the protein described to

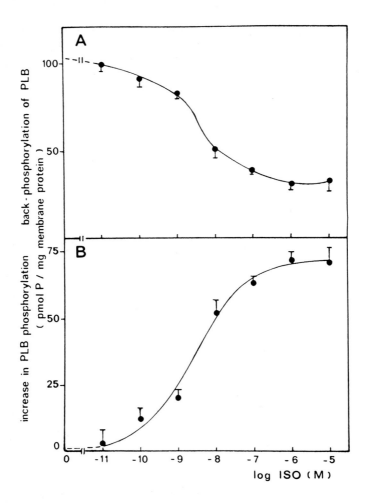

Fig. 4. Dose-response relationship of isoprenaline on phosphorylation of phospholamban in the intact rat heart.
(A) $^{32}P_i$ incorporation into phospholamban obtained *in vitro*
by back-phosphorylation of membranes derived from control
hearts was 102.1 ± 3.7 pmol/mg of protein (n=17), as given
in table 2. After exposure of hearts to increasing concentrations of isoprenaline, $^{32}P_i$ incorporation declined in a
dose-dependent manner. (B) P_i incorporation into phospholamban occuring in response to isoprenaline in the intact heart
during perfusion. These values are the calculated differences of the mean value of $^{32}P_i$ incorporation into phospholamban in membrane fractions from control hearts and the values
obtained in preparations from each individual heart stimulated with isoprenaline. Each value is the mean from 3-5 individual hearts; the error bars represent 1 S.E.M.
(Karczewski et al. 1990)

be localized in the sarcolemma (Presti et al. 1985), was
found to be diminished too. Tab. 2 summarizes the quantitative data obtained by back-phosphorylation of cardiac fractions from control and isoprenaline-treated hearts. The values for phospholamban, the 15 kDa protein and troponin I
represent the amount of cAMP-PrK-phosphorylatable sites in
the respective proteins that are not phosphorylated in the
intact heart but are accessible to exogenously added cAMP-
PrK *in vitro*. The phosphate incorporation expressed in pmol
of phosphate / mg of membrane or supernatant protein decreases significantly after exposure to isoprenaline. This decrease in the topping-up value for the proteins studied is
completed within 30 seconds after the onset of isoprenaline
perfusion of the hearts.

When hearts were perfused with increasing concentrations
of the stimulating agent, the decrease in $^{32}P_i$ incorporation
occured in a dose-dependent manner as shown for phospholamban in Fig. 4A. Thus the back-phosphorylation technique gives quantitative data on graduated responses of cAMP-depen-
dend protein phosphorylation in intact cardiac tissue stimulated with the ß-adrenergic agent. The differences in phosphate incorporation in vitro in preparations from stimulated
hearts relative to control perfusion are equal to the preceding phosphorylation of the protein in the intact heart.
Therefore these differences are used for calculating the
amount of phosphate incorporated into the protein by cAMP-
PrK *in vivo*. Plotting these values versus the concentration
of isoprenaline, a typical dose-response curve is obtained
(Fig. 3B). By using mathematical models values for isoprenaline concentrations producing halfmaximal phosphorylation
($K_{0.5}$ value) of the proteins *in vivo* were calculated. This
is useful in order to compare dose-response relationships
for the phosphorylation of different proteins (Tab. 3).

When the dose-response curves to isoprenaline are compared, there is in the intact heart about a 20-fold higher
sensitivity of phosphorylation of troponin I towards the
ß-adrenergic stimulation compared with phosphorylation of

Table 3. Calculated values for isoprenaline concentrations producing half-maximal phosphorylation ($K_{0.5}$ value) of phospholamban, the 15 kDa protein and troponin I. (Karczewski et al. 1990)

	$K_{0.5}$ value (nM)
phospholamban	2.94 ± 0.04
15 kDa protein	4.46 ± 0.24
troponin I	0.13 ± 0.01

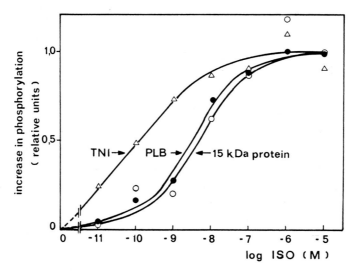

Fig. 5. Dose-response relationships of isoprenaline on phosphorylation of phospholamban (PLB), the 15 kDa protein and troponin I (TNI) in the intact rat heart. The values for phosphorylation of each phosphoprotein studied represent means from preparations of 3-5 individual hearts and are expressed relative to the calculated maximal response of phosphorylation obtained by fitting the dose-response curves to a non-linear model. (Data taken from Karczewski et al. 1990)

membrane-bound phospholamban and the 15 kDa protein (Tab. 3, Fig. 5). These data may be explained by assuming a functional compartmentation of the cAMP signal cascade in the heart (Brunton et al. 1981; England and Shahid 1987; Karczewski et al. 1990).

CONCLUSIONS

Reversible phosphorylation of relevant proteins appears to be an essential step in the regulation of cardiac contractiltity by hormones and drugs. To demonstrate the physiological occurence of these reactions in the intact heart, the $^{32}P_i$-prelabelling method has been extensively used. This technique needs high activities of $^{32}P_i$ in the perfusion medium to achieve high specific labelling of the intracellular ATP. Acurate quantitative measurements rely upon attainment and maintenance of isotopic equilibrium. No indication is given of which protein kinases are responsible for the phosphorylation of the protein under investigation.

The back-phosphorylation does not require radioactive labelling of the whole heart. The specific activity of $[\gamma-^{32}P]$ATP can be adjusted according to the needed sensitivity of the assay.

The presented standardized back-phosphorylation procedure allows the quantitative detection of phosphorylation of proteins catalysed specifically by cAMP-PrK. This is of importance in order to study primary responses in phosphorylation of selected proteins induced by activation of second messenger systems as demonstrated for ß-adrenergic stimulation thought to be mediated through cAMP.

The back-phosphorylation technique is suitable to study dose-response relationships of phosphorylation of relevant proteins, as shown for cardiac phospholamban and troponin I, which can be usefull to obtain information on functional compartmentation of the cell.

By using other types of protein kinases, as e.g. cGMP-dependent protein kinase, Ca^{2+}/calmodulin-dependent protein kinase and protein kinase C, further progress can be expected in the understanding of the interplay of intracellular signalling pathways involving protein phosphorylation.

REFERENCES

Brunton LL, Hayes JS, Mayer SE (1981) Functional compartmentation of cyclic AMP and protein kinase in heart. Adv Cyclic Nucl Res 14:391-398

England PJ (1976) Studies on the phosphorylation of the inhibitory subunit of troponin during modification of contraction in perfused rat heart. Biochem J 160:295-304

England PJ, Shahid M (1987) Effects of forskolin on contractile responses and protein phosphorylation in the isolated perfused rat heart. Biochem J 246:687-695

Forn J, Greengard P (1978) Depolarizing agents and cyclic nucleotides regulate the phosphorylation of specific neuronal proteins in rat cerebral cortex slices. Proc Natl Acad Sci USA 75:5195-5199

Holtzhauer M (1987) An enzyme immunoassay for phospholamban. J Clin Chem Clin Biochem 25:799-804

Karczewski P, Vetter R, Holtzhauer M, Krause E-G (1986) Indirect technique for the estimation of cAMP-dependent and Ca^{2+}/calmodulin-dependent phospholamban phosphorylation state in canine heart in vivo. Biomed Biochim Acta 45: 227-231

Karczewski P, Bartel S, Krause E-G (1987) Isoproterenol induces both cAMP- and calcium-dependent phosphorylation of phospholamban in canine heart in vivo. Biomed Biochim Acta 46:433-439

Karczewski P, Bartel S, Krause E-G (1990) Differential sensitivity to isoprenaline of troponin I and phospholamban phosphorylation in isolated rat hearts. Biochem J 266: 115-122

Krebs EG, Beavo JA (1979) Phosphorylation-dephosphorylation of enzymes. Annu Rev Biochem 48:923-959

LePeuch CJ, Haiech J, Demaille JG (1979) Concerted regulation of cardiac sarcoplasmic reticulum calcium transport by cyclic adenosine monophosphate-dependent and calcium-calmodulin-dependent phosphorylations. Biochemistry 18: 5150–5157

Lindemann JP, Jones LR, Hathaway DR, Henry BG, Watanabe AM (1983) ß-adrenergic stimulation of phospholamban phosphorylation and Ca^{2+}-ATPase activity in guinea pig ventricle. J Biol Chem 258:464–471

Manning DR, DiSalvo J, Stull JT (1980) Protein phosphorylation: quantitative analysis in vivo and in intact cell systems. Mol Cell Endocrinol 19:1–19

Movsesian MA, Nishikawa M, Adelstein RS (1984) Phosphorylation of phospholamban by calcium-activated, phospholipid-dependent protein kinase. J Biol Chem 259:8029–8032

Presti CF, Jones LR, Lindemann JP (1985) Isoproterenol-induced phosphorylation of a 15-kilodalton sarcolemmal protein in intact myocardium. J Biol Chem 260:3860–3867

Ray KP, England PJ (1976) Phosphorylation of the inhibitory subunit of troponin and its effect on the Ca^{2+} dependence of cardiac myofibril ATPase activity. FEBS Lett 70:11–16

Tada M, Katz AM (1982) Phosphorylation of the sarcoplasmic reticulum and sarcolemma. Annu Rev Physiol 44:401–423

Tada M, Kirchberger MA, Katz A (1975) Phosphorylation of a 22,000-dalton component of the cardiac sarcoplasmic reticulum by adenosine 3',5'-monophosphate dependent protein kinase. J Biol Chem 250:2640–2647

Wollenberger A, Ristau O, Schoffa G (1960) Eine einfache Technik der extrem schnellen Abkühlung größerer Gewebestücke. Pflügers Arch 270:399–412

THERMAL TRANSITIONS IN CARDIAC TROPONIN AND ITS SUBUNITS

L.A. Morozova, V.L. Shnyrov, N.B. Gusev*, E.A. Permyakov
Institute of Biophysics
USSR Academy of Sciences
Pushchino, Moscow Region
USSR

Introduction

Cardiac troponin (Tn) takes part in the Ca^{2+}-regulation of the cardiac muscle contraction. It consists of three subunits : troponin C (TnC)-Ca^{2+}-binding protein, troponin I (TnI)-inhibitor of actomyosin ATPase and troponin T (TnT)-tropomyosin binding protein. However, it is still far from clear which changes occur in the conformation of subunits upon their interaction with each other. The parameters of thermal transitions of proteins reflect the changes in their structure. In the present work we studied the thermal denaturation of Tn subunits and their binary and ternary complexes in the presence and absence of Ca^{2+} by means of intrinsic fluorescence and microcalorimetry.

Methods

Tn and its subunits were isolated from bovine heart muscles (Tsukni and Ebashi, 1973; Perry and Cole, 1974; Wilkinson, 1974). The formation of the binary complexes was checked by native electrophoresis at room temperature. The experiments were performed on a differential adiabatic scanning microcalorimetry DACM-4 (USSR) and on a laboratory made spectrofluorimeter. The fluorescence was registered from the first surface of the cuvette. Since TnC has no tryptophan residues, the excitation of tryptophan fluorescence at 297 nm permits to monitor separately the unfolding of the tryptophan containing

* Moscow State University,
  Moscow USSR

NATO ASI Series, Vol. H 56
Cellular Regulation by Protein Phosphorylation
Edited by L. M. G. Heilmeyer, Jr.
© Springer-Verlag Berlin Heidelberg 1991

components (TnI and TnT) in the complexes. The excitation of both tryptophan and tyrosine fluorescence at 280 nm for Tn complex was also used. The heating rate was 1 K/min.

Results and discussion

The Ca^{2+}-binding stabilizes the TnC structure shifting its symmetrical heat sorption peak from 65°C up to 82°C (Table 1). At the same time a cooperative denaturation transition of the individual TnI and TnT (Morozova et al. 1988) and their binary complex TnIT has not been registered up to 100°C, neither by calorimetric nor fluorescence methods. A small inclination of the temperature dependence of fluorescence intensity at 360 nm (or relative quantum yield) from a common fluorescence quenching curve at temperatures above 40°C, indicates only the presence of an uncooperative structural rearrangement in TnT and TnIT upon heating. So, it is reasonable to conclude that the individual TnI, TnT and their binary complex TnIT have no stable, highly ordered structure in solution. In contrast, the complexes TnIC and TnTC exhibit a micro-calorimetric cooperative thermal transition (Fig. 1) both in the presence and absence of Ca^{2+}, and the position of the

Table 1

Denaturation temperatures (T_d) and calorimetric enthalpies (ΔH_{cal}) of the troponin thermal denaturation.

Protein	2 mM CaCl$_2$		3 mM EDTA	
	T_d, °C	ΔH_{cal} KJ/mole	T_d, °C	ΔH_{cal} KJ/mole
TnC	82	460	65	420
TnIC	103	130	55	190
TnTC	113	–	51	190
Tn	57	230	51	220

partial heat sorption peaks of these complexes are shifted
drastically to the higher temperatures in the presence of Ca^{2+}
and to the lower temperatures in the absence of Ca^{2+} in
comparison with the individual TnC (Table 1). At the same
time, the temperature dependence of tryptophan fluorescence

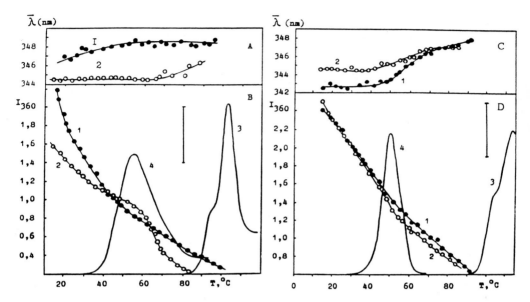

Fig. 1. Thermal denaturation of TnIC (A,B) and TnTC (C,D).
A,C - fluorescence spectrum position; B,D - fluorescence intensity at 360 nm (1,2) and partial heat
sorption (3,4). 1,3 - 2 mM $CaCl_2$; 2,4 - 3 mM EDTA.
Excitation at 297 nm; 10 mM Hepes, pH 7. 1 mg/ml of
each of the protein. Vertical bar is 100 J/K.kg.

intensity at 360 nm of TnT in TnTC complex only slightly
deflect from a common fluorescence quenching curve at 50-70°C
both in the presence and absence of Ca^{2+}. It correlates with a
concomitant long wavelength shift of the spectrum. A similar
effect on the plots of the fluorescence intensity versus
temperature of TnI in TnIC complex was detected at 60-80°C in
the presence of Ca^{2+} and at 45-70°C in the absence of Ca^{2+},
nevertheless it does not coincide with a long wavelength shift
of the spectrum. So, the fluorescence data indicate that TnI
and TnT in the correspondent binary complexes undergo the
Ca^{2+}-dependent structural transition upon heating, however

these transitions cannot be considered as cooperative ones. It is worthy to note that the values of transition enthalpy, ΔH_{cal}, for the binary complexes are rather low (Table 1). ΔH_{cal} was calculated taking into account the masses of both proteins but the main contribution into the heat sorption, perhaps gives TnC. So, it seems that the temperature shift of the heat sorption peaks of the binary complexes in comparison with individual TnC reflects mainly the stabilization or destabilization of TnC structure in the presence or absence of Ca^{2+}, respectively.

In the case of the whole Tn both calorimetric heat sorption peaks and fluorescence effects, such as the peculiarities on the temperatures dependencies of the tryptophan and tyrosine fluorescence intensity and the long wavelength shift of the

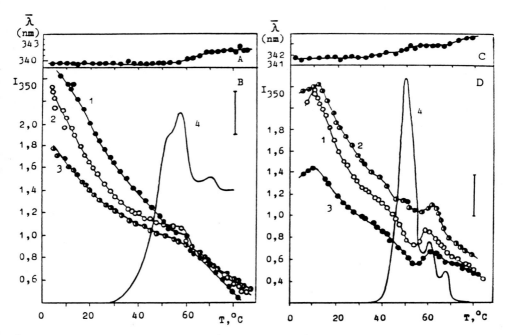

Fig. 2. Thermal denaturation of Tn in the presence of 2 mM $CaCl_2$ (A,B) and in the presence of 3 mM EDTA (C,D). (A,C) - fluorescence spectrum position; B,D - fluorescence intensity at 360 nm (1,2), 310 nm (3) and partial heat sorption (4). Excitation at 297 nm (1) or at 280 nm (2,3). 10 mM Hepes, pH 7. Protein concentration: 1 mg/ml. Vertical bar is 50 J/K.kg.

spectrum, coincide well with each other in temperature scale (Fig. 2). These data indicate the presence of the cooperative unfolding transitions in all subunits of Tn. It seems that the flexible structure of individual TnI and TnT allows them to adjust to each other and to TnC and to gain more stable folded conformation. It can be clearly seen from the complex curves of the temperature dependencies of both heat capacity and tryptophan and tyrosine fluorescence intensities that Tn unfolding passes through several stages corresponding to the melting of the distinct domains. However, it is impossible to ascribe the definite domain to the individual subunit. The Ca^{2+}-binding by Tn induces the complex structural rearrangements in it as it is followed from the changes of the shape of the heat sorption curves and the fluorescence intensities temperature dependencies. Nevertheless, the thermal transition occurs almost in the same temperature region both for Ca^{2+}-loaded and apo-Tn. So, the Ca^{2+}-binding does not cause such pronounced stabilization of Tn structure compared to TnC.

The Tn system is a good model for investigation of the conformational changes in protein upon complex formation.

References

Morozova L.A., Gusev N.B., Shnyrov V.L., Permyakov E.A. (1988) Investigation of the physicochemical properties of troponin I an T from cardiac and skeletal muscles by methods of protein fluorescence and calorimetry (in Russian), Biokhimiya 53 : 531-540

Perry S.V., Cole H.A. (1974) Phosphorylation of troponin and the effect of the interactions between the components of the complex. Biochem. J. 141: 733-743

Tsukui R., Ebashi S. (1973) Cardiac troponin. J. Biochem. 7: 1119-1121

Wilkinson J.M. (1974) The preparation and properties of the components of troponin B. Biochim. Biophys. Acta 359 : 379-388.

DYNAMIC PHOSPHORYLATION OF A SMALL CHLOROPLAST PROTEIN EXHIBITING SO FAR UNDESCRIBED LABELLING PROPERTIES

J. Soll, A. Steidl, I. Schröder
Fachrichtung Botanik der Universität des Saarlandes
D-6600 Saarbrücken

A 19 kDa phosphoprotein from mixed envelope membranes of spinach chloroplasts with extreme labelling kinetics has been characterized. Its localization between the inner and the outer envelope membrane can be deduced by the differential labelling between intact and broken chloroplasts (Table 1), (Soll and Bennett 1988, Soll et al. 1989).

Table 1
Differential labelling of proteins from intact and lysed chloroplasts. Phosphorylation was done at 10 nM ATP at 4 °C for 30 sec. Values are expressed in dpm μg chlorophyll^{-1} x min^{-1}

	intact chloroplasts	lysed chloroplasts	ratio
thylakoid/LHCP	53	125	0.424
stroma a	12	110	0.109
stroma b	11	100	0.11
outer envelope (86 kDa)	63	16	3.9
intermembrane space (64 kDa)	98	54	1.8
19 kDa protein	2340	375	6.24

If the 19 kDa protein is indeed localized in the envelope lumen, intact and broken chloroplasts should differ in their labelling kinetics. Intact chloroplasts still contain residual, endogenous ATP, this means that during labelling of intact chloroplasts two ATP pools exist, with different specific activity which is encountered by envelope membrane proteins and another with low specific activity which is encountered by proteins inside the chloroplast. In intact chloroplasts the outer envelope 86 kDa protein, the 64 kDa and the 19 kDa protein were labelled much earlier and stronger than the stromal and thylakoid phosphoproteins.

The function of the interenvelope space is unknown; the dynamically phosphorylated 19 kDa protein has been purified from mixed envelope membranes (Table 2) and characterized. It seems reasonable that it participates in a signal transduction process. The first purification step was a mild sonication followed by anion exchange chromatography on DEAE cellulose of the supernatant (Table 2). Most of the protein eluted at 125 mM NaCl. Active protein fractions were pooled and purified further on a hydroxylapatite column from which it could be eluted at 60 mM phosphate buffer pH 7.6.

Table 2
Purification of spinach envelope 19 kDa protein

Step	Volume ml	Protein μg	Total units	Specific activity units/mg	Reco- very %	Purifi- cation fold
Envelope membranes	1	6950	63.2	9.1	100	1
Sonication supernatant	0.89	1510	55	36.4	87	4.6
DEAE chromato- graphy	5.6	8.9	28.5	3200	45	352
Hydroxyl- apatite	4.5	0.54	4.4	8161	7	897

1 unit equals 1 fmol ^{32}P incorporated into the 19 kDa protein from $[\gamma-^{32}P]$-ATP x min^{-1}

The 19 kDa protein shows an extreme affinity for ATP and GTP as demonstrated by the low K_m values of 8 nM and 5 nM for ATP and GTP respectively (Fig. 1 A). The phosphorylation, that is trichloracetic acid or acetone precipitable, is dependent on the presence of divalent cations (Mg^{2+} and Mn^{2+}) (Fig. 1 B).

The cation Ca^{2+} has no effect. ADP and GDP inhibit phosphorylation (Fig. 1 D). The optimal pH for phosphorylation is in the range between pH 7 and pH 9 (Fig. 1 C). The pI of the phosphorylated enzyme has the value 6.2, whereas the pI of the non phosphorylated enzyme is 6.3.

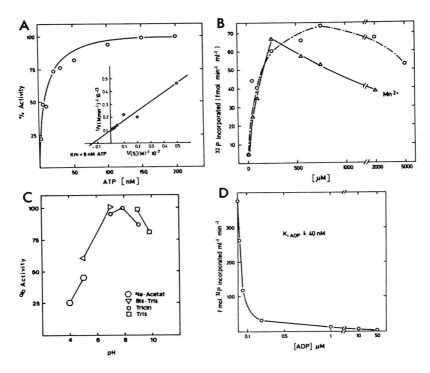

Fig. 1) Characterization of the phosphorylation reaction of the partially purified 19 kDa protein. A) Determination of the Km value for ATP. B) Influence of divalent cations. C) pH dependence. D) The phosphorylation is inhibited by ADP.

The molecular weight of the phosphorylated protein was estimated by SDS gel electrophoresis and found to be 18.8 kDa (Fig. 2 A). The phosphoryl turnover is extremely rapid, as deduced from a pulse-chase experiment. If the protein was labelled in the presence of 8 nM $[\gamma-^{32}P]$-ATP for 60 sec and 10 μM cold ATP was added at this time point, 90 % of the labelled phosphoryl-groups in the protein are turned over within 15 sec.

The determination of the phosphorylated amino acid residue demonstrated that no hydroxylated amino acid was phosphorylated, firstly the phosphate bond was labile to acidic conditions;

secondly after acid hydrolysis of the protein and high voltage electrophoresis no radioactivity was detectable in P-Ser, P-Thr or P-Tyr (Soll et al. 1989). Extraction of the phosphorylated protein by chloroform methanol at pH 1 resulted in no detectable label in the organic solvent phase, but the total radioactivity was still bound to the protein. Exposure of the phosphoprotein to hydroxylamine or pyridine buffered in acetate showed a concentration dependent base catalyzed enhancement of the hydrolysis rate (Fig. 2 B) and excluded most likely aspartate and glutamate as phosphorylgroup acceptor, as those are not susceptible to pyridine treatment (Sabato and Jencks 1961,Hokin et al. 1965). The label was also labile at alkaline pH (Stelte and Witzel 1986). At the moment it seems most likely that we deal with a lysine or histidine phosphate. The phosphorylation of the 19 kDa protein is inhibited by TNP-ATP and by erythrosin (Fig. 2 C,D).

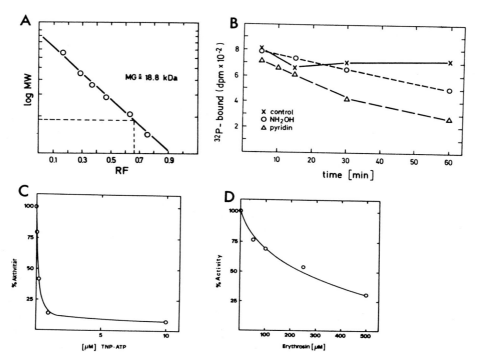

Fig. 2) A. Determination of the molecular weight of the phosphorylated form of the purified 19 kDa protein by SDS-PAGE B. Time course of hydrolysis of the phosphate bond in 1 M acetate buffer pH 5.5 in the presence of 0.1 M hydroxylamin or pyridin. C. Inhibition of the ^{32}P-incorporation by TNP-ATP. D. Inhibition of 19 kDa protein phosphorylation by erythrosine

The purified 19 kDa protein did not show significant ATPase activity (not shown). These findings are corroborated by results (Table 3), which demonstrate the effect of various ATPase inhibitors on the phosphorylation of the 19 kDa protein (Sze et al. 1987, Serrano 1988, Sze and Randall 1987). From further experiments it seems likely that ^{32}p-incorporation into the 19 kDa protein is due to autophosphorylation.

Table 3
Inhibition of 19 kDa protein phosphorylation by various substrates. The purified protein was phosphorylated by [γ-^{32}P]-ATP in the presence of different effectors. A minimum of five different effector concentrations was used in every case.

effector	max.concentration	% inhibition
NaN$_3$	10 mM	0
NaF	20 mM	0
NaF/AlCl$_3$	10 mM/50 μM	0
ortho vanadate	0.5 mM	29
molybdate	2.0 mM	0
nitrate	10 mM	0
oligomycin	0.5 mM	0
DCCD[1]	0.5 mM	0
ouabain	125 μM	0
dihydroxyacetonephohate	2 mM	60
NaCl	150 mM	50

1) The purified protein was preincubated with DCCD for to 2h.

Acknowledgements:
This study was supported by the DFG (SFB 184).

Literature
Hokin, L.E., Sastry, P.S., Galsworthy, P.R. & Yoda, A. (1965)Proc.Natl.Acad.Sci. USA 54, 177-184.
Sabato, G.D. & Jencks, W.P. (1961) J.Am.Chem.Soc. 83, 4393-4400.
Serrano, R. (1988) Biochim.Biophys. Acta 947, 1-28.
Soll, J. & Bennett, J. (1988) Eur.J.Biochem. 175, 301-307.
Soll, J., Berger, V. & Bennett, H.J. (1989) Planta (Berl.) 177, 393-400.
Stelte, B. & Witzel, H. (1986) Eur.J.Biochem. 155, 121-124.
Sze, H., Randall, S.K., Kaestner, K.H. & Lai, S. (1987) in Plant Membranes: Structure, Function Biogenesis, pp. 195-207, eds.C.Leaver & H. Sze, Alan Liss Inc.
Sze, H. & Randall, S.K. (1987) J.Biol.Chem. 262, 7135-7141.

IDENTIFICATION OF PHOSPHORYLATION SITES IN THE NICOTINIC ACETYLCHOLINE RECEPTOR BY EDMAN DEGRADATION AND MASS SPECTROSCOPY LC/MS AND LC/MS/MS

Werner Schroeder[1], Helmut E. Meyer[2], Thomas Covey[3], Klaus Buchner[1], Klaus Eckart[4] and Ferdinand Hucho[1]

[1] Institut für Biochemie
Freie Universität Berlin
Thielallee 63
1000 Berlin 33
Germany

INTRODUCTION

The nicotinic acetylcholine receptor (nACHR) is a heteropentameric glycoprotein transducing signals at cholinergic synapses. Its quaternary structure is $\alpha_2\beta\gamma\delta$ with the α-subunits containing the binding sites for cholinergic agonists and antagonists. Signal transduction occurs upon opening of an ion channel by agonist binding. This channel has been identified as being formed by homologous helical sequences contributed by the subunits (Helix M2 model; Hucho et al. 1989). The nACHR is a phosphoprotein. The physiological role of phosphorylation of the receptor is not clear. Phosphorylation/dephosphorylation has been implied in subunit assembly during development (Changeux 1981) or rapid receptor regulation (desensitization; Huganir et al. 1987). The latter hypothesis has been questioned when slow desensitization was shown to occur in nACHR from which phosphorylation sites have been removed by site directed mutagenesis (Hoffmann et al. 1989). Here we show that at least the δ-subunit is stably and not transiently phosphorylated. This finding supports a role of phosphorylation in long term events rather than in short term regulation.

Three protein kinases have been shown to be able to phosphorylate the nACHR from Torpedo californica electric tissue, a cAMP dependent protein kinase (PKA; Huganir et al. 1984), PKC (Safran et. al. 1987) and a tyrosine kinase (Huganir et al. 1984). They have been predicted to phosphorylate the receptor subunits at a total of seven potential sites (Huganir et al. 1987). Most of the sites are predicted to be phosphorylated only on the basis of the substrate specificity of the respective kinase. By direct localization through protein sequencing only γSer 353 and δSer 361 were shown to be phosphorylated. So far all attempts to localize

---

[2] Institut f. Physiol. Chemie , Ruhruniversität , Bochum , Germany
[3] SCIEX, Division of MDS Health group Ltd., 55 Glen Cameron road, Thornhill, Ontario, Canada
[4] MPI f. Experim. Medizin, 3400 Göttingen, Germany

NATO ASI Series, Vol. H 56
Cellular Regulation by Protein Phosphorylation
Edited by L. M. G. Heilmeyer, Jr.
© Springer-Verlag Berlin Heidelberg 1991

phosphorylation sites in the primary structure addressed ^{32}P-phosphate groups introduced in vitro by means of added kinase. In the present investigation we could show that the δ-subunit in our receptor preparation is already phosphorylated to a high stoichiometry even in the absence of phosphatase inhibitors.

RESULTS

We isolated the receptor as described previously (Schiebler et al. 1978) and separated the subunits by preparative gel electrophoresis. First we determined the total number of phosphate groups in various subunits. We found as previous authors up to four phosphate groups in the δ-subunit (3.7 ± 0.22; Schroeder et al. 1991). This value was obtained by the method of Ames (1966). Further we determined the content of phosphoserine residues by derivatizing the δ-subunit with thioethane according to the method of Meyer et al. (1986). In this method the phosphorylated serine residues were converted to S-ethyl-cysteine (SEC) by ß-elimination/addition mechanism which is easily detectable in the amino acid analysis and stable during Edman degradation. The amount of SEC in the amino acid analysis was in several determinations 2.5-3 Mol/Mol δ-subunit. In addition we determined the yield of derivatization with phosphokemptide. Calibration was performed with authentic SEC and with the appropriate internal standard.

We localized phosphoserine residues in the δ-subunit by Edman sequencing. For this we digested the separated δ-subunit with endoproteinase Lys-C and Glu-C, separated the peptides by HPLC (Vydac C18) and screened each fraction for phosphoserine containing peptides by OPA amino acid analysis with the method of Meyer et al. (for chromatogram see Schroeder et al. 1991).

One major peak containing SEC was found in the Lys-C digest. Edman degradation revealed the sequence Leu_{357}-Lys_{368} but unexpectedly this peptide was found to be phosphorylated not only in position $δSer_{361}$ but also in position $δSer_{362}$. Cleavage of purified δ-subunit with endoproteinase Glu-C yielded another SEC containing peptide starting at position Tyr_{372}. This contains Ser_{377}, a site predicted to be phosphorylated by PKC. This site is not detectable in the Lys-C digest because of its N-terminal position. The SEC method does not work with N-terminal phosphoserine residues because upon ß-elimination of the phosphate group pyruvate is formed and no thioethane can be added.

We tried to identify the protein kinase responsible for the phosphorylation of the novel phosphorylation site $δSer_{362}$ by in vitro phosphorylation of the synthetic peptide Leu_{357}-Lys_{368} with purified PKC and commercial PKA catalytic subunit. It was shown previously that PKA phosphorylates the δ-subunit at position Ser_{361} (Yee et al. 1987) and that PKC phosphorylates a synthetic peptide $δAsp_{354}$-Ile_{367} at an unidentified site (Safran et al. 1987).

Phosphorylation of our synthetic peptide Leu_{357}-Lys_{367} with PKA yielded in the HPLC two components, the monophosphorylated peptide phosphorylated in position Ser_{361} and the diphosphorylated peptide phosphorylated at position Ser_{361} and Ser_{362}. Phosphorylation with PKC yielded only one derivatized peptide phosphorylated in position δSer_{362} (Schroeder et al. 1991).

Identification of phosphopeptides by LC/MS and LC/MS/MS

In addition to this investigation we tried to identify phosphopeptides of the δ-subunit of the nACHR by ion spray LC/MS and LC/MS/MS on a Sciex API III triple quadrupole mass spectrometer. The MS was connected on-line with an HPLC. First we tested with phosphokemptide (LRRAS(P)LG) the stability of serine phosphoester residues (Fig. 1).

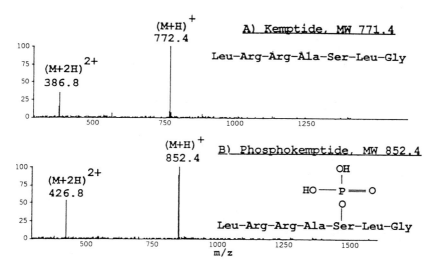

Fig.1 Ion evaporation mass spectrum of 40 picomoles of kemptide and phosphokemptide.

Subsequently we attempted to identify phosphopeptides in the Lys-C digest of the nACHR δ-subunit. Each peak of the total ion current (profile of the eluant detected by MS, Fig. 2) was subjected to further mass analysis (Schroeder et al. 1990). The only peptides of the δ-subunit matching the masses of 1512 and 1432 Da detected in the mass spectrum of one the HPLC peaks is the mono and doubly phosphorylated δ-peptide LRRSSSVGYISK (Fig. 3). In the mass spectrum of a second HPLC peak (Fig. 4) we identified the mono and doubly charged ions corresponding to the masses of 1207.5 (1205) and 1091.5 (1092) Da. Again only the phosphorylated Lys-C peptides SRSELMFEK ($\delta377$-$\delta385$) and AQEYFNIK ($\delta369$-$\delta376$) match these masses. The former peptide contains the predicted PKC substrate site and the latter the tyrosine kinase site.

Further investigations by ion spray LC/MS/MS show that phosphopeptides with a phospho-serine residues can be identified in the daughter ion spectrum by the neutral loss of H_3PO_4 which occurs upon fragmentation in the collision cell (shown with kemptide and phosphokemptide, Fig. 5). A neutral loss function in the tandem MS was developed to screen for all peptides which give rise to a loss of H_3PO_4 (from the doubly charged parent ion loss of 49 Da was observed and not 98 Da). By this method we identified again the mono and doubly phosphorylated δ-peptide Leu_{357}-Lys_{367}. The total ion current of these two peptides indicated that the monophosphorylated peptide is a minor component.

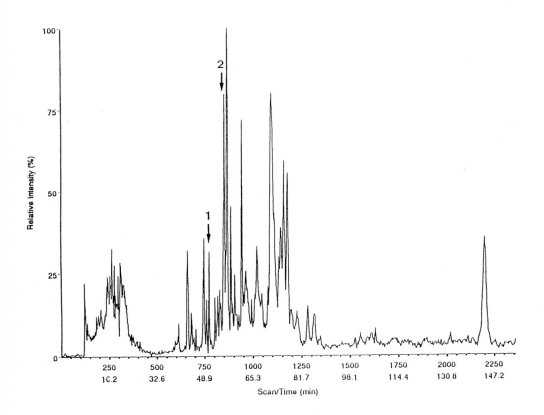

Fig.2 Total ion current of all masses. Lys-C digest of the nicotinic acetylcholine receptor (δ-subunit); amount injected 300 pmol, on-line from HPLC (ABI Aquapore C18, 1 mm x 10 cm), flow rate 40 μL/min, linear gradient: 0-5 min 100 % A, 5-160 min to 70 % B, A: H_2O, 0.1% TFA, B: acetonitrile, 0.1 % TFA. Arrows indicate peptides later identified as phosphopeptides (see Fig. 3 and 4)

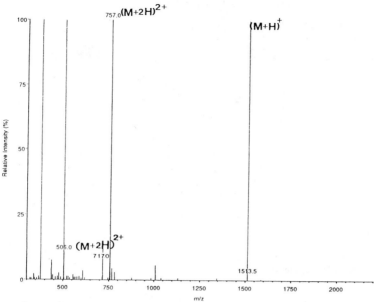

Fig.3 Mass spectrum of peak 1 of Fig. 2. The spectrum shows mono and doubly charged molecular ions of peptide δL_{357}-K_{368}. The molecular mass corresponds to the mass of the peptide plus two phosphate groups. The minor component $(M+2H)^{2+}$ 717.0 represents the same but monophosphorylated peptide. The other peaks were not identified.

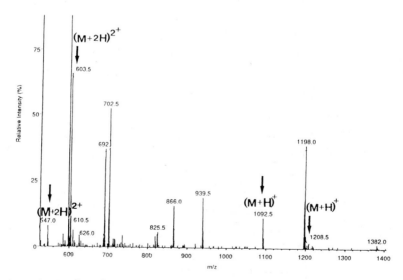

Fig.4 Mass spectrum of peak 2 of Fig. 2. Two phosphorylated peptides were identified: m/z 1092.5 represents the mono charged molecular ion of the phosphorylated peptide δA_{369}-K_{376}. m/z 547 represents the corresponding doubly charged ion. These peaks indicate that the δ-subunit was phosphorylated at tyrosine 372. m/z 12o8.5 and 603.5 represent the mono and doubly charged ions of δS_{377}-K_{385} plus one phosphate group. The peptide identified comprises the site phosphorylated by PKC.

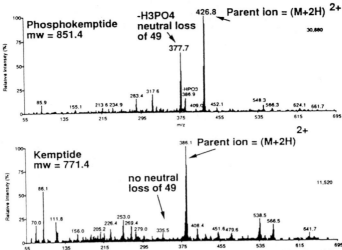

Fig.5 Daughter ion sprectrum of phosphokemptide and kemptide obtained by collision induced fragmentation. The fragmentation of phosphokemptide is dominated by the loss of phosphate.

CONCLUSION

We have shown by two independent methods (Edman sequencing and MS) that the predicted phosphorylation sites of the nACHR δ-subunit are stably (and not transiently) phosphorylated to near stoichiometry. This refers to the sites phosphorylated by PKA, PKC and tyrosine kinase. Phosphorylation of the latter was deduced from the difference in the number of phosphate groups obtained with the Ames and SEC method. The stably phosphorylated tyrosine kinase site was furthermore identified by MS. A hitherto unpredicted phosphorylation in position δSer$_{362}$ was detected both by Edman sequencing of the SEC derivatized peptide and by MS. Both methods seem to be complementary. Concerning the possible function of the receptor phosphorylation these findings support a developmental rather than a short term regulatory role.

Acknowledgements: We gratefully acknowledge H.Bayer's expert technical assistance. This work was supported by grants from the Deutsche Forschungsgemeinschaft (SFB 312) and the Fonds der Chemischen Industrie.

REFERENCES
Ames BN (1966) Methods Enzymol. 8:115-118
Changeux JP (1981) Harvey Lect. 75:85-254
Hoffmann P, Choi KL, Kienker P, Yellen, G and Huganir RL (1989) Society Neuroscience
    Abstracts Vol. 15:85-254
Hucho F and Hilgenfeld R (1989) FEBS Lett. 257:17-23
Huganir RL and Greengard P (1983) Proc. Natl. Acad. Sci. 80:1130-1134
Huganir RL, Miles K and Greengard P (1984) Proc. Natl. Acad. Sci. 81:6968-6972
Huganir RL and Greengrad P (1987) TIPS 8:472-477
Meyer HE, Hoffmann-Posorske E, Korte H and Heilmeyer LMG (1986) FEBS Lett. 204:61-66
Safran A, Sagi-Eisenberg R, Neumann D and Fuchs S (1987) J. Biol. Chem. 262:10506-10510
Schiebler W and Hucho F (1978) Eur. J. Biochem. 85:55-63
Schroeder W, Covey T and Hucho F (1990) FEBS Lett. 273:31-35
Schroeder W, Meyer HE, Buchner K, Bayer H and Hucho F (1991) Biochemistry, in press
Yee GH and Huganir RL (1987) J. Biol. Chem. 262:16748-16753

A SENSITIVE COLORIMETRIC ASSAY
FOR PROTEIN PHOSPHATASE ACTIVITY

T.P. Geladopoulos, T.G. Sotiroudis[1] and A.E. Evangelopoulos

Institute of Biological Research, The National Hellenic Research Foundation,
48 Vassileos Constantinou Avenue, Athens 116 35, Greece

Phosphorylation of proteins is a posttranslational modification of major significance in cellular regulation (Krebs & Beavo 1979). Protein kinases catalyze the addition to, and phosphatases the removal of phosphate groups from serine, threonine, tyrosine and other amino acid residues (Boyer & Krebs, 1986). We present here a simple and sensitive colorimetric assay for protein phosphatase activity based on the determination of released Pi by an improved malachite green procedure (Baykov et al. 1988). The method is described in detail elsewhere (Geladopoulos et al. 1991).

There is a number of methods for measuring Pi release in protein phosphatase assays (Weller 1979). The most sensitive technique involves acid precipitation of a protein substrate containing bound [32P]phosphate and determination of the released radioactivity in the supernatant. This procedure is handicapped by restrictions for working with radioactivity and it cannot distinguish between the release of acid-soluble ^{32}P-labeled peptides through proteolysis and the release of free [32P]Pi by the phosphatase. On the other hand in non-radioisotopic phosphatase assays, the measurement of released Pi, using the formation of molybdenum blue by the reduction of phospho-molybdate, although simple, lacks sensitivity (Buss & Stull 1983).

Highly sensitive assays for Pi have relied upon formation of a colored complex between the acidified molybdate and the dye malachite green (Itaya & Ui 1966). The sensitivity of this procedure is 30 times greater than the widely used Fiske-SubbaRow method (Bus & Stull 1983). Recently a modified malachite green procedure has been reported (Baykov et al. 1988). This

---

[1] To whom correspondence should be addressed

NATO ASI Series, Vol. H 56
Cellular Regulation by Protein Phosphorylation
Edited by L. M. G. Heilmeyer, Jr.
© Springer-Verlag Berlin Heidelberg 1991

method takes advantage of the high solubility and stability of the malachite green in the presence of 6N H_2SO_4. Furthermore, all necessary reagents are combined in one concentrated solution making the assay more sensitive and convenient (Baykov *et al.* 1988). In order to extend this sensitive method for Pi determination as a protein phosphatase assay, we had to overcome the interference due to protein precipitation that usually occurs.

In the experiments presented in this paper the malachite green method for Pi determination of Baykov *et al.* (1988) was adapted to measure protein phosphatase activity.

Deproteinization and Protein Stabilization

A. H_2SO_4 and $HClO_4$ Treatment

We used several proteins and protein containing biological materials as standards in order to establish methods suitable for deproteinization or protein stabilization in acid solution prior to the malachite green Pi assay. All proteins examined (albumin, protamine, phosphorylase *b*, phosvitin, and brain homogenate), were either removed and/or stabilized by treating the sample with H_2SO_4 (0.25 N), or with $HClO_4$ (3% w/w), so that no interference with subsequent Pi determination was observed. Deproteinization with sulfuric or perchloric acid was performed by adding 1 volume of 2.25 N H_2SO_4 or 70% (w/w) $HClO_4$ to 8 or 22 volumes of phosphatase reaction mixture, respectively. After standing for 10 min at $0°C$, the mixture was centrifuged for 3 min in an Eppendorf (type 5414S) centrifuge and the supernatant was removed for phosphate determination (see below). In the original method (Baykov *et al.* 1988) the color reaction is completed within 10 min. However, the increase of acidity in the final Pi assay mixture drastically reduces the rate of color development (more than tenfold). In our effort to overcome the retardation effect of overacidification we found that an increase in the final ammonium molybdate concentration up to 3% (w/v), highly accelerates color formation, so that the reaction is almost complete within 1 min, even in presence of excess H_2SO_4 or $HClO_4$.

A mixture of casein peptides (prepared by trypsinolysis of casein) was also soluble in the final Pi assay system, up to 0.1 mg/ml, without affecting color formation.

B. SDS/KCl Treatment

Taking into account that there is always the possibility of acid hydrolysis of labile phosphate during deproteinization with strong acids, we established an alternative procedure for protein removal or stabilization at neutral pH. When we first denature the proteins with SDS (2.5%) at neutral pH and then remove the excess of detergent by precipitation with KCl (1M) (SDS gives intense color reaction with malachite green reagent), proteins examined do not interfere with Pi determination (phosvitin was the only exception).

Phosphate Determination

The malachite green reagent was prepared and Pi was determined as described by Baykov *et al.* (1988) except that the final reagent solution contained 3% (w/v) ammonium molybdate.

Standard plots obtained in the presence of the mentioned proteins after treatment with H_2SO_4, $HClO_4$ or with SDS/KCl were essentially identical with those obtained in the absence of protein. Moreover, standard curves obtained by any of the above treatments were linear up to 4 nmol Pi.

Application of the method to the determination of protein phosphatase activity

Based on the previous results, the malachite green micromethod was applied in order to determine protein phosphatase activities. Using the H_2SO_4-treatment procedure, we were able to follow the alkaline phosphatase-catalyzed release of Pi from phosvitin.

Furthermore, it was also possible to determine glycogen phosphorylase phosphatase activity in freshly prepared bovine brain extracts. Standard phosphorylase phosphatase activity measurements were performed by determining the release of trichloroacetic acid-soluble radioactivity from ^{32}P-labelled phosphorylase *a* (Shenolikar & Ingebritsen 1984). The comparison of the time course of dephosphorylation determined colorimetrically with that monitored by following the release of [32P]Pi, clearly showed that the results obtained by both methods are almost the same. To our knowledge this is the first non-radioactive procedure applied for monitoring the enzymatic release of Pi from phosphorylase *a*, a large homodimer with only one phosphorylated residue in each polypeptide chain (97.400 Da).

In conclusion, using the most sensitive colorimetric procedure for Pi determination reported so far (Baykov *et al.* 1988, Van Veldhoven & Mannaerts 1987), we were able to establish a method for determination of protein phosphatase activities which is non-radioactive, simple, non-expensive and sensitive enough to detect 0.2 nmol Pi/sample. This method permits to use very low concentrations of substrate and can be applied to a variety of phosphorylated substrates. It also allows the determination of Pi released by phosphatase from soluble, low Mr phosphopeptides in an one step procedure and most important to assay the enzymatic dephosphorylation of biologically important protein-bound endogenous phosphate when it is not possible to find suitable conditions or the particular protein kinase necessary for the radiolabeling of the specific site(s).

Finally, we must outline certain disadvantages of our method: (a) Multisite dephosphorylation of polyphosphorylated protein substrates would complicate the interpretation of kinetic data. In this respect, the use of monophosphorylated peptides or protein substrates is advantageous.
(b) The use of complex biological samples as a source of protein phosphatase, may produce large blank values because of enzymatic Pi release from endogenous phosphoproteins or from free and protein-associated (noncovalently) phosphate-containing compounds. In this case, one must select an appropriate method for the efficient removal of such phosphocompounds.

REFERENCES

Baykov AA, Evtushenko OA, Avaeva SM (1988) Anal Biochem 171:266-270

Boyer PD, Krebs EG (eds) (1986) The Enzymes, vol XVII, Academic Press Inc

Buss JE, Stull JT (1983) Meth Enzymol 99:7-14

Geladopoulos TP, Sotiroudis TG, & Evangelopoulos AE (1991) Anal Biochem *(in press)*

Itaya K, Ui M (1966) Clin Chim Acta 14:361-366

Krebs EG, Beavo JA (1979) Ann Rev Biochem 48:923-959

Shenolikar S, Ingebritsen TS (1984) Meth Enzymol 107:102-129

Van Veldhoven PP, Mannaerts GP (1987) Anal Biochem 161:45-48

Weller M (ed.) (1979) Protein Phosphorylation, Pion Limited, London

III. STRUCTURE - FUNCTION RELATIONSHIP

PROTEIN KINASE STRUCTURE & FUNCTION: cAMP-DEPENDENT PROTEIN KINASE

S. S. Taylor, W. Yonemoto, W. R. G. Dostmann, D. L. Knighton, J. M.
Sowadski, F. W. Herberg, J. A. Buechler and Y. Ji-Buechler
Department of Chemistry, 0654
University of California, San Diego
9500 Gilman Drive
La Jolla, CA 92093-0654

Protein phosphorylation is a major mechanism for regulation in
eukaryotic cells, and the protein kinases represent a large and very
diverse family of enzymes. Nearly all major metabolic pathways are
regulated at some step by phosphorylation. In addition, protein kinases
play critical roles in mitogenesis, in cell cycle events, and in many types
of oncogenesis. One of the first protein kinases to be discovered was
cAMP-dependent protein kinase (cAPK) (Walsh DA, et al., 1968). In the
intervening decades, the family has grown to well over 100. These
enzymes are complex and differ in terms of their size, subunit structure,
subcellular localization, and mechanism of activation. cAPK, however,
remains as one of the simplest. Furthermore, despite the diversity of
the kinases, all share a conserved catalytic core that is included within
the free catalytic (C) subunit of cAPK (Hanks SK, et al., 1988). Thus, the
C-subunit can serve as a framework for the entire family (Figure 1).

    With the exception of the oncoproteins, most protein kinases are
maintained in an inhibited state in the absence of an activating signal.
In the case of the C-subunit, inactivation is achieved by association with
a regulatory (R) subunit. The inactive holoenzyme is an R_2C_2 tetramer.
cAMP binds with a high affinity to the R-subunit thus promoting
dissociation into an $R_2(cAMP)_4$ dimer and 2 free and active C-subunits.
It is this mechanism of activation that contributes significantly to the
simplicity of cAPK, for, unlike the other kinases, the major regulatory
entity can be readily removed and studied independently (for reviews
see (Beebe SJ and Corbin JD, 1986;Bramson HN, et al., 1984;Taylor SS,
et al., 1990)).

NATO ASI Series, Vol. H 56
Cellular Regulation by Protein Phosphorylation
Edited by L. M. G. Heilmeyer, Jr.
© Springer-Verlag Berlin Heidelberg 1991

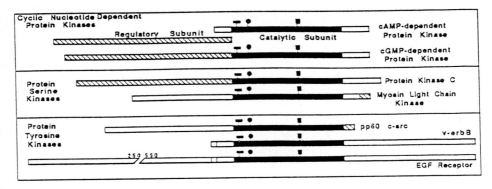

Figure 1: Sequence alignment of protein kinases. The conserved catalytic core is shown in black. Invariant residues are Asp (■), Lys (●), and Gly 50, 52, and 55 (−). M = N-myristoyl. Stippled = membrane spanning segment. Regulatory regions (crosshatched) are not conserved.

Understanding the relationship between structure and function requires a number of diverse techniques. Protein chemistry has provided insights into specific functional sites and residues. Some of the key features associated with the C-subunit are summarized in Figure 2 and are discussed in detail in two recent reviews (Taylor SS, et al., 1990;Taylor SS, et al., 1990). Localization of the ATP binding site near the amino-terminus, for example, initially was based on affinity labeling with fluorosulfonyl benzoyl 5'-adenosine (FSBA) which modified Lys72 (Zoller MJ and Taylor SS, 1979). Differential labeling with a group specific reagent can provide a more global picture of the protein, but can also be used to localize functional sites. This approach was utilized for the C-subunit by labeling with acetic anhydride, a small lysine-specific reagent that should have access to solvent-accessible side chains. Group specific labeling with acetic anhydride confirmed that the N-terminal region of the C-subunit was protected by MgATP (Buechler JA, et al., 1989). Lys47 and Lys72, in particular, were very reactive in the absence of MgATP, but protected from modification with acetic anhydride in the presence of MgATP. A conserved glycine-rich segment (Gly50 through Gly55) preceding Lys72 is also an important feature of the ATP binding site (Taylor SS, et al., 1990).

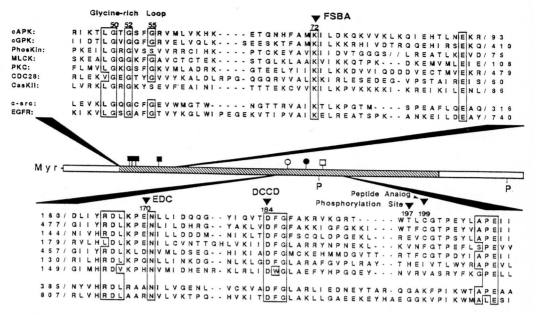

Figure 2: Functional sites in the C-subunit and sequence alignment with some other known protein kinases. Myristoylation (Myr) and phosphorylation sites (P) are indicated. Other sites include Gly 50, 52, and 55 (■■) as well as Lys 72 (■) labeled with FSBA, Asp 184 labeled with EDC (O), and Cys 199 (□), affinity labeled with a peptide analog. Highly conserved sequences are indicated also (●) (Figure taken from Taylor, Yonemoto, and Buechler **Annu. Rev. Biochem.**).

In order to identify any carboxyl groups that may lie close to the ATP binding site, the C-subunit was treated with a hydrophobic carbodiimide, dicyclohexyl carbodiimide (DCCD). This led to the identification of an essential reactive carboxyl group, Asp184, and furthermore established that Asp184, in the absence of MgATP, can be cross-linked readily to Lys72 (Buechler JA and Taylor SS, 1988;Buechler JA and Taylor SS, 1988). Both of these residues, as well as the glycine-rich loop, are highly conserved features of all protein kinases (Hanks SK, et al., 1988;Taylor SS, et al., 1990;Taylor SS, et al., 1990).

Since peptide recognition differs for each kinase, specific residues that line the substrate binding pocket very likely differ for each kinase although the folding of the polypeptide chain will, in general, be conserved. Furthermore, it should be emphasized that the substrate is relatively large so that peptide recognition will certainly involve several spatially distinct sites. In the case of the C-subunit , the general consensus sequence for recognition is Arg-Arg-X-Ser (Bramson HN, et al., 1984). Affinity labeling showed that Cys199, Gly125, and Met127 are all close to the peptide binding site (Bramson HN, et al., 1982;Miller WT and Kaiser ET, 1988), while group-specific labeling with a water soluble carbodiimide identified several carboxyl groups that may contribute to peptide recognition: Glu170, Asp328, and Glu332 (Buechler JA and Taylor SA, 1990). In general, these residues associated with peptide recognition are not highly conserved in other kinases. An overall summary of the active site region showing some of the residues thought to lie near the substrates is shown in Figure 3.

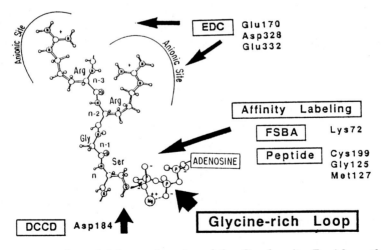

Figure 3: Proposed model for active site of the C-subunit. Residues thought to be in close proximity are indicated.

While chemical approaches enable us to map out functional sites, additional techniques are required in order to fully understand a molecule this size. On the one hand, a high resolution crystal structure is essential. In addition, however, recombinant techniques and having prokaryotic expressions systems for both the R- and the C-subunit (Saraswat LD, et al., 1986;Slice LW and Taylor SS, 1989) are invaluable tools for interpreting that structure.

Recombinant Approaches: The cloning of the $C\alpha$-subunit opened the way for using recombinant methods to probe its structure and function (Uhler MD, et al., 1986). While many questions can be answered by overexpressing mutant and native proteins in cultured cells, we chose to develop prokaryotic expression systems that would enable us to purify significant amounts of protein. After trying many different vectors and strains of *E. coli*, expression of active and soluble enzyme was eventually achieved using a Pt7-7 vector where the gene encoding the C-subunit is placed behind a T7 polymerase promoter (Slice LW and SS Taylor, 1989). The amount of soluble protein can be increased by carrying out the induction at low temperatures (Yonemoto W, et al., 1990). This recombinant enzyme, rC, is kinetically indistinguishable from the mammalian enzyme; however, it is more labile to heat denaturation. Both the free rC-subunit and the corresponding holoenzyme have a T_m that is approximately 5° lower than the corresponding mammalian enzyme (Yonemoto W, et al., 1990).

The mammalian C-subunit is known to undergo two types of posttranslational modification, myristoylation at the N-terminus (Carr SA, et al., 1982) and phosphorylation (Shoji S, et al., 1979). The rC-subunit lacks the myristoyl moiety since *E. coli* does not express an N-myristoyl transferase (NMT). In order to determine the role of myristoylation, the C-subunit was coexpressed with NMT. This allowed us to reconstruct a eukaryotic posttranslational modification system in *E. coli*, and the rC-subunit expressed in the presence of NMT was stoichiometrically modified at its N-terminus with myristic acid. The myristoylated rC-subunit was kinetically identical to the non-myristoylated enzyme; however, the addition of the acyl group stabilized the enzyme. The T_m of the myristoylated C-subunit is nearly identical to the mammalian C-subunit (Duronio RJ, et al., 1990;Yonemoto W, et al., 1990).

The rC-subunit is also phosphorylated and since expression of an inactive form of the coenzyme is not phosphorylated, it is clear that the autophosphorylations are autocatalytic (Yonemoto W, et al., 1990). For example, replacing the essential lysine, Lys72, with His or Arg leads to a 2-3 order of magnitude loss in enzymatic activity, and this protein, in contrast to the native rC-subunit, shows very little phosphorylation when expressed in the presence of $[^{32}P_i]$. The two sites that are normally phosphorylated in the mammalian C-subunit, Thr197 and Ser338, are phosphorylated in the rC-subunit. In addition, Ser10 and Ser149 are autophosphorylated. Ser10 is a site of reversible autophosphorylation in the mammalian C-subunit (Yonemoto and Taylor, unpublished results).

Other mutations are now being introduced into the C-subunit in order to probe specific amino acid residues thought to be important for substrate and MgATP recognition and for catalysis. In addition, deletions are being introduced at the C-terminus and at the N-terminus to better define the true core that is essential for function.

Biophysical Approaches: The progress in solving the crystal structure is presented in the Knighton abstract. While the chain tracing is not yet complete, the shape of the molecule is clear. There are two general lobes with a cleft between. The smaller lobe contains primarily β-structure while α-helices dominate in the larger lobe. A 20 residue peptide inhibitor, derived from the heat stable protein kinase inhibitor (Cheung H-C, et al., 1986), and MgATP both bind in the cleft between the two lobes.

Several types of conformational changes are associated with substrate binding to the C-subunit. Neutron scattering showed that the free C-subunit had a radius of gyration (R_g) of 20.0Å. In the presence of the 20 residue inhibitor peptide (PKI:5-24) and MgATP the R_g was 18.5Å, indicating a significant compacting of the molecule (Parello J, et al., 1990). Peptide alone was sufficient to induce most of this change in R_g while MgATP alone had no effect. Hence we propose that substrate binding causes the cleft to close. This is also consistent with the crystallographic results since the free C-subunit crystallizes in a cubic form while the ternary complex yields a hexagonal crystal (Knighton DR, et al., 1990).

While the neutron scattering indicates global changes in overall conformation that are induced primarily by peptide binding, intrinsic protein fluorescence can be used to measure more subtle and localized changes in conformation. In contrast to the neutron scattering results, binding of peptide causes no changes in the intrinsic tryptophan fluorescence of the C-subunit, while the binding of MgATP causes a significant quenching of fluorescence (Driscoll and Taylor, unpublished results). Thus there are at least two separate probes that are sensitive enough to detect different types of conformational changes in the C-subunit. These will be particularly important as we begin to characterize mutant forms of the C-subunit.

Inhibition of the Catalytic Subunit: There are two known inhibitors of the C-subunit: the regulatory (R)-subunits (Taylor SS, et al., 1990) and the heat stable protein kinase inhibitor (PKI) (Walsh DA, et al., 1990). The mechanism by which each inhibits the C-subunit is remarkably similar. Each contains an inhibitory domain that resembles a peptide substrate. This region then occupies the peptide binding site thus preventing other substrates from binding (Taylor SS, et al., 1990). The primary difference between the two mechanisms is that the inhibition by the R-subunit can be reversed by the binding of cAMP while the inhibition by PKI appears to be irreversible.

Unlike the C-subunit, the R-subunits have a well-defined domain structure as summarized in Figure 4. Limited proteolysis provided the first clues about this domain structure. More recently, mutagenesis of the recombinant R-subunit has enabled us to extend this model by generating numerous deletion mutants (Ringheim GE, et al., 1988). There are at least four unique gene products in the R-subunit family (Clegg CH, et al., 1988;Jahnsen T, et al., 1986;Lee DC, et al., 1983;Scott JD, et al., 1987), but all have conserved the same overall domain structure. Two tandem gene duplicated cAMP binding domains lie at the C-terminus. cAMP binds with positive cooperativity and induces conformational changes that cause dissociation of the subunits. A model for the folding of the polypeptide chain in each cAMP-binding domain has been proposed based on the crystal structure of the catabolite gene activator protein (CAP) in E. coli which shows extensive sequence similarities with the cAMP-binding domains of the R-subunit

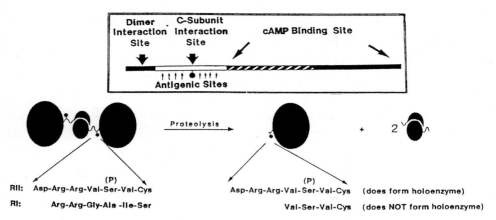

Figure 4: Model and domain structure of the R-subunit. The overall domain structure of the R-subunit is summarized in the inset. A model for the asymmetric R-subunit is shown below.

(Weber IT, et al., 1987). The proposed model for cAMP-binding site A is shown in Figure 5. The cAMP binds in a β-barrel with the adenine ring interacting with a long helix. Two highly conserved features of this binding site are an Arg (Arg209) that interacts with the exocyclic oxygens of cAMP and a Glu (Glu200) that hydrogen bonds to the 2'-OH of the ribose ring.

The R-subunit is an asymmetric dimer, and the major sites for dimer interaction lie at the N-terminus. Approximately 90-100 residues from the N-terminus lies the inhibitory domain also referred to as a "hinge" region because it is labile to proteolysis in the absence of C-subunit. This segment contains a substrate-like sequence. Proteolytic cleavage up to the arginines in this sequence generates a monomeric R-subunit that still binds to C with a high affinity in a cAMP-dependent manner. Cleavage just beyond the two arginines, however, results in a fragment that no longer associates with the C-subunit (Weldon SL and Taylor SS, 1985). This inhibitory region differs in the two families of R-subunits. The type II R-subunits contain an autophosphorylation site here while the type I R-subunits contain a pseudophosphorylation site (Figure 5). The type II R-subunit can be autophosphorylated in the holoenzyme complex (Rosen OM and Erlichman J, 1975). The type I holoenzyme binds MgATP with a high affinity (10$\underline{nM}$) in contrast to a K_m of 10$\underline{mM}$ for the free C-subunit and this increases the K_a(cAMP) by 5-10 fold (Hofmann F, et al., 1975). If Ala97 in the R^I-subunit is replaced with Ser, the R^I-subunit also can be autophosphorylated by an intramolecular event in the holoenzyme providing compelling evidence that this segment of the R-subunit occupies the actual peptide binding site in the holoenzyme complex (Durgerian S and Taylor SS, 1989).

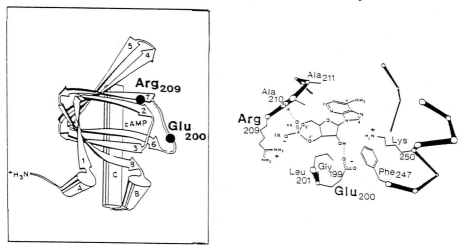

Figure 5: Model of cAMP-binding site A based on the crystal structure of CAP.

This mechanism of inhibition via a pseudosubstrate region has been proposed now for several other protein kinases including protein kinase C (House C and Kemp BE, 1987), the calcium calmodulin-dependent protein kinase II (Kwiatkowski AP and King MM, 1989), and myosin light chain kinase (Ikebe M, et al., 1987). However, in none of these cases is the evidence as clear as it is for cAMP-dependent protein kinase. Thus, once again, this simple kinase, cAPK, serves as a framework for a mechanism of inhibition that appears to be widespread in the protein kinase family.

The other known inhibitor of the C-subunit is PKI (Walsh DA, et al., 1990). Like the R^I-subunits, it contains a pseudosubstrate site that is localized at the amino-terminus. A 20 residue peptide, PKI(5-24), from this region inhibits the C-subunit with a K_i of 2$\underline{nM}$ (Cheung H-C, et al.,

1986). It is this peptide that was cocrystallized with the C-subunit. So far there is no known mechanism for releasing the C:PKI complex.

Microinjection: While it is essential to understand the chemical structure of the C and R subunits, if one is to truly appreciate how the enzyme functions at the molecular level, one also wants to ultimately understand how the enzyme functions and is regulated within the cell. One such approach to understanding cellular function is microinjection. In the case of cAPK, this approach is quite feasible since large amounts of protein are readily available. With good expression systems in *E. coli* in hand for both subunits, mutant proteins also can be generated easily.

One of the persistent questions regarding the C-subunit is its role in the regulation of gene expression and, in addition, its location in the cell. Many genes are known to be regulated by cAMP. These genes are typically preceded by a cAMP responsive element. The protein that binds to this element, the CRE binding protein or CREB, is regulated by phosphorylation by the C-subunit and the C-subunit alone is sufficient to induce the expression of genes preceded by a CRE (Gonzalez GA, et al., 1989;Grove JR, et al., 1987). A critical question now is whether the C-subunit is, at some point, translocated to the nucleus and whether nuclear localization of the C-subunit is necessary for regulating gene expression.

In an effort to answer this question, C-subunit was labeled with fluoresceine isothiocyanate under conditions where activity and capacity to form holoenzyme were retained. This fluorescent C-subunit (FITC-Cat) was then microinjected into rat fibroblast cells under a variety of conditions (Meinkoth JL, et al., 1990). When C-subunit alone was microinjected, it migrated first to a perinuclear location and then to the nucleus where it remained for up to 3 hours. In contrast, when holoenzyme formed with FITC-Cat was microinjected, it remained in the cytoplasm and showed a filamentous distribution. Only in the presence of 8-Br-cAMP sufficient to cause holoenzyme dissociation did FITC-Cat migrate to the nucleus. When the R-subunit was labeled with fluoresceine isothiocyanate and microinjected, it remained in the cytoplasm showing a distribution that mimicked the holoenzyme. When holoenzyme formed with FITC-R was microinjected, the fluorescence remained in the cytoplasm both in the presence and absence of 8-Br-cAMP. A mutant form of the C-subunit containing a Gly1Ala replacement and incapable of being myristoylated at the N-terminus also migrated to the nucleus following microinjection. Hence, myristoylation is not a requirement for nuclear translocation.

 <u>Conclusions:</u> We have tried to stress the importance of using multiple approaches to understand kinase structure and function. All these pieces of information, ranging from a high resolution crystal structure to microinjection of isolated subunits, are important. Hopefully, by superimposing this information, we shall begin to understand how this enzyme functions both at the molecular level and at the physiological level.

REFERENCES

Beebe SJ and Corbin JD (1986) Cyclic Nucleotide-Dependent Protein Kinases. The Enzymes: Control by Phosphorylation Part A,Vol. XVII eds. P.D. Boyer and E. G. Krebs Academic Press, Inc.New York

Bramson HN, Kaiser ET and Mildvan AS (1984) Mechanistic Studies of cAMP-Dependent Protein Kinase Actions. CRC Crit. Rev. Biochem. 15:93-124

Bramson HN, Thomas N, Matsueda R, Nelson NC, Taylor SS and Kaiser ET (1982) Modification of the Catalytic Subunit of Bovine Heart cAMP-Dependent Protein Kinase with Affinity Labels Related to Peptide Substrates. J. Biol. Chem. 257:10575-10581

Buechler JA and Taylor SA (1990) Differential Labeling of the Catalytic Subunit of cAMP-Dependent Protein Kinase with a Water-Soluble Carbodiimide: Identification of Carboxyl Groups Protected by MgATP and Inhibitor Peptides. Biochemistry 29:1937-1943

Buechler JA and Taylor SS (1988) Dicyclohexyl carbodiimide crosslinks two essential residues at the active site of the catalytic subunit of cAMP-dependent protein kinase: Asp 184 and Lys 72. Biochemistry 28:2065-2070

Buechler JA and Taylor SS (1988) Identification of Asp 184 as an Essential Residue in the Catalytic Subunit of cAMP-Dependent Protein Kinase. Biochemistry 27:7356-7361

Buechler JA, Vedvick TA and Taylor SS (1989) Differential Labeling of the Catalytic Subunit of cAMP-dependent Protein Kinase with Acetic Anhydride. Biochemistry 28:3018-3024

Carr SA, Biemann K, Shuji S, Parmalee DC and Titani K (1982) n-Tetradecanoyl in the NH2 Terminal Blocking Group of the Catalytic Subunit of the Cyclic AMP-Dependent Protein Kinase from Bovine Cardiac Muscle. Proc. Natl. Acad. Sci. USA 79:6128-6131

Cheung H-C, van Patten SM, Smith AJ and Walsh DA (1986) An Active Twenty-Amino-Acid-Residue Peptide Derived from the Inhibitor Protein of the Cyclic AMP-dependent Protein Kinase. Biochem. J. 231:655-661

Clegg CH, Codd GG and McKnight GS (1988) Genetic Characterization of a Brain-specific Form of the Type I Regulatory Subunit of cAMP-dependent Protein Kinase. Proc. Natl. Acad. Sci. USA 85:3703-3707

Durgerian S and Taylor SS (1989) The Consequences of Introducing an Autophosphorylation Site into the type I regulatory subunit of cAMP-dependent Protein Kinase. J. Biol. Chem. 264:9807-9813

Duronio RJ, Jackson-Machelski E, Heuckeroth RO, Olins P, Devine CS, Yonemoto W, Slice LW, Taylor SS and Gordon JI (1990) Protein N-myristoylation in *Escherichia coli*: Reconstitution of a Eukaryotic Protein Modification in Bacteria. Proc. Natl. Acad. Sci. USA 87:1506-1510

Gonzalez GA, Yamamoto KK, Fischer WH, Karr D, Menzel P, Biggs III W, Vale WW and Montminy MR (1989) A Cluster of Phosphorylated Sites on the Cyclic AMP-regulated Nuclear Factor CREB Predicted by Its Sequence. Nature 337:749-752

Grove JR, Price DJ, Goodman HM and Avruch J (1987) Recombinant Fragment of Protein Kinase Inhibitor Blocks Cyclic AMP-Dependent Gene Transcription. Science 238:530-533

Hanks SK, Quinn AM and Hunter T (1988) The Protein Kinase Family: Conserved Features and Deduced Phylogeny of the Catalytic Domains. Science 241:42-52

Hofmann F, Beavo JA, Bechtel PJ and Krebs EG (1975) Comparison of Adenosine 3':5'-Monophosphate-Dependent Protein Kinases from Rabbit Skeletal and Bovine Heart Muscle. J. Biol. Chem. 250:7795-7801

House C and Kemp BE (1987) Protein Kinase C Contains a Pseudosubstrate Prototope in Its Regulatory Domain. Science 238:1726-1728

Ikebe M, Stepinska M, Kemp BE, Means AR and Hartshorne DJ (1987) Protoelysis of Smooth Muscle Lyosin Light Chain Kinase. J. Biol. Chem. 260:13828-13834

Jahnsen T, Hedin L, Kidd VJ, Beattie WG, Lohmann SM, Walter V, Durica J, Schulz TZ, Schlitz E, Browner M, Lawrence CB, Goldman D, Ratoosh SL and Richards JS (1986) Molecular cloning, cDNA structure, and regulation of the regulatory subunit of type II cAMP-dependent protein kinase from rat ovarian granulosa cells. J. Biol. Chem. 261(26):12352-12361

Knighton DR, Xuong N-h, Taylor SS and Sowadski JM (1990) Crystallization Studies of cAMP-Dependent Protein Kinase.

Cocrystals of the catalytic subunit with a twenty amino acid peptide inhibitor and MgATP diffract to 3.0Å resolution. J. Mol. Biol. Submitted:

Kwiatkowski AP and King MM (1989) Autophosphorylation of the Type II Calmodulin-dependent Protein Kinase is Essential for Formation of a Proteolytic Fragment with Catalytic Activity. Implications for Long-term Synaptic Potentiation. Biochemistry 28:5380-5385

Lee DC, Carmichael DF, Krebs EG and McKnight GS (1983) Isolation of a cDNA Clone for the Type I Regulatory Subunit of Bovine cAMP-Dependent Protein Kinase. Proc. Natl. Acad. Sci. USA 80:3608-3612

Meinkoth JL, Ji Y, Taylor ST and Feramisco JR (1990) Dynamics of the Cyclic AMP-Dependent Protein Kinase Distribution in Living Cells. Proc. Natl. Acad. Sci. USA 87:9595-9599

Miller WT and Kaiser ET (1988) Probing the Peptide Binding Site of the cAMP-Dependent Protein Kinase by Using a Peptide-Based Photoaffinity Label. Proc. Natl. Acad. Sci. USA 85:5429-5433

Parello J, Timmins PA, Sowadski JM and Taylor SS (1990) Major Conformational Changes in the Catalytic Subunit of cAMP-Dependent Prortein Kinase Induced by the Binding of a Peptide Inhibitor: A Small-Angle Neutron Scattering Study in Solution. J. Mol. Biol. In press

Ringheim GE, Saraswat LD, Bubis J and Taylor SS (1988) Deletion of cAMP-binding Site B in the Regulatory Subunit of cAMP-dependent Protein Kinase Alters the Photoaffinity Labeling of Site A. J. Biol. Chem. 263:18247-18252

Rosen OM and Erlichman J (1975) Reversible Autophosphorylation of a Cyclic 3':5'-AMP-Dependent Protein Kinase from Bovine Cardiac Muscle. J. Biol. Chem. 250:7788-7794

Saraswat LD, Filutowics M and Taylor SS (1986) Expression of the Type I Regulatory Subunit of cAMP-Dependent Protein Kinase in Escherichia coli. J. Biol. Chem. 261:11091-11096

Scott JD, Glaccum MB, Zoller MJ, Uhler MD, Helfan DM, McKnight GS and Krebs EG (1987) The Molecular Cloning of a Type II Regulatory Subunit of the cAMP-dependnet Protein Kinase from Rat Skeletal Muscle and Mouse Brain. Proc.Natl. Acad. Sci. USA 84:5192-5196

Shoji S, Titani K, Demaille JG and Fischer EH (1979) Sequence of Two Phosphorylated Sites in the Catalytic Subunit of Bovine Cardiac

Muscle Adenosine 3':5'-Monophosphate-Dependent Protein Kinase. J. Biol. Chem. 254:6211-6214

Slice LW and Taylor SS (1989) Expression of the Catalytic Subunit of cAMP-dependent Protein Kinase In *Escherichia coli*. J. Biol. Chem. 264:20940-20946

Taylor SS, Buechler JA and Knighton DR (1990) cAMP-Dependent Protein Kinase: Mechanism for ATP: Protein Phosphotransfer. CRC. Crit. Rev. Biochem. 1-42

Taylor SS, Buechler JA and Yonemoto Y (1990) cAMP-dependent Protein Kinase: Framework for a Diverse Family of Regulatory Enzymes. Annu. Rev. Biochem. 59:971-1005

Uhler MD, Carmichael DF, Lee DC, Chivia JC, Krebs EG and McKnight GS (1986) Isolation of cDNA Clones for the Catalytic Subunit of Mouse cAMP-Dependent Protein Kinase. Proc. Natl. Acad. Sci. USA 83:1300-1304

Walsh DA, Angelos KL, Van Patten SM, Glass DB and Garetto LP (1990) The Inhibitor Protein of the cAMP-Dependent Protein Kinase. CRC Crit. Rev. Biochem. 43-84

Walsh DA, Perkins JP and Krebs EG (1968) An Adenosine 3',5'-Mono-Phosphate-Dependent Protein Kinase from Rabbit Skeletal Muscle. J. Biol. Chem. 243:3763-3765

Weber IT, Steitz TA, Bubis J and Taylor SS (1987) Predicted Structures of the cAMP Binding Domains of the Type I and II Regulatory Subunits of cAMP-Dependent Protein Kinase. Biochemistry 26:343-351

Weldon SL and Taylor SS (1985) Monoclonal Antibodies as Probes for Functional Domains in cAMP-Dependent Protein Kinase II. J. Biol. Chem. 260:4203-4209

Yonemoto W, McGlone MM, Slice LW and Taylor SS (1990) Prokaryotic Expression of the Catalytic Subunit of cAMP-dependent Protein Kinase. Methods Enzymol. In press:

Zoller MJ and Taylor SS (1979) Affinity Labeling of the Nucleotide Binding Site of the Catalytic Subunit of cAMP-Dependent Protein Kinase using p-Fluorosulfonyl-[14C]-Benzoyl 5'-Adenosine: Identification of a Modified Lysine Residue. J. Biol. Chem. 254:8363-8368

EXPRESSION IN *E. COLI* OF MUTATED R SUBUNITS OF THE cAMP-DEPENDENT PROTEIN KINASE FROM *DICTYOSTELIUM DISCOIDEUM.*

Marie-Noëlle SIMON, Sandrine GAZEAU, Olivier PELLEGRINI and Michel VERON
Unité de Biochimie Cellulaire
Institut Pasteur
75724 PARIS Cedex

ABSTRACT

The cDNA coding for the regulatory subunit of the cAMP-dependent protein kinase from *Dictyostelium discoideum* was cloned in an *E. coli* expression vector. Oligonucleotide-directed mutagenesis was used to introduce single amino-acid substitutions in either cAMP binding site A (Gly to Glu at position 135) or site B (Gly to Glu at position 261). Analysis of the mutated R subunits showed that both single mutants retain high affinity cAMP binding activity (Kd = 20 nM) while cAMP did not bind to the protein carrying substitutions in both sites. The results show that the regulatory subunit from *Dictyostelium* contains two high affinity cAMP binding sites and that cAMP binding to one site is sufficient to activate the holoenzyme.

INTRODUCTION: Upon food starvation, the amoebae from *Dictyostelium discoideum* embark on a development program leading to the formation of a fruiting body composed of two cell types. Extracellular cAMP plays a crucial role in the differentiation as a chemoattractant coordinating movement of the amoebae during aggregation and in the control of the expression of many developmentally regulated genes through cAMP specific membrane receptors (Devreotes, 1989). In addition, cAMP plays a role at the intra-cellular level through the cAMP dependent protein kinase (Simon *et al.*, 1989; Harwood *et al.*, manuscript in preparation). The predicted primary structure of the R_D protein deduced from its cDNA sequence indicates that it does not carry the N-terminal domain responsible for the association of R subunits in dimers (Mutzel *et al.*, 1987a). This explains why the

NATO ASI Series, Vol. H 56
Cellular Regulation by Protein Phosphorylation
Edited by L. M. G. Heilmeyer, Jr.
© Springer-Verlag Berlin Heidelberg 1991

holoenzyme of *Dictyostelium* PKA is an RC dimer as opposed to the typical R_2C_2 structure of other PKAs. Although the sequence of *Dictyostelium* PKA contains two in tandem cAMP binding motifs homologous to those of bovine RI or RII, the stoichiometry of cAMP binding to R has previously been reported to be 1mol/mol of R subunit (de Wit *et al.*, 1982; de Gunzburg *et al.*, 1984). In order to determine whether the two cAMP binding sites of R_D are functional, we have used site-directed mutagenesis to inactivate one or the other of the putative cAMP binding sites.

MATERIALS AND METHODS: *Materials, Vectors and site directed mutagenesis*: The pMS501 vector was constructed by cloning the cDNA 2.1 (Mutzel *et al.*, 1987a) coding for R in pMS5 (Simon *et al.*, 1988). The protein expressed from pMS501 contains one substitution and three additional residues at its N-terminus compared with the *Dictyostelium* R subunit, due to vector construction. For *in vitro* mutagenesis, the *Eco*RV-*Pst*I fragment of the cDNA was subcloned in M13tg131 (P.L. Biochemicals). The single stranded DNA was used as a template for *in vitro* mutagenesis according to Kunkel (1985). The mutated clones were screened by sequencing using the sequenase 2.0 kit from USB. The double mutant was constructed by the same method with the DNA fragment mutated in site A as template for mutagenesis of the site B. The 640 bp *Eco*RV-*Pst*I DNA fragments containing the mutation were purified, ligated back in the large 4600 bp *Eco*RV-*Pst*I fragment from pMS501 and transformed into *E. coli* TG1. The mutations were confirmed by the presence of new *Mbo*II sites created by the mutations. The proteins expressed from these vectors were designed as RmutA, RmutB and RmutAB.

Expression of the R subunits: The transformed strains were grown at 37°C in LB medium supplemented with 50µg/ml ampicillin. When the culture reached a density of 0.6 at 600 nm, expression of the protein was induced by IPTG to a final concentration of 0,2mM. The proteins were analyzed by western blotting using a polyclonal antiserum against the R subunit as described (Mutzel *et al.*, 1987b). Protein concentrations were measured by the technique of Bradford (1976) using γ-globuline as protein standard.

Partial purification of the C subunit from Dictyostelium: Crude extract of differentiated amoebae from strain AX2-2 was prepared as previously described (de Gunzburg *et al.*, 1981). In order to dissociate the holoenzyme, 5mM of cAMP was added to the extract before chromatography on Q-sepharose. The free C subunit was eluted by a 0-0.6M linear salt gradient. The fraction containing the catalytic activity was adjusted to 4M NaCl and adsorbed on Phenyl-Sepharose CL-4B resin (P.L. Biochemicals). After successive washing with 4M and 3M NaCl, the C subunit was eluted by an inverse salt gradient (3-0M NaCl). Fractions containing catalytic activity were pooled and dialysed.

Activity assays: cAMP binding activity was measured in the presence of 0.5 μM ^{3}H-cAMP by the technique of filtration on nitrocellulose filters and the activity of the C subunit from *Dictyostelium* was measured using Kemptide as substrate (Mutzel *et al.*, 1987b). Reconstitution of the holoenzyme was made by adding increasing amount of partially purified R subunits to a fixed amount of C subunit. Residual catalytic activity was measured in absence and in the presence of 0.5mM cAMP.

RESULTS AND DISCUSSION: In the mammalian RI subunit, substitution of Gly 200 (site A) by a glutamic acid or Gly 324 (site B) by an aspartate abolishes cAMP binding on the mutated site (Kuno *et al.*, 1988, Correll *et al.*, 1989). Two synthetic 20 mers oligonucleotides (GTAGTTTTGAAGAATTAGCT for site A and ATTACTTTGAAGAATTGCA for site B) were used for *in vitro* mutagenesis to selectively change Gly 135 and Gly 261 by a glutamic acid (figure 1).

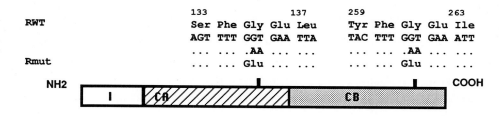

figure 1: **Localization of the amino-acid substitutions.** R$_D$ subunit is represented as composed of three functional domains: I: domain of interaction with the C subunit, A and B: sequences homologous to cAMP binding domains of mammalian RI and RII subunits.

The expression of the R protein from the different strains, analyzed by SDS-PAGE and immunolabeling, showed that recombinant R$_D$ as well as the mutant forms are stably expressed in *E. coli* (data not shown). As shown in figure 2, cAMP binding activity is observed in crude bacterial extracts expressing RmutA or RmutB protein.

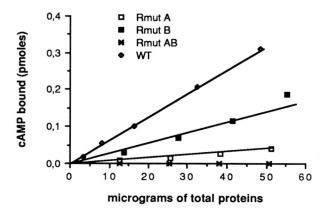

Figure 2: Measurement of cAMP binding activity in crude bacterial extracts expressing wild type or mutated recombinant R subunits. The levels of R subunit in the extracts were monitored by western blot analysis to verify that the mutants are expressed in the same amount than the wild type recombinant protein.

cAMP binding to RmutB is approximatively 50% of that measured with the wild type recombinant protein. cAMP binding to RmutA is also clearly demonstrated although it is lower than the binding observed with RmutB. The millipore filtration technique used for this assay has been shown to underestimate binding to one site in mammalian R subunits (Ogreid and Dokesland, 1980) which could explain the 1mol/mol stoichiometry first reported (de Gunzburg *et al.*, 1984; de Wit *et al.*, 1982). It is also possible that western blot analysis did not detect small differences of expression of RmutA and RmutB in *E.coli*. The protein carrying amino acid substitutions in both site A and B showed no detectable specific cAMP binding indicating that the cAMP binding on RmutA and RmutB occured on the residual non-mutated site. The Kd of the cAMP binding on RmutA and RmutB are comparable to those of the recombinant protein (Kd=20nM) showing that both site A and site B possess a high affinity for cAMP.

The mutated R subunits were analyzed for their ability to reconstitute inactive holoenzymes in the presence of *Dictyostelium* C subunit. Mutations in the cAMP binding sites did not prevent interaction between the catalytic and the regulatory subunits since addition of partially purified RmutA, RmutB or RmutAB to *Dictyostelium* C subunit inhibits its catalytic activity (data not shown). Most of the kinase activity can be recovered by addition of cAMP to the holoenzymes formed with RmutA and RmutB. This indicates that only one functional cAMP binding site is required for activation of the holoenzyme in agreement with previous results described for the mammalian protein. Addition of cAMP has no effect on the inhibition of catalytic activity by RmutAB as expected from the lack of its binding.

In conclusion, these data provide evidence that the R subunit from *Dictyostelium discoideum* PKA has two high affinity cAMP binding sites. Further experiments will show whether these two sites can be distinguished by their kinetic properties which could allow differential sensibility of the PKA to cAMP at different steps of development.

REFERENCES

Bradford MM (1976) A rapid and sensitive method for the quantitation of microgram quantities of protein utilizing the principe of protein-dye binding. Anal Biochem 72: 248-254.

Correll LA, Woodford TA, Corbin JD, Mellon PL, McKnight GS (1989) Functionnal characterization of cAMP-binding mutations in type I protein kinase. J Biol Chem 262: 9668-9673.

de Gunzburg J, Part D, Guiso N, Veron M (1984) An unusual Adenosin 3',5'-Phosphate dependent Protein Kinase from *Dictyostelium discoideum*. Biochem. 23: 3805-3812.

Devreotes, P. (1989) *Dictyostelium discoideum:* A model system for cell-cell interactions in development. *Science* 245, 1054-1058.

de Wit RJW., Arents JC, van Driel R (1982) Ligand binding properties of the cytoplasmic cAMP-binding protein of *Dictyostelium discoideum*. FEBS Lett. 145: 150-154.

Kemp BE, Graves DJ, Benjamini E, Krebs EG (1977) Role of multiple basic residues in determining the substrate specificity of cAMP-dependent protein kinase. J Biol Chem 252: 4888

Kunkel TA (1985) Rapid and efficient site-specific mutagenesis without phenotypic selection. Proc Natl Acad Sci USA 82: 488-492

Kuno T, Shuntoh H, Sakaue M, Saijoh K, Takeda T, Fukuda K, Tanaka, C. (1988) Site-directed mutagenesis of the cAMP binding sites of the recombinant type I regulatory subunit of cAMP-dependent protein kinase. Biochem Biophys Res Commun 153: 1244-1250

Mutzel R, Lacombe M.-L, Simon M-N, de Gunzburg J, Veron M. (1987a) Cloning and cDNA sequence of the regulatory subunit of cAMP-dependent protein kinase in *Dictyostelium discoideum*. Proc Natl Acad Sci USA 84: 6-10.

Mutzel R, Simon M-N, Lacombe M-L, Veron M, (1987b) Expression and properties of the regulatory subunit of *Dictyostelium* cAMP-dependent Protein Kinase encoded by λgt11 cDNA Clones. Biochem 27: 481-486.

Ogreid D, Dokesland SO (1980) Protein kinase II has two distinct binding sites for cAMP, only one of which is detectable by the conventional membrane filtration method. FEBS LETTERS 121: 340-344.

Simon M-N, Mutzel R, Mutzel H, Veron M (1988) Vectors for expression of truncated coding sequences in *E. coli*. Plasmid 19: 94-102.

Simon M-N, Driscoll D, Mutzel R, Part D, Williams J, Veron M (1989) Overproduction of the regulatory subunit of the cAMP dependent protein kinase blocks the differentiation of *Dictyostelium discoideum*. Embo J 8: 2039-2043.

Expression of the γ-subunit of Phosphorylase b Kinase in *Escherichia coli*

Sarah Cox and L N Johnson

Laboratory of Molecular Biophysics

University of Oxford

South Parks Road

Oxford OX1 5PB,

U.K.

ABSTRACT

The γ-subunit of phosphorylase b kinase has been expressed in *Escherichia coli*. It is produced in inclusion bodies and can be solublised and refolded to an active species. The protein has been partially purified and shown to have properties similar to those of the γ-subunit prepared from holoenzyme by HPLC.

Inrtoduction

Phosphorylase kinase is a multisubunit protein of approximate Mr 1.3×10^6. It consists of a tetramer of four subunits $(\alpha\beta\gamma\delta)_4$. The γ-subunit is the active protein kinase subunit, while the α,β and δ-subunits are regulatory. The α and β–subunits confer control of catalytic activity by phosphorylation and proteolysis and the δ-subunit, which is identical to calmodulin, confers control by calcium. The γ-subunit makes up only 13% of the total protein mass. Although it can be separated from the other subunits by denaturation followed by either reverse phase HPLC (Crabb and Heilmeyer, 1984) or size exclusion chromatography (Reimann *et. al.,*1984), prohibitively large quantities of the holoenzyme would be required to produce quantities for structural studuies. To obtain the amounts required for crystallographic analysis we have developed a system for expression of the protein in *E.coli.*

Materials and Methods

Vector construction: A 2.1 kb cDNA encoding the entire γ-subunit was provided in the bluescript plasmid (Stratagene, USA) by PTW Cohen (da Cruz e Silva and Cohen,1987). An Nde1 site (CATATG) was introduced at the initiator methionine ATG codon by site directed mutagenesis using the Eckstein method (Amersham International, UK). The full coding sequence for the γ-subunit was introduced into the pTacTac vector obtained as the kind gift of M Browner UCSF (Browner *et.al.,* 1990) using standard molecular biology techniques. Full details of the construction will be reported elsewhere.

Optimisation of γ-subunit expression in *E.coli.*: Strains of *E.coli*, MC1061, DH1, BMH, JM105, JM109, NM554 and CAG629 were transformed with the pTacTac vector containing the γ-subunit. *E.coli* were grown to early log phase ($OD_{600} \approx 0.4$) at 37, 30 or 22°C in LB media containing 50μg/ml ampicillin or 50μg/ml ampicillin and 12μg/ml tetracyclin (CAG629). Expression was induced by adding isopropyl β-D-thiogalactopyranoside (IPTG) to 0.2 or 1 mM. Growth was continued and cells harvested at time intervals for analysis of expression by western blotting.

Western Blotting: Monoclonal antibody mAb88 (a kind gift from D.Wilkinson and G. Carlson , Tenessee) raised against phosphorylase kinase holoenzyme reacts specifically with the γ-subunit and was used for western blot analyses throughout. Proteins were separated by 12% SDS-PAGE (Laemmli U.K., 1970) and transferred to nitrocellulose paper (Schleicher and Schuell, FRG) using a 2110 Multiphore II (Pharmacia-LKB, Sweden). The nitrocellulose was blocked in borate pH 8.3, 0.1M NaCL, 1% BSA (buffer B) for 40 min followed by incubation with mAb88 at 1/1000 in buffer B for 2 hours or O/N. The nitrocellulose was then washed twice in buffer B for 10 min and incubated with rabbit anti-mouse IgG-alkaline phosphatase conjugate at 1/500 in buffer B for 1-2 hours and washed as before. The blot was developed in 75mM Tris-HCl pH 9.6, 4mM $MgCl_2$, 0.005% nitroblue tetrazolium, 0.0025% 4-bromo 5-chloro indoyl phosphate for 1-3 hours (all chemicals were from Sigma, U.K.).

Inclusion body preparations: 2 l of culture (CAG629 containing pTacTac-γ) following overnight induction were harvested by centrifugation at 5000xg for 15 min and resuspended in 1/20 volume of 100 mM Tris-HCl pH 7.5, 1 mM EDTA and protease inhibitors (0.1mM PMSF, 1μM leupeptin, 10μg/ml BPTI and 0.7μg/ml pepstatin). Cells were lysed by adding lysozyme at 27μg/ml for 10min on ice , 0.1% deoxycholate for 10 min on ice and 10μg/ml deoxy-ribonuclease, 10mM $MgCl_2$ for 30 min on ice. Inclusion bodies were pelleted by centrifugation at 10,000 x g for 15 min and washed twice with 0.1% triton X-100 and twice with distilled H_2O. All solutions were pre-treated with protease inhibitors as above.

Activity Assays: *Phosphorylase b*. Incorporation of ^{32}P into phosphorylase b was carried out in 50μl incubations at 30°C containing 50μM phosphorylase b, 50mM β-glycerophosphate pH 8.2, 10mM $MgCl_2$, 1mM EDTA, 0.5mM EGTA, 1mM DTT, 1.2mM ATP, 10μCi [γ^{32}P]-ATP and 0.6mM $CaCl_2$. The reactions were started by adding enzyme or bacterial extract. Samples were analysed by autoradiography. *Peptide assay*. Routinely activity was assesed by incorporation of ^{32}P into a peptide substrate corresponding to residues 4-18 of phosphorylase. Incubations were as above but contained 2mM peptide instead of phosphorylase. Reactions were terminated by precipitation of protein with 12% TCA (with carrier 2mg/ml BSA) on ice for 30 min and centrifugation at 10,000xg for 5 min. Peptide was recovered by spotting 10μl samples onto P81 filter paper (Whatman International, U.K.) and excess ATP removed by 4 x 5 min washes in 75 mM phosphoric acid. Cherencov counting was in a 1121 Minibeta scintillation counter (LKB). The peptide for this assay was synthesized in house on an ABI 430 peptide synthesizer using f-moc chemistry.

Results and Discussion

Expression of γ-subunit in *E.coli*

Preliminary studies on the expression of the γ-subunit of phosphorylase kinase in *E.coli* using a heat inducible dual origin vector, pMG (Celltech U.K.; Wright *et. al.,* 1986), resulted in the production of γ-subunit in inclusion bodies. Activity was not detected after solublisation and refolding of this material by a variety of methods. The pTacTac system, in which protein expression is under the control of the chemical inducer IPTG, has been used to produce phosphorylase b in a soluble form (Browner *et. al.,* 1990). Soluble protein was only obtained when inductions were done at low temperatures and low inducer concentrations. In an attempt to produce γ-subunit in a soluble form the cDNA was subcloned into the pTacTac vector and a range of *E.coli* strains, temperature and inducer conditions screened for expression.

γ-subunit was expressed in seven *E.Coli* strains MC1061, DH1, BMH, JM105, JM109, NM554 and CAG629. The protein was expressed in all strains but, in six, accumulated only to low levels even after growth for 20 hours in the presence of IPTG (not shown). In the heat-shock and protease deficient strain of *E.coli*, CAG629, protein accumulated, after an overnight induction, to up to 4mg/l as judged by commassie staining and western blotting as compared to holoenzyme control (fig. 1).

Fractionation of soluble and insoluble material indicated that γ-subunit was produced in inclusion bodies (fig 1). Growth at the lower temperatures of 30 or 22°C and/or induction at lower IPTG (0.2mM) concentrations did not prevent inclusion body formation and no soluble protein was detected (not shown).

Figure1:
Expression of
γ-subunit in
CAG629. Lane 1 markers;2-4 vector only; 5-7 vector with γ-subunit gene; 2,5 whole cells after o/n induction;3,6: soluble protein; 4,7: insoluble protein;8 phosphorylase kinase holoenzyme; (a) coomassie (b) western blot.

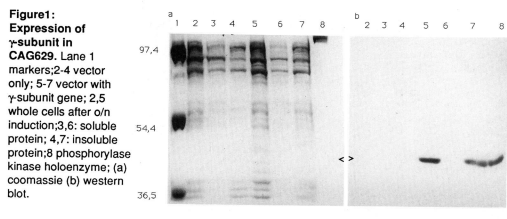

Inclusion bodies were prepared as described from 2 l of culture grown at 30°C, solublised in 8M urea, 50mM Tris-HCl pH 8.5, 4mM DTT and refolded by dilution into refolding buffer(RB: 50mM β-glycerophosphate pH 8.2, 2mM EDTA, 1mM EGTA, 1mM DTT, 10% glycerol) as described by Chen *et. al.* (1989) for the refolding of CAT-γ-subunit fusion protein produced in *E.coli.* Crude solublised inclusion body material containing γ-subunit incorporated ^{32}P into phosphorylase b (fig. 2) while inclusion bodies from CAG629 carrying pTacTac containing an alternative coding sequence did not. The specific activity of the crude material against peptide substrate was found to be 20nM ATP/mg/min.

Figure 2: 32 P Incorporation into
phosphorylase b. Incubations contained 1.5mM γ^{32}P-ATP, 50 mM glycerophosphate pH 8.2, 2mMEDTA, 0.6 mM CaCl$_2$, 10mM MgCl$_2$, 0.1 mM ßME, 0.1mM phosphorylase b and 0/50, 5/50 or 10/50 µl refolded inclusion bodies. 1-3 with γ-subunit,4-6 no γ-subunit. The figure is an autoradiograph.

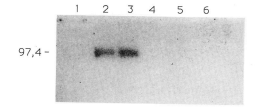

Partial Purification of recombinant γ-subunit

Figure 3 shows the fractionation of refolded γ-subunit by DEAE cation exchange chromatography, This resulted in separation of the enzyme into two species. Active protein eluted at 0.1M NaCl while the majority of immunoreactive material eluted at 0.2 M NaCl and showed no activity towards peptide substrate. This suggested that only a proportion of the γ-subunit present had refolded correctly following dilution from urea.

It has been observed for a number of proteins produced in inclusion bodies that treatment with ion-exchange resins before removal of denaturant resulted in improved refolding properties (Darby and Creighton, 1990). It was suggested that this was due to the removal of a poly-anionic species of unknown nature which comlexed tightly with misfolded proteins in inclusion bodies. Recombinant γ-subunit was subjected to SP-sepharose chromatography in 8M urea. Material was loaded at pH3.5 (0.1M phosphoric acid, 4mMDTT), washed at pH5.5 (0.1M citrate-phosphate, 4mMDTT) and eluted at pH 8.0 (50mM Tris-HCl, 4mMDTT).

Figure 3: DEAE-Cellulose fractionation of γ-subunit. Lane 1 refolded inclusion bodies; 2: unbound; 3: 0.1M NaCl eluate; 4: 0.2M NaCl eluate; 5: 0.5 M NaCl eluate DEAE was equilibrated in R buffer (see Methods) and eluted with R buffer and NaCl as indicated. (a) coomassie, (b) western blot.

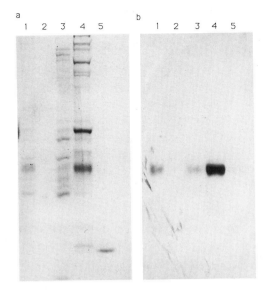

After refolding this material still showed anomalous behaviour on DEAE. Inactive γ-subunit, eluted from DEAE at 0.2M NaCl, regained some activity after dialysis into 8M urea and then dialysis back into refolding buffer (not shown).

Characterisation of recombinant γ-subunit

Partially purified γ-subunit from DEAE chromatography assayed by following ^{32}P incorporation into peptide gave a linear time dependency up to 20 min (fig 4a).

Figure 4: Characterisation of γ-subunit kinase activity. Time course of incorporation or ^{32}P into peptide substrate (a); Ca^{2+}dependence of peptide phosphorylation (b); calmodulin dependence of peptide phosphorylation (c). All reactions contained 140µg/ml of partially purified γ-subunit, calmodulin was added to 60µg/ml in the presence or absence of Ca^{2+} as indicated.

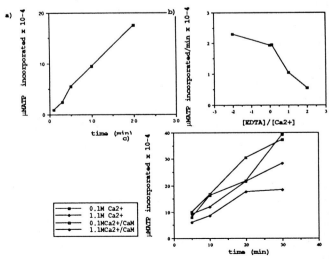

The ratio of activity pH6.8/ pH8.2 was 0.7 ±0.2. This is compared with a ratio of 0.5

-0.6 obtained for HPLC purified γ-subunit (Kee and Graves, 1986). Unlike holoenzyme, γ-subunit is insensitive to Ca^{2+} at μM concentrations (Kee and Graves, 1987). Recombinant γ-subunit was also insensitive to μM Ca^{2+} but was inhibited by mM Ca^{2+} (fig 4b). Whether this is due to competition with Mg^{2+} in MgATP, or interaction of Ca^{2+} at divalent metal binding sites on the γ-subunit (Kee and Graves, 1987) has not been determined. Calmodulin has been reported to stimulate the activity of HPLC purified γ-subunit in the presence of $1\mu M$ Ca^{2+} to 1.4 fold. No effect of calmodulin either in the presence or absence of μM Ca^{2+} on recombinant γ-subunit was observed (fig 6b). However calmodulin partially rescued the inhibition by mM Ca^{2+}.

Preliminary experiments with calmodulin coupled to agarose beads indicated that recombinant γ-subunit interacted with calmodulin. The lack of observed effect of calmodulin on the recombinant protein may be due to the use of the small peptide as substrate instead of phosphorylase.

Conclusions

The γ-subunit of phosphorylase kinase has been expessed in *E.coli* to a level suitable for the production of quantities of material for crystallographic studies. The protein accumulates in inclusion bodies and approximately 10% of the protein produced can be recovered in an active form by solublisation and refolding. We are now persueing proceedures for purification of the active protein.

Acknowlegements: This work was supported by SERC and MRC under the auspices of the Oxford Center for Molecular Sciences. Thanks to Patricia Cohen for the γ-subunit gene, Deborah Wilkinson and G. Carlson for the monoclonal antibody, Michelle Browner for the vector, Rob Greist for the peptide synthesis and to Tony Rees for much helpful advice.

References
Browner MF, Rasor P, Tugendreich S and Fletterick RJ (1990) Temperature Sensitive Production of Rabbit Muscle Glycogen Phosphorylase in *Eschericia coil. Protein Engineering* submitted.
Chen LR, Yuan CJ, Somasekhar G, Wejksnora P, Peterson JE, Myers AM, Graves L, Cohen PTW, da Cruz e Silva and Graves DJ (1989) Expression and Characterisation of the γ-subunit of Phosphorylase Kinase. *Biochem. Biopys. Res. Commun.* **161** 746-753.
Crabb JW and Heilmeyer LMGJr (1984) High Performance Liqiud Cromatography Purification and Sructural Characterisation of the Subunits of Phosphorylase b Kinase. *J Biol Chem* **259** 6346-6350.
da Cruz e Silva and Cohen PTW (1987)Isolation and sequence analysis of a cDNA clone encoding the entire catalytic subunit of phosphorylase kinase. *FEBS Lett.* **220** 36-42.
Darby NJ and Creighton TE (1990) Folding Proteins. *Nature* **344** 715-716.
Kee SM and Graves DJ (1986) Isolation and Properties of the Active γ-subunit of Phosphorylase Kinase. *J Biol Chem* **261** 4732-4737.
Kee SM and Graves DJ (1987) Properties of the γ-subunit of Phosphorylase Kinase. *J Biol Chem* **262** 9448-9453.
Laemmli UK. (1970) Cleavage of structural proteins during the assembly of the head of bacteriophage T4. *Nature* **227** 680-685.
Reimann EM, Titani K, Ericsson LH, Wade RD, Fischer EH and Walsh KA (1984) Homology of the γ-subunit of Phosphorylase Kinase with cAMP-Dependent Protein Kinase. *Biochemistry* **23** 4185-4192.
Wright EM, Humphreys GO and Yarrenton GT (1986) Dual origin plasmids containing an amplifiable ColE1 ori; temperature-controlled expression of cloned genes.*Gene* **49** 311-321.

CONFORMATIONAL AND SHAPE CHANGES ASSOCIATED WITH cAMP- DEPENDENT PROTEIN KINASE

Friedrich W. Herberg and Susan S. Taylor

Department of Chemistry
University of California, San Diego
9500 Gilman Drive
La Jolla, CA 92093-0654 USA

INTRODUCTION

Cyclic-AMP dependent protein kinase (cAPK) was one the first discovered protein kinases (Walsh et al., 1968) and is one of the best understood biochemically of the protein kinases (Taylor et al., 1990). cAPK is composed of both regulatory (R) and catalytic (C) subunits. In the absence of cAMP the subunits associate with a high affinity to form an inactive tetrameric (R_2C_2) holoenzyme. cAMP binds to the R-subunit promoting the dissociation of the complex into a dimeric R-subunit and two monomeric active C-subunits which transfer the γ-phosphate of ATP to a serine or threonine of a peptide or protein substrate (Beebe and Corbin, 1986).

The enzyme undergoes a number of conformational changes, the most obvious being the dissociation of the holoenzyme complex caused by cAMP-binding. However, more subtle changes in conformation are also observed. For example neutron scattering measurements show that a 20 residue peptide, PKI (5-24), derived from the heat stable protein kinase inhibitor, PKI (Van Patten et al.,1986) causes the molecule to compact from 20 Å to 18.5 Å (Parello et al., 1990). Circular dichroism also indicates that there is a significant conformational change associated with the binding of shorter peptides (Reed et al., 1985)

Analytical gel chromatography (Ackers, 1975), combined with new resin materials which can withstand high back pressures (FPLC), is a rapid and highly sensitive technique for monitoring conformational changes such as those described above. Using this method very small changes in the Stokes' radius (R_s) can be measured.

MATERIALS AND METHODS

All gel filtration experiments were carried out with a Pharmacia FPLC-instrument using columns of different sizes packed with various resins. Protein was monitored at 280 nm in a flow detector. Prepacked Pharmacia Superose 12 10/30 and Superdex 75 10/30 columns were run at room temperature with flow rates between 0.5 and 1 ml/min. Samples were injected in a volume of 100-200 μl in small zone experiments (Ackers, 1975). All buffers and samples were filtered through Nylon-66 0.22 μm filters (Rainin Instrument Co). The running buffer used for gel filtration contained at least 150 mM KCl and 25 mM $PO_4\,HPO_4$ or 20 mM MOPS, pH 7.0. ß-mercapthoethanol (0.5 mM) or DTT (5 mM) were added directly prior to use.

Blue dextran (Sigma) was used to determine the void volume (V_0) and tryptophan (Sigma) of the columns for the internal volume (V_i). The partition coefficient, σ, was calculated using the equation: $\sigma = (V_r - V_0)/(V_i - V_0)$, where V_r is the retention volume. Ribonuclease, α-chymotrypsinogen, ovalbumine and bovine serum albumine (Pharmacia), ß-lactoglobulin and aldolase (Sigma) were used as standards to determine the Stokes' radius (R_s).

Both the regulatory and the catalytic subunit were overexpressed in *E. coli* as described previously (Saraswat et al., 1986; Slice and Taylor 1989). Site directed mutagenesis was carried out using the Kunkel method (Kunkel, 1985). Activity assays of the C-subunit were performed according to Cook et al. (1982).

cAMP-free R-subunit was prepared as follows. The R-subunit was unfolded by incubating the protein for 2 hours at room temperature in the running buffer containing 5 M urea. cAMP and protein were separated then with Pharmacia NAP 10 columns. Immediately after this separation, a buffer exchange against running buffer was performed using again NAP 10 columns.

RESULTS AND DISCUSSION

Measurements were made using several different columns and resins in order to find the most sensitive system for detecting changes in the conformation of each subunit. From 5 resins tested the best results were obtained using a Superose 12 (10/30, prepacked) column for the R-subunit or holoenzyme and a Superdex 75 (10/30, prepacked) column for the C-subunit, PKI and PKI(5-24). Under optimum conditions the Stokes' radius could be measured with a precision of 0.2 Å with as little as 40 pmoles (in a volume of 200 μl) of enzyme.

Catalytic Subunit: The Stokes' radius of mammalian C-subunit was determined to be 26.1 Å, in good agreement with previous independent measurements, and also consistent with the C-subunit being a relatively spherical molecule (Zoller et al., 1978). Although the recombinant enzyme has a specific activity and an electrophoretic mobility identical to the mammalian enzyme, it does not contain a myristic acid at the N-terminus in contrast to the mammalian enzyme (Carr et al., 1982). The recombinant C-subunit shows an R_s of 27.1 Å; a myristoylated form of this enzyme (Duronio et al., 1990) still has a R_s of 27.1 Å.

Protein	partition coefficient [σ]	Stokes' radius (R_S) [Å]	Previous measurements [Å] #
native r-C-subunit	0.2995	**27.1** ±0.2	ND
mammalian C-Subunit	0.3069	**26.1** ±0.2	26.1
myristoylated r-C-subunit	0.3005	**27.0** ±0.2	ND
native recomb. RI-Subunit	0.1827	**43.2** ±0.3	43.8
Holoenzyme [R_2C_2]	0.1702	**46.0** ±0.8	53.8

Table I   Hydrodynamic properties of native recombinant (r-) and mammalian proteins (# Zoller et al., (1978))

The effect of peptide binding on the Stokes' radius of the C-subunit was determined. High affinity binding of the I-form (Walsh et al., 1990) of the heat stable protein kinase inhibitor protein (PKI) to the C-subunit causes the R_s to increase from 27.2 Å to 31.1 Å. This increase is larger than the expected calculated value and may suggest that the PKI:C-subunit complex has significant dimensional asymmetry. Previous work indicated that MgATP is essential for high affinity binding of PKI to the C-subunit (Walsh et al., 1990). However, the stability of the C-subunit:PKI-complex at 0.2 μM or higher is independent of MgATP. At this protein concentration which is a 1000-fold higher than the K_d for PKI, MgATP is not required for binding of PKI. Further experiments with labelled compounds will show if an ATP effect is detectable at lower protein concentrations.   PKI does not compete well for the C-subunit when it is part of a holoenzyme complex.

Using   PKI(5-24), an inhibitory segment of PKI, an increase in Stokes' radius of 0.3 Å was observed at protein concentrations in the range of 0.2 μM. MgATP is required to see this change. In contrast, neutron scattering experiments done in collaboration with J. Parello, CNRS Montpellier, show that the radius of gyration (R_g) of the C-subunit decreases from 20 Å to 18.5 Å after the binding of PKI(5-24) and decreases to 18.0 Å after the addition of PKI(5-24) and MgATP (Parello et al., 1990). These experiments do not necessarily contradict each other because measurements of the R_s include the hydration shell of the protein whereas the radius of gyration, R_g does not. Nevertheless there are conformational changes associated with the binding of PKI(5-25) especially when MgATP is present.

Regulatory Subunit: In contrast to the C-subunit, the R-subunit shows a significant degree of dimensional asymmetry (frictional coefficient, $f/f_o = 1.47$; Zoller et al., 1979). A R_s of 43.2 Å was measured for the R^I subunit which is consistent with earlier measurements (table I) The addition of C-subunit does not significantly increase the R_s.

The R-subunit contains two tandem and homologous cAMP binding sites (site A and site B) at the carboxy terminus. Each site has an essential arginine which probably binds to the exogenic oxygen of cAMP (Weber et al. 1987). Replacement of each Arg (Arg 209 in site A and/or Arg 333 in site B) with Lys causes a signifi-cant decrease in the cAMP binding affinity. Purified w-type R-subunit typically contains 2 moles of bound cAMP per monomer. The following mutant R-subunits R209K, R333K and R209K;R333K were therefore used to determine the effect that cAMP binding has on the R_s. Each mutation effectively leaves the mutated site unoccupied with cAMP. As shown in Table II the protein which is fully occupied with cAMP has the largest R_s. A decrease in R_s (-0.5 to -0.8 Å) is measured when either of the cAMP binding sites is mutated. When both sites are mutated, the R_s is reduced by 1.8 Å. This value is larger than the simple additive values for each single mutation. A similar result was obtained when each of the R-subunits was artificially stripped of cAMP with urea (Table II). The results therefore indicate that, in contrast to the C-subunit, the binding of ligands increases the R_s of the R-subunit and that the binding of cAMP to each site contributes to the overall conformational change.

Protein	Partition coefficient [σ]	Stokes' radius (R_s) [Å]	σ + 50 μM cAMP	R_s [Å] + 50 μM cAMP
r-R^I-subunit w-type	0.183	43.2 ±0.20	0.183	43.2 ±0.15
urea unfolded	0.201	41.0 ±0.40		
r-R^I-subunit mutant R333 K	0.187	42.7 ±0.30	0.187	42.7 ±0.25
urea unfolded	0.198	41.4 ±0.35		
r-R^I-subunit mutant, R209K	0.189	42.4 ±0.20	0.183	43.2 ±0.25
urea unfolded	0.201	41.1 ±0.30		
r-R^Isubunit mutant, R209K;R333K	0.195	41.5 ±0.30	0.192	41.8 ±0.25

Table II: Hydrodynamic properties of mutant forms of the R^I-Subunit

LITERATURE

G. K. Ackers (1975) Molecular Sieve Methods of Analysis In: The Proteins, 3rd Edition. Vol. 1 (Neurath, H., and Hill, R. eds.):Academic Press New York

S. J. Beebe and J. D. Corbin (1986) Cyclic Nucleotide-Dependent Protein Kinases. The Enzymes: Control by Phosphorylation Part A,Vol. XVII eds. P.D. Boyer and E. G. Krebs Academic Press, Inc. New York.

S. A Carr, .Biemann, K.Shuji, S.Parmalee, D. C. Titani, K. (1982) n-Tetradecanoyl in the NH2 Terminal Blocking Group of the Catalytic Subunit of the Cyclic AMP-Dependent Protein Kinase from Bovine Cardiac Muscle. Proc. Natl. Acad. Sci. USA 79: 6128-6131

P.F. Cook,.Neville, M. E.Vrana, K. E.Hartl, F. T.Roskoski, Jr., R. (1982) Adenosine Cyclic 3',5'-Monophosphate Dependent Protein Kinase: Kinetic Mechanism for the Bovine Skeletal Muscle Catalytic Subunit. Biochemistry 21: 5794-5799

R. J.Duronio, Jackson-Machelski, E.Heuckeroth, R. O.Olins, P.Devine, C. S.Yonemoto, W.Slice, L. W.Taylor, S. S. Gordon, J. (1990)I.Protein N-myristoylation in *Escherichia coli* : Reconstitution of a Eukaryotic Protein Modification in BacteriaProc. Natl. Acad. Sci. USA 87: 1506-1510

T.A. Kunkel (1985) Rapid and efficient site-specific mutagenesis without phenotypic selection Proc. Natl. Acad. Sci. USA 82: 488-492

J. Parello, P. A. Timmins, J. M. Sowadski and S. S. Taylor (1990) Major Conformational Changes in the Catalytic Subunit of cAMP-Dependent Protein Kinase Induced by the Binding of a Peptide Inhibitor: A Small-Angle Neutron Scattering Study in Solution. J. Mol. Biol. submitted

J. Reed, .Kinzel, V Kemp, B.E. Cheung, H.-C Walsh, D.A. (1985) Circular Dichroic Evidence for an Ordered Sequence of Ligand/Binding Site Interactions in the Catalytic Reaction of the cAMP-Dependent Protein Kinase Biochem. 24: 2967-2973

L. D. Saraswat, M. Filutowics and S. S. Taylor (1986) Expression of the Type I Regulatory Subunit of cAMP-Dependent Protein Kinase in Escherichia coli. J. Biol. Chem. 261:11091-11096

L. W. Slice and S. S. Taylor (1989) Expression of the Catalytic Subunit of cAMP-dependent Protein Kinase In *Escherichia coli.* J. Biol. Chem. 264:20940-20946

S. S. Taylor, J. A. Buechler and Y. Yonemoto (1990) cAMP-dependent Protein Kinase: Framework for a Diverse Family of Regulatory Enzymes. Ann. Rev. Biochem. 59:971-1005

S. M. Van Patten, W. H. Fletcher and D. A. Walsh (1986) The Inhibitor Protein of the cAMP-Dependent Protein Kinase-Catalytic Subunit Interaction. J. Biol. Chem. 261:5514-5523

D. A. Walsh, K. L. Angelos, S. M. Van Patten, D. B. Glass and L. P. Garetto (1990) The Inhibitor Protein of the cAMP-Dependent Protein Kinase. CRC Crit. Rev. Biochem. 43-84

I. T.Weber, Steitz, T. A.Bubis, J.Taylor, S. S. (1987) Predicted Structures of the cAMP Binding Domains of the Type I and II Regulatory Subunits of cAMP-Dependent Protein Kinase Biochem 26: 343-351

D. A. Walsh, J. P. Perkins and E. G. Krebs (1968) An Adenosine 3',5'-Mono-Phosphate-Dependent Protein Kinase from Rabbit Skeletal Muscle. J. Biol. Chem. 243:3763-3765

M.N. Zoller, Kerlavage, A. R.Taylor, S. S. (1979) Structural Comparisons of cAMP-Dependent Protein Kinases I and II from Porcine Skeletal Muscle J. Biol. Chem. 254: 2408-2412

CRYSTALLOGRAPHIC STUDIES OF THE CATALYTIC SUBUNIT OF cAMP-DEPENDENT PROTEIN KINASE

Daniel R. Knighton‡, Jianhua Zheng‡, Victor A. Ashford§, Susan S. Taylor‡, Nguyen-huu Xuong§‡¶, and Janusz M. Sowadski||

University of California, San Diego
9500 Gilman Drive
La Jolla, CA 92093-0654

INTRODUCTION

The cAMP-dependent protein kinase (reviewed in Beebe and Corbin, 1986) is a tetrameric enzyme, probably ubiquitous in eukaryotic cells, composed of two regulatory and two catalytic (C) subunits. Upon binding cAMP the regulatory dimer releases two monomeric C-subunits, which then use MgATP to phosphorylate serine or threonine residues found typically in the sequence Arg-Arg-X-Ser/Thr in target proteins. Protein phosphorylation is a well-known mechanism for regulating protein function (reviewed in Krebs, 1985), and a large number of proteins have been found whose activity the C-subunit regulates this way, including many from the glycolytic and gluconeogenetic pathways (Krebs, 1985).

In addition to thus being a central cell regulatory protein, the C-subunit generates additional interest from its membership in a large family of protein kinases possessing a common ATP:protein phosphotransferase activity and likely a common structural basis for that activity. These other protein kinases possess short segments of sequence identity to the C-subunit over a 250-residue range; spacing between these segments is also conserved (Hanks et al., 1988). That the C-subunit and these other protein kinases share a common domain possessing the protein kinase activity is further suggested by specific chemical labeling of the C-subunit (Zoller and Taylor, 1979) and two other members of the

---

‡   Department of Chemistry
§   Department of Biology
¶   Department of Physics
|| Department of Medicine

NATO ASI Series, Vol. H 56
Cellular Regulation by Protein Phosphorylation
Edited by L. M. G. Heilmeyer, Jr.
© Springer-Verlag Berlin Heidelberg 1991

family (Kamps, et al., 1984; Russo et al., 1985) at an invariant lysine residue using the reactive ATP analog fluorosulfonyl benzoyl 5'-adenosine. Such labeling destroys protein kinase activity, and the essential nature of this lysine for normal enzymatic and biological activity has been confirmed for several members using site-directed mutagenesis (Chen et al., 1987; Chou et al., 1987; Kamps and Sefton, 1986; Snyder et al., 1985). The importance of determining the crystal structure of the cAMP-dependent protein kinase catalytic subunit thus derives from the insight that will be gained into the structural basis for C-subunit function and into the structural basis for protein kinase function of roughly one hundred known relatives (Hanks et al., 1988) for whom no crystal structures have been reported.

MATERIALS & METHODS

Porcine heart C-subunit was purified according to Nelson and Taylor (1983). Recombinant murine C-subunit was prepared as described in Slice and Taylor (1989). Porcine heart C-subunit crystals were prepared using hanging-drop vapor diffusion as described in Knighton et al (1990). Recombinant murine C-subunit crystals were prepared as described in Zheng et al (submitted Methods Enzymol., 1990) and Zheng et al (submitted J. Mol. Biol, 1990). PKI(5-24), sequence TTYADFIASGRTGRRNAIHD-NH2, was synthesized on an Applied Biosystems Model 430A peptide synthesizer. All crystal characterization and diffraction measurements were done using the area detector facilities of the NIH Resource for Protein Crystallographic Data Collection at UCSD, La Jolla, California (Xuong et al., 1985).

RESULTS AND DISCUSSION

Since the initial monoclinic form (Sowadski et al., 1985), four crystal forms have been grown. The first two, a cubic and a hexagonal form, were grown from porcine heart C-subunit (Knighton et al., 1990). The cubic form represents the apoenzyme, but the hexagonal represents a ternary complex with MgATP and PKI(5-24), the active portion of the heat-stable cAMP-dependent protein kinase inhibitor protein (Cheng et al., 1986). These two forms were grown from crystallization conditions differing only by the addition of MgATP and PKI(5-24) in molar ratio 5:20:1.1:1 for Mg:ATP:PKI:C to the cubic condition. Because this resulted in a different crystal form, the hexagonal form may represent a different conformation of the C-subunit than that of the apoenzyme. This hypothesis receives support from the low-angle neutron scattering experiments of Parello et al. (1990), which showed that the ternary complex of a murine recombinant C-subunit (Slice and Taylor, 1989) (rC) with MgATP and PKI(5-24) has a radius of gyration 1.5 Å less than that of free enzyme, and by the earlier circular dichroism studies of Reed, et al. (Reed and Kinzel, 1984; Reed and Kinzel, 1984; Reed et al., 1985), that found major secondary structure changes in the C-subunit upon binding substrates.

After growth of the cubic and hexagonal forms, two orthorhombic crystal forms were obtained using rC, which differs from porcine C-subunit primarily in that it lacks an N-terminal myristic acid moiety (Slice and Taylor, 1989). These two crystal forms both represent complexes of rC. The first of the pair is analagous to the hexagonal form in that it represents a ternary complex of MgATP:PKI(5-24):rC, while the second is a binary complex of PKI(5-24):rC. The space groups are identical, and their cell dimensions are nearly close enough to be considered isomorphous. Both of these crystal forms diffract to at least 2.7 Å.

Structure solution has been attempted on all 4 forms, but only the attempt on the binary rC:PKI complex has succeeded. The methodology used was isomorphous replacement (Blundell and Johnson, 1976). Several heavy-atom derivatives have been obtained, the most useful of which has been a two-site Hg derivative. The positions of the Hg atoms in this derivative are 24.5 Å apart. Because the C-subunit has two free cysteine residues, and because Hg reagents preferentially bind to thiols, it is likely that Cys 199 and Cys 343 have been mercurated. The inter-Hg distance is in rough agreement with the lower-limit inter-cysteine distance of approximately 30 Å determined by fluorescence resonance energy transfer (First et al., 1989).

A difference Fourier synthesis using the phases from the binary complex has revealed the location of the ATP in the orthorhombic ternary complex form. The location of the central Hg site coincides with the ATP location, and since it is known that Cys 199 is close to the ATP site (Bramson et al., 1982), the locations of Cys 199 and Cys 343 have probably been established.

Structural results available from model-building work are still too preliminary to describe in detail, but model building is continuing in a 2.7-Å electron density map. When completed, this model will allow rapid solution of the crystal structures of the hexagonal and orthorhombic ternary complexes, which will reveal details of the MgATP substrate binding and perhaps of the N-terminal myristic acid.

REFERENCES

Beebe SJ and Corbin JD (1986) Cyclic Nucleotide-Dependent Protein Kinases. In: P.D. Boyer and E. G. Krebs (eds) The Enzymes: Control by Phosphorylation Part A,Vol. XVII Academic Press, Inc. New York

Blundell TL and Johnson, LN (1976) Protein Crystallography. Academic Press, Inc. New York

Bramson HN, Thomas N, Matsueda R, Nelson NC, Taylor SS and Kaiser ET (1982) Modification of the Catalytic Subunit of Bovine Heart cAMP-Dependent Protein Kinase with Affinity Labels Related to Peptide Substrates. J. Biol. Chem. 257:10575-10581

Chen WS, Lazar CS, Poenie M, Tsien RY, Gill GN and Rosenfeld MG (1987) Requirement for intrinsic protein tyrosine kinase in the immediate and late actions of the EGF receptor. Nature 328:820-823

Cheng H-C, Kemp BE, Pearson RB, Smith AJ, Misconi L, Van Patten SM and Walsh DA (1986) A Potent Synthetic Peptide Inhibitor of the cAMP-Dependent Protein Kinase. J. Biol. Chem. 261:989-992

Chou CK, Dull TJ, Russell DS, Gherzi R, Lebwohl AU and Rosen OM (1987) Human Insulin Receptors Mutated at the ATP-binding Site Lack Protein Tyrosine Kinase Activity and Fail to Mediate Postreceptor Effects of Insulin. J. Biol. Chem. 262:1842-1847

First EA, Johnson DA and Taylor SS (1989) Fluorescence Energy Transfer between Cysteine 199 and Cysteine 343: Evidence for MgATP dependent Conformational Change in the Catalytic Subunit of cAMP-Dependent Protein Kinase. Biochemistry 28:3606-3613

Hanks SK, Quinn AM and Hunter T (1988) The Protein Kinase Family: Conserved Features and Deduced Phylogeny of the Catalytic Domains. Science 241:42-52

Kamps MP and Sefton BM (1986) Neither Arginine Nor Histidine Can Carry Out the Function of Lysine-295 in the ATP-binding Site of p60src. Mol. Cell. Biol. 6:751-752

Kamps M , Taylor SS and Sefton BM (1984) Oncogenic Tyrosine Kinases and cAMP-Dependent Protein Kinases Have Homologous ATP Binding Sites. Nature 310:589-592

Knighton DR, Xuong N-h, Taylor SS and Sowadski JM (1990) Crystallization Studies of cAMP-Dependent Protein Kinase. Cocrystals of the catalytic subunit with a twenty amino acid peptide inhibitor and MgATP diffract to 3.0Å resolution. Submitted: J. Mol. Biol.

Krebs EG (1985) The Phosphorylation of Proteins: A Major Mechanism for Biological Regulation. Biochem. Soc. Trans. 13:813-820

Nelson N and Taylor SS (1983) Selective Protection of Sulfydryl Groups in cAMP-Dependent Protein Kinase II. J. Biol. Chem. 258:10981-10987

Parello J, Timmins PA, Sowadski JM and Taylor SS (1990) Major Conformational Changes in the Catalytic Subunit of cAMP-Dependent Prortein Kinase Induced by the Binding of a Peptide Inhibitor: A Small-Angle Neutron Scattering Study in Solution. Submitted: J. Mol. Biol.

Reed J and Kinzel V (1984) Ligand Binding Site Interaction in Adenosine Cyclic 3', 5'-Monophosphate Dependent Protein Kinase Catalytic Subunit: Circular Dichroic Evidence for Intramolecular Transmission of Conformational Change. Biochemistry 23:968-973

Reed J and Kinzel V (1984) Near- and Far-Ultraviolet Circular Dichroism of the Catalytic Subunit of Adenosine Cyclic 5'-Monophosphate Dependent Protein Kinase. Biochemistry 23:1357-1362

Reed J, Kinzel V, Kemp BE, Cheung H-C and Walsh DA (1985) Circular Dichroic Evidence for an Ordered Sequence of Ligand/Binding Site Interactions in the Catalytic Reaction of the cAMP-Dependent Protein Kinase. Biochemistry 24:2967-2973

Russo MW, Lukas T, Cohen S and Staros JV (1985) Identification of Residues in the Nucleotide Binding Site of the Epidermal Growth Factor Receptor/Kinase. J. Biol. Chem. 260:5205-08

Slice LW and Taylor SS (1989) Expression of the Catalytic Subunit of cAMP-dependent Protein Kinase In *Escherichia coli*. J. Biol. Chem. 264:20940-20946

Snyder M, Bishop JM, McGrath J and Levinson A (1985) A Mutation in the ATP-binding Site of pp60v-src Abolishes Kinase Activity, Transformation, and Tumorigenicity. Mol. Cell. Biol. 5:1772-1779

Sowadski JM, Xuong N-h, Anderson D and Taylor SS (1985) Crystallization Studies of cAMP-Dependent Protein Kinase: Crystals of Catalytic Subunit Suitable for High Resolution Structure Determination. J. Mol. Biol. 182:617-620

Xuong N-h, Sullivan D, Nielsen C and Hamlin R (1985) Use of the Multiwire Area Detector Diffractometer as a National Resource for Protein Crystallography. Acta Cryst. B41:267-269

Zheng J-h, Knighton DR, Parello J, Taylor SS and Sowadski JM (1990) Crystallization of the Catalytic Subunit of cAMP-Dependent Protein Kinase. In press: Meth. Enzymol.

Zheng J-h, Knighton DR, Xuong N-h, Parello J, Taylor SS and Sowadski JM (1990) Crystallization Studies of the Catalytic Subunit of cAMP-Dependent Protein Kinase: Crystals of Murine Recombinant Catalytic Subunit and a Mutant, Cys 343 Æ Ser, Diffract to 2.7Å Resolution. Submitted: J. Mol. Biol.

Zoller MJ and Taylor SS (1979) Affinity Labeling of the Nucleotide Binding Site of the Catalytic Subunit of cAMP-Dependent Protein Kinase using p-Fluorosulfonyl-[14C]-Benzoyl 5'-Adenosine: Identification of a Modified Lysine Residue. J. Biol. Chem. 254:8363-8368

6-PHOSPHOFRUCTO 2-KINASE/FRUCTOSE 2,6-BISPHOSPHATASE: KINETIC CHANGES INDUCED BY PHOSPHORYLATION

Authors: VENTURA, F., ROSA J.L., AMBROSIO, S., GIL, J., TAULER, A. and BARTRONS, R.

Unitat de Bioquímica. Departament de Ciències Fisiològiques. Universitat de Barcelona

INTRODUCTION:

Fructose 2,6-bisphosphate (Fru-2,6-P_2) is a potent stimulator of 6-phosphofructo 1-kinase which has been identified in all eukaryotic cells. Its synthesis and breakdown are catalyzed by 6-phosphofructo 2-kinase (PFK-2) and fructose 2,6-bisphosphatase (FBPase-2), respectively. These two activities belong to separate domains of the same homodimeric protein (Van Schaftingen, 1987; Pilkis & EL-Maghrabi, 1988). Differences between tissues in the PFK-2/FBPase-2 activity ratio, kinetic and antigenic properties have suggested the existence of several PFK-2/FBPase-2 isoenzymes (Hue & Rider, 1987; Taniyama et al., 1988). The liver "L" and the muscle "M" isozymes have a common sequence of 438 aminoacids and differ at the N-terminus. In the "L" type, the divergent sequence is 32 residues long and contains the serine phosphorylated by cAMP-dependent protein kinase (PKA). The "M" type has a divergent sequence of 10 residues, containing a protein kinase C (PKC) target (Lively et al.,1988; Darville et al.,1989). The bovine heart "H" isozyme has a well-conserved sequence of 530 aminoacids, differing only at the N and C-terminus. Interestingly, it has phosphorylation targets for both PKA and PKC at the C-terminus (Sakata & Uyeda, 1990).

We have now studied the brain, heart, muscle and liver PFK-2/FBPase-2, comparing their kinetic properties and the influence of phosphorylation by both cyclic AMP-dependent protein kinase and protein kinase C.

NATO ASI Series, Vol. H 56
Cellular Regulation by Protein Phosphorylation
Edited by L. M. G. Heilmeyer, Jr.
© Springer-Verlag Berlin Heidelberg 1991

RESULTS AND DISCUSSION:

PFK-2 from brain, heart, liver and muscle were purified by polyethylene glycol fractionation and Mono-Q high performance liquid chromatography. The physical behaviour of PFK-2 differed between brain and heart activities respect to the liver and muscle counterparts. Brain and heart PFK-2 activities were eluted at 0.2 M NaCl, whereas liver and muscle were eluted in two peaks at about 0.4 M NaCl (Fig 1). The second peak eluted from Mono-Q column represents a phosphorylated form of the enzyme, since incubation of the fraction corresponding to the second peak with alkaline phosphatase and reaplication into the column leads to its disappearence.

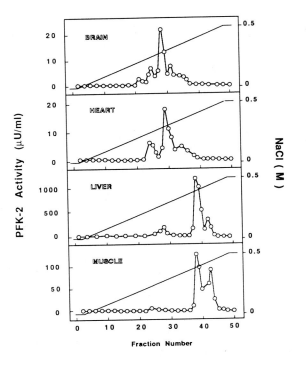

Fig 1.- ELUTION PROFILES OF BRAIN, HEART, LIVER AND MUSCLE PFK-2 FROM MONO-Q COLUMN

Polyethilene glycol fractions (6-21%) of rat tissues were applied to Mono-Q column (1ml). Fractions of 0.5 ml were collected after addition of the specific elution buffer. Pooled fractions contained: 0.55, 0.30, 3.4 and 0.53 mU/mg of protein for brain, heart, liver and muscle respectively. Other two experiments were done with similar results.

All PFK-2 and FBPase-2 activities were co-eluted in the same fractions although their activity ratios are quite different. Enzymes from brain and heart resemble more a kinase than a phosphatase whereas the opposite occurs for the muscle type. The purified enzymes also showed distinct responses to allosteric modulators such as citrate and glycerol 3-phosphate (Table I). PFK-2 from liver and heart are phosphorylated by PKA (Van

Phosphorylation effects on PFK-2/FBPase-2

Properties	Tissues			
	Liver	Muscle	Heart	Brain
PFK-2/FBPase-2 activity ratio	1 - 4	0.2	8 0	7 0
Phosphorylation by PKA	+	-	+	N.D.
Activity changes	+	-	+ / -	-
Phosphorylation by PKC	-	N.D	+	N.D.
Activity changes	-	-	+ / -	-
Glycerol 3-P inhibition	+	-	-	-
Citrate inhibition	+	+	+ +	+ +

Table I .- **PHOSPHORYLATION EFFECTS ON PFK-2/FBPase-2**

Data were taken from (Rider & Hue, 1986; Hue & Rider, 1987; Kitamura **et al.**, 1988 and present results). Abbreviations: **+**, detectable effects; **-**, no effects; **N.D.**, not done.

Schaftingen, 1987; Pilkis & El-Maghrabi, 1988; Kitamura **et al.**, 1988). However, their activity is clearly modified only for the liver isozyme, resulting in an inhibition of the kinase and activation of the phosphatase activities (Van Schaftingen, 1987; Pilkis & El-Maghrabi, 1988). In the other hand, heart PFK-2/FBPase-2 is phosphorylated by PKA and PKC (Rider & Hue, 1986; Kitamura **et al.**, 1988). This phosphorylation does not alter the activity of the enzyme (Rider & Hue, 1986), although Kitamura **et al.**, (1988) have shown slightly activations by phosphorylation. The phosphorylated serines in the heart PFK-2 are located near the C-terminus (Sakata & Uyeda, 1990), whereas the phosphorylation site of the liver isozyme is known to be located near the N-terminus (Lively **et al.**, 1988; Darville **et al.**, 1989). We have analyzed the effects of phosphorylation by PKC on the PFK-2 activity of brain, heart, liver and muscle isoforms. We found no changes in the activities in any of them, either at saturating or subsaturating concentration of substrates (Table I).

The similarity of the mRNA for the three isozymes (Sakata & Uyeda, 1990; Darville **et al.**, 1989) suggests that they arise from the same gene by using alternative splicing. The "M" isozyme differs from the "L" form only in the first exon. This substitution yields a change in the phosphorylation targeting. The cDNA sequence of bovine heart isozyme (Sakata & Uyeda, 1990) suggests that very likely this form has been originated by another alternative splicing or arises from a close related gene. In the first case, the 1 and 14 exons would be changed (Fig. 2). Northern blot analysis of total RNA from brain showed identical pattern than that of heart RNA and clearly different from those of liver and muscle (not shown).

All these results suggest that the brain contains an isoenzymatic form related to the heart and different from the liver and muscle isozymes.

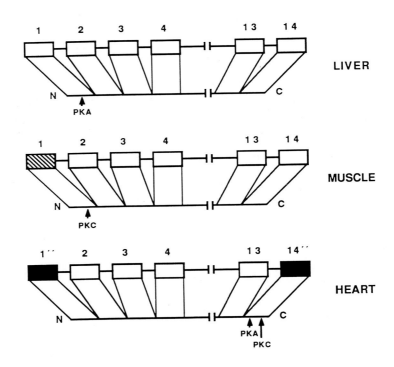

Fig 2.- **HYPOTHESIZED ALTERNATIVE SPLICING OF PFK-2/FBPase-2 ISOZYMES.**

Alternative splicing of the first exon for liver and muscle isozymes and the first and fourteenth exons for the heart isozyme are shown. Arrows point to the phosphorylation targets for these isozymes.

REFERENCES

Darville, M.I., Crepin, K.M., Hue, L. and Rousseau, G.G. (1989) " 5' flanking sequence and structure of a gene encoding rat 6-phosphofructo 2-kinase/fructose 2,6-bisphosphatase", Proc. Natl. Acad. Sci. USA, **86**, 6543-6547

Hue, L. and Rider, M.H. (1987) " Role of fructose 2,6-bisphosphate in the control of glycolysis in mammalian tissues", Biochem. J. **245**, 313-324

Kitamura, K., Kangawa, K., Matsuo, H. and Uyeda, K. (1988) "Phosphorylation of myocardial Fructose 6-phosphate 2-kinase/fructose 2,6-bisphosphatase by cAMP-dependent protein kinase and protein kinase C", J.Biol. Chem. **263**, 16796-16801

Lively, M.O., El-Maghrabi, M.R., Pilkis, J., D'Angelo, G., Colosia, A.D., Ciavola, J.A., Fraser, B.A. and Pilkis, S.J. (1988) "Complete amino acid sequence of rat liver 6-phosphofructo 2-kinase/fructose 2,6-bisphosphatase" J. Biol. Chem. **263**, 839-849

Pilkis, S.J. and El-Maghrabi, M.R. (1988) "Hormonal regulation of hepatic gluconeogenesis and glycolysis", Ann. Rev. Biochem. **57**, 755-783

Rider, M.H. and Hue, L. (1986) " Phosphorylation of purified bovine heart and rat liver 6-phosphofructo 2-kinase by protein kinase C and comparison of the fructose 2,6-bisphosphatase activity of the two enzymes", Biochem. J. **240**, 57-61

Sakata, J. and Uyeda, K. (1990) "Bovine heart fructose 6-phosphate 2-kinase/fructose 2,6-bisphosphatase: Complete amino acid sequence and localization of phosphorylation sites", Proc. Natl. Acad. Sci. USA **87**, 4951-4955

Taniyama, M., Kitamura, K., Thomas, H., Lawson, J.W.R. and Uyeda, K. (1988) " Isozymes of Fructose 6-phosphate 2-kinase/fructose 2,6-bisphosphatase in rat and bovine heart, liver and skeletal muscle", Biochem. Biophys. Res. Commun. **157**, 949- 954

Van Schaftingen, E. (1987) " Fructose 2,6-bisphosphate", Adv. Enzymology **59**, 315-395

ENZYMES INVOLVED IN THE REVERSIBLE PHOSPHORYLATION IN MICROVESSELS OF THE BRAIN

U. Dechert, M. Weber-Schaueffelen, S. Lang-Heinrich and E. Wollny
Institut für Biochemie
TH Darmstadt
Petersenstraße 22
6100 Darmstadt
F.R.G.

Introduction

The Blood-Brain Barrier (BBB) of vertebrates is located in the microvessels of the brain. The endothelial cells coating these capillaries are sealed together by tight junctions to restrict any uncontrolled diffusion of water soluble substances across the BBB. The transport of polar substances including nutrients and waste products into and out of the brain is mediated by specific transport mechanisms. Brain capillaries have been shown to be very active in the reversible protein phosphorylation which led to the hypothesis of a rapid control of brain microvessel functions by protein kinases and phosphatases (Pardridge et al., 1985). There are hints that all groups of second messenger dependent protein kinases are present in the endothelial cells of these vessels (Oláh et al., 1988).

Results

Here we describe three enzymes involved in the reversible phosphorylation in capillaries of the porcine brain. These proteins comprise a phosphoprotein phosphatase, a second messenger independent protein kinase and a phospholipase C. The physical and enzymatic data of these enzymes are compared to proteins derived from other tissues and species.

THE PROTEIN KINASE

The protein kinase has a molecular mass of 97,000 as judged by SDS-gelelectrophoresis. The kinase activity was determined in a

NATO ASI Series, Vol. H 56
Cellular Regulation by Protein Phosphorylation
Edited by L. M. G. Heilmeyer, Jr.
© Springer-Verlag Berlin Heidelberg 1991

kinase assay followed by gel electrophoresis and subsequent autoradiography as published by Dechert et al. (1989). Antibodies were raised against the enzyme and used for the screening of a cDNA library.

CONSTRUCTION OF A CDNA LIBRARY

The RNA was isolated from porcine brain microvessels and Poly(A)$^+$-RNA was prepared by oligo-(dT)-cellulose chromatography. The cDNA library was constructed in a λ-ZAP XR vector following the instruction manuals ZAP-cDNA synthesis kit and Uni-ZAP XR cloning kit, Fa. Stratagene, La Jolla, CA.

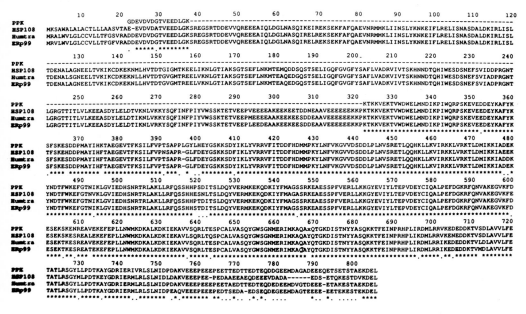

Figure 1 Protein sequence alignment of the porcine protein kinase (PPK) versus HSP108 (chicken cell cultures), Humtra (human tumor cells) and ERp99 (murine endoplasmin). Asterisks indicate a match across all sequences, while conservative substitutions are marked by dots. The aminoterminal sequence data of the PPK (22-37) were derived by Edmann degradation.

SCREENING OF THE LIBRARY

The λ-ZAP bacteriophage recombinant library was screened using the antibodies mentioned above as described in the manufactures instructions (PicoBlue Immunoscreening kit, Fa. Stratagene). The

97kD-protein kinase aminoacid sequence derived from the cDNA sequence (484 aminoacids) was compared to SwissProt release 15.0
and PIR-NBRF release 26.0 Protein Sequence Library databases using the FASTA program on a VAX 8530. The best scoring sequences
were ERp99 (A29317), Humtra (A30935, S06189) and HSP108 (Hhch08)
in case of the protein database search. As well as a nucleic
acid database search the protein database search revealed a high
degree of similarity between the protein kinase and the HSP´s
(95%(human) – 88%(chicken) identity resp.) as shown in fig.1.

HISTOCHEMISTRY

Histological studies using the antibodies mentioned above,
showed that the enzyme is not localized within the nuclei of
capillary endothelial cells but rather within the cell membranes
or the cytosol (fig.2a and b).

THE PHOSPHOPROTEIN PHOSPHATASE

The phosphatase was purified and characterized as published by
Weber et al 1987. The enzyme which shows an apparent molecular
mass of 56kD on SDS PAGE is strongly inhibited by addition of
inhibitor 2. Therefore the enzyme is a type 1 phosphatase according to Cohen´s nomenclature. Metal ions as manganese, magnesium or calcium ions did not show any effect on the phosphatase whereas zinc ions inhibit the enzyme in micromolar concentrations. The physical and enzymatic data of the protein are
very similar to a phosphatase derived from rabbit reticulocytes
as described earlier (Tipper et al., 1986).

THE PHOSPHOLIPASE C

The phospholipase C has an apparent molecular mass of 57kD as
judged by SDS PAGE as well as Western blot analysis. Using the
antibodies raised against the protein we identified the corresponding cDNA in a brain capillary derived expression cDNA library in bacteriophage λgt11. The screening and sequencing procedures were performed according to the protein kinase described
above. 396 nucleotides of the coding region (30% of the total
protein sequence) could be determined and aligned to existing
sequence data. The results indicate a high degree of similarity
(89%) of the brain microvessel derived enzyme as compared to the
phospholipase C from guinea pig uterus.

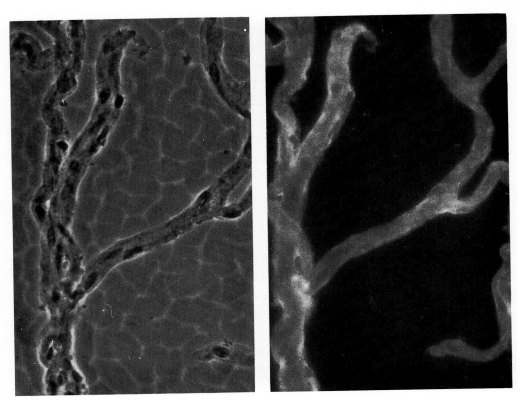

Figure 2a Phasecontrast microscopy of porcine brain microvessels. Amplification: 600-fold.
Figure 2b Immunohistochemistry: porcine brain microvessels stained with specific anti-97K-PK antibodies and fluoresceine labeled second antibodies (goat-anti-rabbit-IgG). Amplification: 600-fold.

Discussion

Three enzymes involved in the reversible protein phosphorylation in brain microvessels where isolated and characterized. The protein kinase seems to be a member of a class of heat shock proteins (HSP 90). The cDNA and aminoacid data which are available show that the kinase may be identical to three heat shock proteins derived from chicken, mouse and human. This result is surprising because an enzymatic activity never was attributed to a class of heat shock proteins. Recently however (Flaherty et al., 1990) could show that a proteolytic fragment of HSP 70 has the identical three dimensional structure as hexokinase as deduced by X-ray crystallography. The sequence data of the HSP´s shown in fig.1 indicate certain sequence motifs that are required or common for protein kinases. The sequence stretch from position 196 to 201 for example is characteristic for a nucleotide binding site. The histological data suggest that the kinase is an amphitrophic protein located both in the membranes and cytoplasm of the endothelial cells. The phosphoprotein phosphatase and phospholipase C seem to be identical with or at least related to corresponding enzymes derived from various tissues and species. The phospholipase C as a key enzyme in the signal transduction pathways mediated by diacylglycerol and inositoltriphosphate as second messengers may proof to be useful in further studies of the kinase C. This enzyme is known to be involved in a variety of regulatory processes in many kinds of tissues including brain microvessels.

ACKNOWLEDGMENT: We wish to thank Prof.Dr.H.Zimmermann, Zool.Inst., J.W.Goethe Universität, Frankfurt for providing the histological data.

References

Dechert U,Weber M,Weber-Schaeuffelen M and Wollny E (1989) Isolation and Partial Characterization of an 80,000-Dalton Protein Kinase from the Microvessels of the Porcine Brain. J Neurochem 53:1268-1273

Flaherty KM,DeLuca-Flaherty C and McKay DB (1990) Three-dimensional structure of the ATPase fragment of a 70K heat-shock cognate protein. Nature 346:623-628

Oláh Z,Novák R,Dux E and Joó,F (1988) Kinetics of Protein Phosphorylation in Microvessels Isolated from Rat Brain:Modulation by Second Messengers.J Neurochem 51:49-56

Pardridge WM,Yang J and Eisenberg,J (1985) Blood-Brain Barrier Protein Phosphorylation and Dephosphorylation. J Neurochem 45:1141-1147

Tipper J,Wollny E,Fullilove S,Kramer G and Hardesty B (1986) Interaction of the 56,000-Dalton Phosphoprotein Phosphatase from Reticulocytes with Regulin and Inhibitor 2. J Biol Chem 261:7144-7150

Weber M,Mehler M and Wollny E (1987) Isolation and Partial Characterization of a 56,000-Dalton Phosphoprotein Phosphatase from the Blood-Brain Barrier. J Neurochem 49:1050-1056

COVALENT MODIFICATION OF CREATINE KINASE BY ATP: EVIDENCE FOR AUTOPHOSPHORYLATION

W. Hemmer[1], S. J. Glaser[2], G. R. Hartmann[2], H. M. Eppenberger[1] and T. Wallimann[1]

[1]Institute for Cell Biology
  ETH Hönggerberg
  CH-8093 Zürich, Switzerland

[2]Institute for Biochemistry
  Ludwig-Maximilians-Universität
  D-8000 München 2, FRG

INTRODUCTION

Creatine kinase (CK, EC 2.7.3.2) catalyses the reversible reaction:

$$\text{MgATP} + \text{Creatine (Cr)} \leftrightarrow \text{MgADP} + \text{Phosphocreatine (PCr)} + \text{H}^+ \qquad (1)$$

The reaction mechanism has been classified as rapid equilibrium-type with all evidence pointing to a direct, in-line transfer of the phosphoryl group between bound substrates (Kenyon, Reed 1983), without any direct evidence for a covalent phosphoryl-enzyme intermediate. Only one report so far indicated the possible existence of such a covalent phosphoryl-enzyme intermediate (Molnar, Lorand 1960). Rat brain CK (Mahadevan et al., 1984) and recently also chicken brain-type creatine kinase have been shown to be partially phosphorylated (Quest et al. 1990); in the latter case the phosphorylated CK correlates with enzyme species showing altered kinetic parameters with respect to the K_m for PCr.

MATERIALS AND METHODS

All chemicals were of analytical grade quality. $^{32}\text{P-}\gamma\text{-ATP}$ was purchased from Amersham (PB10218), nonradioacative ATP and rabbit M-CK from Boehringer Mannheim. Protein determinations were performed according to Bradford, using Biorad reagent. CK activity was measured by pH-stat (Wallimann et al. 1984).

Autophosphorylation assay: CK was routinely incubated at a concentration of 0.2 mg/ml with 50 nM $^{32}\text{P-}\gamma\text{-ATP}$ in 0.1 M glycylglycine pH 7.5, 0.5 mM EDTA and 2 mM Mg-acetate for 2 h before being analysed by SDS-PAGE. The reaction was started by addition of enzyme to the reaction mixture and stopped by addition of

NATO ASI Series, Vol. H 56
Cellular Regulation by Protein Phosphorylation
Edited by L. M. G. Heilmeyer, Jr.
© Springer-Verlag Berlin Heidelberg 1991

SDS-sample buffer. Gels were electrophoretically blotted onto PVDF membranes and analysed by autoradiography. For quantitation, the protein bands of amidoblack stained blots were excised and analysed by Cherenkov counting or alternatively autoradiograms were scanned with a Computing Densitometer (Molecular Dynamics 300 A).

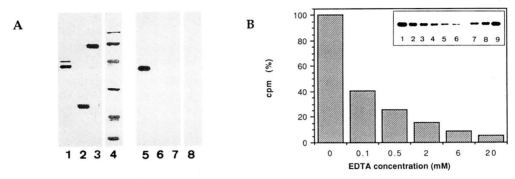

Figure 1): A, CK is autophosphorylated: Identical amounts of rabbit M-CK (lanes 1,5), myokinase (2,6), purified in our laboratory, and BSA (3,7) were incubated under standard assay conditions, analysed by SDS-PAGE, blotted on PVDF (lanes 1 to 4) and autoradiographed.(lanes 5 to 8). MW standards of 97, 66, 43, 31, 21 and 14 kD were loaded in lane 4 (8). Only CK was radioactively labeled (5). B, Inhibition of autophosphorylation by EDTA: Identical amounts of CK were incubated as in A but with increasing amounts of EDTA (see graph) leading to a decrease in autophosphorylation. (Lanes 1-6 of the autoradiogram shown in the inset correspond to the fractions represented by the bars). Lanes 7, 8 and 9 of the inset show that addition of increasing amounts of MgCl2 (0.1, 1 and 5mM in lanes 7, 8 and 9 respectively) lead to reversal of the inhibition of autophosphorylation by EDTA.

RESULTS AND DISCUSSION

Autophosphorylation of creatine kinase: Extensively dialysed creatine kinase, purified from rabbit skeletal muscle, showed incorporation of ^{32}P when incubated with ^{32}P-γ-ATP, whereas other proteins tested, even if they use ATP as a substrate (myokinase), are not radioactively labeled in such assays (Fig.1A). The ATP-mediated modification of CK is of covalent nature since the reaction product is stable in SDS-gels and withstands treatment with 8 M urea, 4 M guanidinium hydrochloride or TCA-precipitation (not shown).

Preliminary experiments with other CK isoenzymes from rabbit and chicken, purified to homogeneity by completetely different methods, all resulted in radioactively labeled CK isoproteins, indicating that this modification is a specific feature of CK. In addition, coincubations of CK together with the proteins shown in Fig.1 or with casein (Sigma) resulted also in specific labeling of the CK band (not shown). These facts make accidental contamination of different CK preparations by a protein kinase a very unlikely explanation for the above observations and strongly indicate that CK can undergo autophosphorylation.

Characterization of creatine kinase autophosphorylation: Like the enzymatic reaction of CK itself, also the autophosphorylation reaction of CK depends on divalent metal ions, for the incorporation of radioactivity can be inhibited by increasing concentrations of EDTA (Fig.1B). This inhibition by EDTA can be reversed by addition of magnesium ions as shown in the inset of Fig.1B. Similar results were obtained with manganese but not with zinc ions (not shown). Furthermore, the reaction is temperature dependent and is inhibited by high salt.

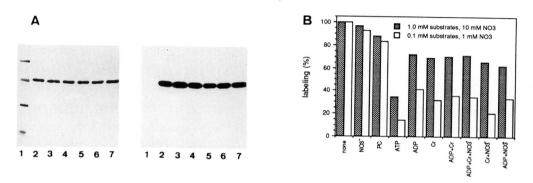

Figure 2): A, Autophosphorylation is not due to a phospho-intermediate of the CK reaction: Identical amounts of CK (left panel shows protein staining) were incubated as described before with 50 nM ^{32}P-γ-ATP. Then the reaction was stopped (lane 2) or one of the substrates of the CK reaction (ATP, PCr, ADP and Cr) (lanes 3-6) was added to a final concentration of 10 mM each and the mixture incubated for an additional 2h. Even if both phosphate acceptor molecules for CK (ADP and Cr) were added together to the incubation mixture (each at 10 mM, lane 7), no reduction of radioactive labeling could be observed (In the case that labeling were due to a phospho-intermediate, a reduction or removal of the signal on the autoradiogram [right panel] would be expected). B, No significant protection by substrates: CK was incubated under standard assay conditions with or without additional substrates, each substrate at 0.1mM and NO_3 at 1mM (open bars) or substrates each at 1mM and NO_3 at 10mM (dashed bars). Even the combination of substrates forming a transition state-analogue complex with CK do not significantly reduce autophosphorylation (see text).

To investigate whether the labeling of CK by ^{32}P-γ-ATP was due to a phospho-intermediate of its enzymatic reaction or not, CK was labeled under standard assay conditions, checked for enzymatic activity by pH stat and then further incubated in the presence of excess of substrates for an additional 2 hours. Activity measurements revealed that CK remained fully active during the incubation time. None of the CK substrates was able to remove the radioactive label significantly even when added at 2 x 10^5- fold excess over ^{32}P-γ-ATP (Fig.2A). Also coincubation with 10 mM Cr plus 10 mM ADP, both possible phosphate acceptors of a hypothetical phospho-intermediate, did not reduce the labeling of CK (Fig.2A, lane 7). In further experiments, the influences of single substrates and substrate combinations, including the one forming a transition state-analogue complex with CK (CK•MgADP•Cr•NO$_3^-$; Milner-White, Watts 1971), were analysed by adding these together with ^{32}P-γ-ATP to the standard incubation mixture (Fig.2B). Addition of cold ATP lead to the strongest inhibition of labeling, ADP and creatine had less and PCr only very little if any effect. The effect of Cr was caused by the reduction of the effective ^{32}P-γ-ATP concentration due to the catalytic conversion of the latter into ADP and ^{32}PCr, whereas the effect of ADP, although less clear, might be explained by allosteric competition. Most interestingly, addition of the combination of substrates able to form a transition state-analogue complex with CK (see above) did not lead to stronger inhibition than addition of single substrates or simple substrate combinations (Fig.2B). This indicates that the observed autophosphorylation is not due to a phospho-intermediate formed during the catalytic CK reaction and that the autophosphorylationsite(s) might be distant to the catalytic site. This was supported by the fact, that treatment of CK with Dinitrofluorobenzene (DNFB), a reagent which reacts specifically with the reactive sulfhydrylgroup of CK (James et al. 1990), inhibits completely enzyme activity, but not the autophosphorylation. (not shown).

In order to examine whether the mechanism of CK autophosphorylation is an

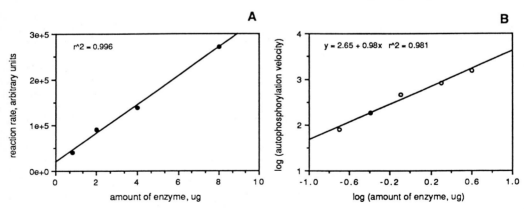

Figure 3): Indications for an intramolecular reaction. CK was incubated either under standard assay conditions (open circles) or in the presence of additional cold ATP at a total concentration of 50 µM (closed circles). A: Autophosphorylation is first-order with respect to enzyme concentration; B: Relationship between log CK autophosphorylation velocity and log enzyme concentration (van't Hoff plot). The slope this diagram (0.97) clearly indicates that the observed autophosphorylation is an intramolecular reaction, for such a plot would have a slope of 1 for an intramolecular or a slope of 2 for an intermolecular reaction.

intra- or intermolecular process, the influence of enzyme concentration on the reaction rate was investigated as it has been described for type II cAMP-dependent protein kinase (Ehrlichman et al. 1983) or protein kinase C (Huang et al. 1986). All experiments indicated that for CK the mechanism is intramolecular (Fig 3) since phosphorylation was first-order with respect to enzyme concentration(Fig.3A) and van't Hoff plots revealed slopes between 0.8 and 1 (Fig.3B, here the slope is 0.97). Theoretically, such a plot would yield slopes of 1 and 2 for an intramolecular and an intermolecular reaction, respectively.

In conclusion, the data presented show that creatine kinase from rabbit skeletal muscle can undergo autophosphorylation. Obviously, this is not due to the formation of a phospho-intermediate during the enzymatic reaction, but nevertheless the mechanism of the autophosphorylation appears to be an intramolecular one. The reaction-rate of the autophosphorylation did neither correlate with nor ultimately depend on the enzymatic activity. The results indicate that CK is able to react with ATP in at least two different ways which would be in agreement with the presence of distinct adeninenucleotide binding sites within a CK monomer as already had been suggested from results with ^{31}P-NMR (Nageswara, Cohn 1981).

REFERENCES

Erlichmann J, Rangel-Aldao R, Rosen OM (1983) Reversible autophosphorylation of type II cAMP-dependent protein kinase: distinction between intramolecular and intermolecular reactions. Meth Enzymol 99:176-186

Huang KP, Chan KFJ, Singh TJ, Nakabayashi H, Huang FL (1986) Autophosphorylation of rat brain Ca2+-activated and phospholipid-dependent protein kinase. J Biol Chem 261: 12134-12140

James P, Wyss M, Lutsenko S, Wallimann T, Carafoli E (1990) ATP binding site of mitochondrial c reatine kinase: Affinity labelling of Asp-335 with CIRATP. FEBS Letters 273: 139-143

Kenyon GL, Reed GH (1983) Creatine kinase: structure-activity relationships. In: Meister A (ed) Adv in Enzymology and Related Areas in Molecular Biology, vol 54. J Wiley & Sons Inc, p 367-425

Mahadevan LC, Whatley SA, Leung TKC, Lim L (1984) The brain isoform of a key ATP-regulating enzyme, creatine kinase, is a phosphoprotein. Biochem J. 222: 139-144

Milner-White EJ, Watts DC (1971) Inhibition of Adenosine 5'-Triphosphate-Creatine Phosphotransferase by Substrate-Anion Complexes. Biochem J 122: 727-740

Molnar J, Lorand L (1960) Phosphoryltransfer with Phosphocreatine or Phosphoenolpyruvate and Adenosinemonophosphate. Fed Proc 19: 260

Nageswara, B. D.; Cohn, Mildret (1981) 31P NMR of enzyme-bound substrates of rabbit muscle creatine kinase. J Biol Chem 256: 1716-1721.

Quest AFG, Soldati T, Hemmer W, Perriard J-C, Eppenberger HM Wallimann T (1990) Phosphorylation of chicken brain-type creatine kinase: effect on a relevant kinetic parameter and gives rise to protein-microheterogeneity in vivo. FEBS Letters 269: 457-464

Wallimann T, Schloesser T, Eppenberger HM (1984) Function of M-line-bound creatine kinase as an intramyofibrillar ATP regenerator at the receiving end of the phosphorylcreatine shuttle in muscle. J Biol Chem 259: 5238-5246

TWO ADJACENT PHOSPHOSERINES IN BOVINE, RABBIT AND HUMAN CARDIAC TROPONIN I

Karin Mittmann, Kornelia Jaquet and Ludwig M.G. Heilmeyer Jr.
Institut für Physiologische Chemie, Abteilung für Biochemie Supramolekularer Systeme, Ruhr-Universität Bochum, Postfach 102148, D-4630 Bochum, FRG

INTRODUCTION

Cardiac troponin I, the inhibitory subunit of troponin, is phosphorylated in mammalian hearts upon β-adrenergic stimulation (England, 1975; Solaro et al., 1976). On rabbit troponin I it is believed to occur at serine-20 (Moir & Perry, 1977). In hormonally stimulated perfused hearts a correlation between phosphorylation of this single amino acid and change in contractility was expected which, however, could not be clearly demonstrated (Westwood & Perry, 1981). Freshly isolated troponin I from both rabbit and bovine heart contains 1.5–1.9 mol of phosphate per mol of protein (Cole & Perry, 1975; Swiderek et al., 1988) indicating that in addition to phosphoserine-20 a second phosphate residue of unknown location is bound in the N-terminal sequence (Moir & Perry, 1981). Recently, two adjacent phosphoserine residues in position 23 and 24 were identified in troponin I from bovine heart (Swiderek et al., 1988).

Skeletal muscle troponin I does not contain the N-terminal region present in the heart isoform. Therefore, cardiac troponin I is an excellent heart specific cell marker which can be detected in blood plasma of patients with acute myocardial infarction (Cummins et al., 1987). Antibodies against phosho- and dephospho troponin I may even allow to analyze the hormonal status of a heart. A prerequisite is the localization of the phosphorylation domain in human troponin I.

The location of a second in vivo phosphorylation site of rabbit cardiac troponin I will be reported requiring a correction of the amino acid sequence. Additionally, the phosphorylation domain of human cardiac troponin I will be shown.

MATERIAL AND METHODS

Freshly excised hearts from rabbit, immediately frozen with liquid nitrogen, were stored at −50°C. Myocardial tissue probes from explantated human hearts were obtained from the Herzzentrum Nordrhein Westfalen, Bad Oeynhausen.

NATO ASI Series, Vol. H 56
Cellular Regulation by Protein Phosphorylation
Edited by L. M. G. Heilmeyer, Jr.
© Springer-Verlag Berlin Heidelberg 1991

Troponin I was purified from both human and rabbit heart according to Syska et al. (1974). The protein-containing fractions were applied onto PD-10 columns (Pharmacia) to remove urea. Troponin I fragments were isolated after endoproteinase Lys-C and trypsin digestion as described previously by Swiderek et al. (1988). Phosphoserine containing peptides were analyzed following S-ethylcysteine modification (Meyer et al., 1987; Swiderek et al., 1988). Protein-bound phosphate was determined by the method of Stull and Buss (1977). Amino acid and sequence analyses were carried out as described previously (Bidlingmeyer et al., 1984; Swiderek et al., 1988). Mass-spectra were recorded on a plasma desorption mass spectrometer in the Biomedical Center of the University of Uppsala. 1 nmol of the phosphopeptide was solubilized in 30 μl 0.1 % (v/v) trifluoroacetic acid. 7 μl of the protein solution were incubated with ethanol for 15 min before the probe was applied to the mass-spectrometer.

RESULTS

To localize phosphoserine residues in rabbit and human troponin I fragments obtained by digestion with lysine-specific proteinase were separated by reversed-phase high-performance liqiud chromatography (Fig. 1). In each chromatogram two peaks, (a and b for rabbit or c and d for human), were found to contain one or two moles of phosphoserine per mol of lysine respectively (table I). Peptides a and b from rabbit cardiac troponin I differ in mass by one phosphate group (Fig. 2). The determined molecular masses of both these peptides (4098 and 4018) agree with the amino acid composition (table I). The calculated molecular mass of the N-terminal Lys-C peptide 1-31 (3444) according to the published rabbit cardiac troponin I sequence (Grand et al., 1976) is by 574 smaller than the determined mass of peptide b (4018). Therefore the amino acid composition as well as the molecular mass indicate that approximately 5 amino acids are present additionally in this peptide.

The isolated N-terminal blocked peptides a, c and d were further digested with trypsin. Peptide aI from rabbit (Fig. 3A), as well as peptide cI from human cardiac troponin I (Fig. 3B) showed a characteristic ratio of approximately 3 mol arginine to 2 mol phosphoserine (table I). The monophosphorylated peptide dI (Fig. 3C) derived from human cardiac troponin I contained only two arginine residues. Following S-ethylcysteine modification all three tryptic phosphopeptides aI, cI and dI were subjected to Edman degradation. During degradation of both bisphosphorylated peptides from rabbit and human cardiac troponin I (Fig. 4A and B) nearly exclusively the phenylhydantoin derivatives of S-ethylcysteine (S-Et-Cys) were found in

151

cycle 3 and 4. Thus, in both the rabbit and the human peptides phosphoserine residues were present at these two positions. In agreement with the amino acid composition the phosphopeptide from human cardiac troponin I showed no alanine residue following the second phosphoserine residue. During Edman degradation of the monophosphorylated peptide *dI* serine was found in cycle 2 and S-ethylcysteine in cycle 3 (Fig. 4C).

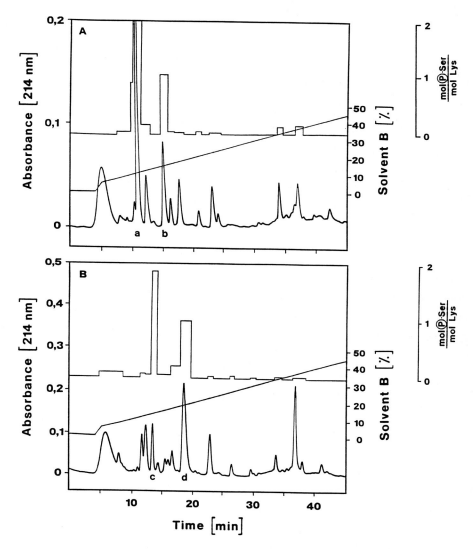

Fig. 1. Separation of endoproteinase Lys-C peptides from rabbit and human cardiac troponin I.
10 nmol of each troponin I from rabbit (A) containing 1.9 mol phosphate per mol protein and human cardiac muscle (B) containing 1.0 mol phosphate were digested with endoproteinase Lys-C. The generated peptides were separated as described previously (Swiderek et al., 1988).

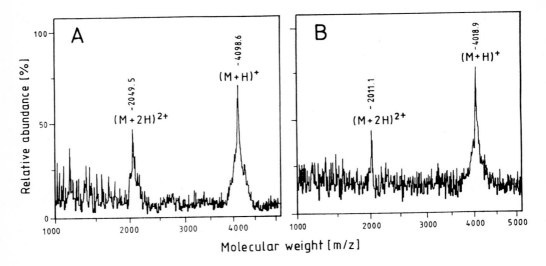

Fig. 2. Plasma desorption mass spectra of the phosphorylated N-terminal peptides *a* and *b* from rabbit cardiac troponin I.
230 pmol of each peptide *a* (A) and peptide *b* (B) were applied. The calculated masses of 4098 and 4018 respectively according to amino acid composition are given in table I.

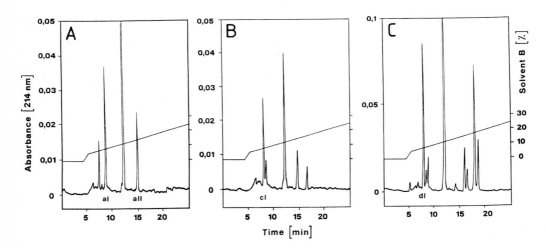

Fig. 3. Separation of the tryptic fragments from the phosphoserine containing troponin I peptide *a* (rabbit), peptide *c* and *d* (human).
1.0 nmol peptide *a* (A), 0.9 nmol peptide *c* (B) and 2.3 nmol peptide *d* (C) were digested with trypsin and the generated fragments were separated as described by Swiderek et al. (1988).

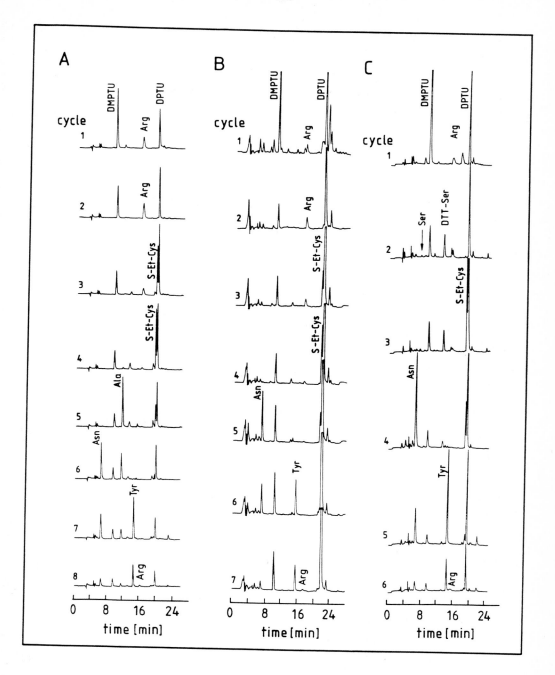

Fig. 4. Sequence analysis of the tryptic peptides *aI*, *cI* and *dI*. Samples were sequenced following derivatization of phosphoserine residues to S–ethylcysteine.
1.2 nmol peptide *aI* (A), 0.2 nmol peptide *cI* (B) and 0.6 nmol peptide *dI* (C) were applied. DMPTU, Dimethylphenylthiourea; DPTU, Diphenylthiourea; DTT–Ser, Dithiothreitol–serine; S–Et–Cys, S–Ethylcysteine.

Table I. Amino acid compositions of phosphoserine containing peptides from troponin I.
The amount of amino acids of the Lys-C peptides *a–d* was related to 1 mol lysine and of the tryptic peptides *aI*, *cI* and *dI* to 1 mol tyrosine. Values were not corrected for destruction and slow liberation. The amino acid composition of the N-terminus 1-31 from rabbit cardiac troponin I containing phosphoserine-20 (Moir & Perry, 1977) was taken from Grand et al. (1976). The calculated and determined mass of the N-terminus are given (compare Fig. 2). n.d.; not determined.

Amino acid	------Rabbit cardiac--------					---Human cardiac---			
	peptide	peptide	peptide	N-terminus		peptide		peptide	
	a	*aI*	*b*	1-36	1-31	*c*	*cI*	*d*	*dI*
Asx	1.8	0.9	1.8	3	3	1.6	0.7	1.2	0.7
Glx	2.3	–	2.4	3	3	1.7	–	1.2	–
Ser	1.0	–	2.2	1	1	1.7	–	2.5	0.8
Ser(P)	2.0	2.1	1.1	2	1	2.1	2.0	1.0	1.2
Gly	1.2	–	1.3	1	1	1.2	–	1.4	–
His	1.1	–	0.9	1	1	1.0	–	1.0	–
Arg	6.4	3.3	5.8	6	5	5.7	3.1	5.9	1.9
Thr	(2.1)	–	(2.0)	1	1	(1.5)	–	(1.4)	–
Ala	9.7	1.3	9.6	10	9	7.4	–	7.6	–
Pro	4.0	–	4.1	4	3	4.8	–	5.1	–
Tyr	2.2	1.0	1.8	2	1	2.0	1.0	2.1	1.0
Val	1.3	–	1.2	1	1	–	–	–	–
Ile	–	–	–	–	–	0.9	–	1.0	–
Lys	1.0	–	1.0	1	1	1.0	–	1.0	–
Mass	4098	n.d.	4018	4098	3444	n.d.	n.d.	n.d.	n.d.

The sequence of the bisphosphorylated peptide derived from rabbit cardiac troponin I is incompatible with the published sequence (Grand et al., 1976). The corresponding tryptic fragment from human cardiac troponin I showed a very similar amino acid sequence. During the preparation of this communication Vallins et al. (1990) published the nucleotide sequence of the human cardiac troponin I gene. The deduced amino acid sequence is identical with the determined sequence of the N-terminus (Fig. 5).

```
                                                                   23 24          37
Bovine cardiac                        AcADRSGGSTAG DTVPAPPPVR RRSSANYRAY ATEPHAK
                                                                   \ /
                                                                   PP

                                                                   20            31
Rabbit cardiac                        AcADESRDA-AG EARPAPA-VR R-S--D-RAY ATEPHAK
                                                                   P

                                                                   22 23          36
Rabbit cardiac, corrected             AcADESRDA-AG EARPAPAPVR RRSSANYRAY ATEPHAK
                                                                   \ /
                                                                   PP

                                                                   22 23          36
Human cardiac   (Ac,A,D/N,G,S,S,D/N,A,A,R) EPRPAPAPIR RRSS-NYRAY ATEPHAK
                                                                   \ /
                                                                   PP
```

Fig. 5. Comparison of the N-terminal phosphorylation sites from rabbit, bovine and human cardiac troponin I.

The amino acid sequences were aligned to maximize homology. The sequence data and phosphorylation sites of troponin I from bovine heart were taken from Leszyk et al. (1988) and Swiderek et al. (1988) and from rabbit heart from Grand et al. (1976) and Moir and Perry (1976). No sequence could be determined for the N-terminal blocked peptide, therefore the amino acid composition is given in parentheses. The N-terminal acetylgroup of rabbit troponin I was determined by plasma desorption mass spectrometry.

DISCUSSION

Due to the correction of the rabbit troponin I sequence all three sequences, that of bovine, rabbit and human, are very similar (Fig. 5). Characteristically, there exists a proline rich region upstream of the phosphorylation domain formed by three arginine and two serine residues. The cluster of helix breaking proline residues may change the structure of the N-terminus in such a way that the following phosphorylation domain can adapt to the catalytic center of protein kinases. Indeed, these two serine residues can be phosphorylated by the cAMP dependent, cGMP dependent protein kinase and protein kinase C (Swiderek et al., 1990). Furthermore, the bound phosphate groups are exposed on the surface of the protein as can be judged from ^{31}P NMR data (Beier et al., 1988).

In vitro phosphorylation of troponin I yields the bisphosphorylated product exclusively (Swiderek et al., 1990) vice versa dephosphorylation the completely dephosphorylated material. The monophosphorylated species are not found (Jaquet & Heilmeyer, unpublished). In contrast, both monophosphorylated forms of troponin I are found in the freshly isolated protein (Swiderek et al., 1988; Swiderek et al., 1990). It indicates that these two serine residues are phosphorylated and dephosphorylated differentially. It explains why no linear correlation between phosphate incorporation and change in contractility was observed in perfused rabbit hearts (Westwood & Perry, 1981).

Heart contractility is influenced by two hormonal signals triggered by α_1 and β adrenergic receptor occupation. The corresponding two signal pathways merge at troponin; maybe, differential phosphorylation of two serine residues on troponin I is the molecular equivalent of signal attenuation observed in heart perfusion studies (Osnes et al., 1989).

ACKNOWLEDGEMENTS

This work was supported by the Deutsche Forschungsgemeinschaft, the DAAD, the Herzzentrum Bad Oeynhausen, the Fonds der Chemie and by the Bundesminister für Forschung und Technologie. We thank Dr. A. Engström for recording the plasma desorption mass spectra, Dr. H.E. Meyer for interpretation of sequence data, P. Goldmann for her excellent technical assistance and S. Humuza for editing assistance.

REFERENCES

Beier N, Jaquet K, Schnackerz K, Heilmeyer LMG Jr (1988) Isolation and characterization of a highly phosphorylated troponin from bovine heart. Eur J Biochem 176: 327–334
Bidlingmeyer BA, Cohen SA, Tarvin TL (1984) Rapid analysis of amino acids using pre-column derivatization. J Chromatogr 336: 93–104
Cole HA, Perry SV (1975) The phosphorylation of troponin I from cardiac muscle. Biochem J 149: 525–533
Cummins B, Auckland ML, Cummins P (1987) Cardiac-specific troponin-I radioimmunoassay in the diagnosis of acute myocardial infarction. Am Heart J 113: 1333–1344
England PJ (1975) Correlation between contraction and phosphorylation of the inhibitory subunit of troponin in perfused rat heart. FEBS Lett 50: 57–60
Grand JA, Wilkinson JM, Mole LE (1976) The amino acid sequence of rabbit cardiac troponin I. Biochem J 159: 633–641
Leszyk J, Dumaswala R, Potter JD, Collins JH (1988) Amino acid sequence of bovine cardiac troponin I. Biochemistry 27: 2821–2827
Meyer HE, Swiderek K, Hoffmann-Posorske E, Korte H, Heilmeyer LMG Jr (1987) Quantitative determination of phosphoserine by high-performance liquid chromatography as the phenythiocarbamyl-S-ethylcysteine. Application to picomolar amounts of peptides and proteins. J Chromatogr 397: 113–121
Moir AJG, Perry SV (1977) The sites of phosphorylation of rabbit cardiac troponin I by adenosine 3':5'-cyclic monophosphate-dependent protein kinase. Effect of interaction with troponin C. Biochem J 167: 333–343
Osnes JB, Aass H, Skomedal T (1989) Adrenoceptors in myocardial regulation: concomitant contribution from both alpha- and beta-adrenoceptor stimulation to the inotropic response. Bas Res Cardiol 84: 9–17

Solaro RJ, Moir AJG, Perry SV (1976) Phosphorylation of troponin I and the inotropic effect of adrenaline in perfused rabbit heart. Nature 262: 615–616

Stull JT, Buss JE (1977) Phosphorylation of cardiac troponin by cyclic adenosine 3':5'-monophosphate-dependent protein kinase. J Biol Chem 252: 851–857

Swiderek K, Jaquet K, Meyer HE, Heilmeyer LMG Jr (1988) Cardiac troponin I, isolated from bovine heart, contains two adjacent phosphoserines. A first example of phosphoserine determination by derivatization to S-ethylcysteine. Eur J Biochem 176: 335–342

Swiderek K, Jaquet K, Meyer HE, Schächtele C, Hofmann F, Heilmeyer LMG Jr (1990) Sites phosphorylated in bovine cardiac troponin T and I. Characterization by ^{31}P-NMR spectroscopy and phosphorylation by protein kinases. Eur J Biochem 190: 575–582

Syska H, Perry SV, Trayer IP (1974) A new method of preparation of troponin I (inhibitory protein) using affinity chromatography. Evidence for three different forms of troponin I in striated muscles. FEBS Lett 40: 253–257

Vallins WJ, Brand NJ, Dabhade N, Butler-Browne G, Yacoub MH, Barton PJR (1989) Molecular cloning of human cardiac troponin I using polymerase chain reaction. FEBS Lett 270: 57–61

Westwood SA, Perry SV (1981) The effect of adrenaline on the phosphorylation of the P light chain of myosin and troponin I in the perfused rabbit heart. Biochem J 197: 185–193

CHARACTERISATION OF THYLAKOID MEMBRANE PROTEIN KINASE BY AFFINITY AND IMMUNOLOGICAL METHODS

Ian R. White, Michael Hodges* and Paul A. Millner.
Department of Biochemistry and Molecular Biology,
University of Leeds,
Leeds,
LS2 9JT,
U.K.

INTRODUCTION

In the presence of light, Mg^{2+} and ATP a subset of thylakoid membrane polypeptides undergo reversible phosphorylation. Two kinetically distinguishable populations of phosphoproteins may be discerned. The most abundant population comprises the light harvesting chlorophyll a/b binding proteins of photosystem II (PSII), termed LHCII's. This modification results in lateral migration of the phosphorylated species away from PSII, in the appressed membrane regions, to the non-appressed, photosystem I (PSI) rich regions. The net effect of this process is to decrease the light harvesting efficiency of PSII. Phospho-LHCII is dephosphorylated by a phosphoprotein phosphatase, allowing diffusion back into the appressed membrane regions and resumption of light harvesting. This process provides a dynamic regulatory mechanism for distribution of light energy between the two photosystems (Allen *et al.*, (1981)). The second population of proteins comprises the 8.3kDa *psbH* gene product, and three other PSII polypeptides *i.e.* reaction centre polypeptides D1 (psbA) and D2 (psbD) and the chlorophyll a binding protein CPa2 (psbC). The consequences of phosphorylation of these proteins are less clear.

Regulation of kinase activity is realised *via* the redox state of plastoquinone (Allen *et al.*, (1981)), and the cytochrome b$_6$f complex (Bennett *et al.*, (1988)), although the mechanisms involved have not been fully elucidated at a molecular level. However, phosphorylation of LHCII, compared with the PSII polypeptides, may be characterised by greater susceptibility to electron transport inhibitors which block electron flow from plastoquinol to the cytochrome b$_6$f complex (Bennett *et al.*, (1988)). Moreover, recent studies have shown that mutants lacking a functional cytochrome b$_6$f complex display impaired phosphorylation of LHCII, but not of PSII

---

* Present Address: Lab. de Physiologie Végétale Moléculaire, Bât. 430, Université de Paris Sud, 91405 Orsay Cedex, France.

NATO ASI Series, Vol. H 56
Cellular Regulation by Protein Phosphorylation
Edited by L. M. G. Heilmeyer, Jr.
© Springer-Verlag Berlin Heidelberg 1991

proteins (Bennett *et al.*, (1988),Gal *et al.*, (1988)).

This evidence, suggesting more than one kinase or regulatory mechanism, is further supported, by fundamental differences in the sites of phosphorylation (Table 1).

Protein	Source	N-terminal Sequence	Reference
LHCII	Pea	MRKSATTKKVASSGSPW-	Cashmore (1984)
psbH	Spinach	ATQTVESSSR-	Westhoff *et al* ., (1986)
psbC	"	Ac-TAILERRGS-	Michel *et al* ., (1988)
psbA	"	Ac-TLFNGTLTLAGR-	" " "
psbD	"	Ac-TIAVGKFTK-	" " "

TABLE 1 N-terminal sequences of the major phosphorylated polypeptides of the thylakoid membrane. Residues phosphorylated *in vivo* are underlined.

Recently we have shown the preference of the LHCII kinase for basic resides N-terminal to the phosphate accepting residues T_6 and T_7, and that peptide analogues with more basic N-terminal flanking regions than the endogenous substrate were better substrates for the kinase *in vitro* (White *et al* ., (1990)). In this communication we present supplementary fluorescence measurements in support of these findings and preliminary data showing the use of immobilised peptide substrate analogues as affinity ligands for kinase purification. We also report immunological studies employing anti-peptide antibodies raised to conserved kinase domains VI (Hanks *et al.*, (1988) and VIII (Lawton *et al* .,1988)).

METHODOLOGY

Thylakoid membranes were prepared from 14 day old *Pisum sativum* (var. Feltham First) seedlings and resuspended in 25mM Tricine.NaOH pH 8.0, 330 mM sorbitol,10 mM $MgCl_2$.

Fluorescence measurements were carried out on thylakoid membranes at a chlorophyll concentration of 100μg/ml. Following 10 min dark adaption, +/- 150 μM peptide, membranes were illuminated for 3 min with white light followed by addition of 1mM ATP.

Peptide affinity columns for kinase isolation and antibody purification were constructed using Sulpholink Coupling Gel (Pierce) and Affigel-10 (BioRad) following the manufacturers instructions.

Antibodies to peptides were raised in rabbits. PKCVI [LIYRDLKPENL] was coupled to BSA, using glutaraldehyde (Doolittle (1986)), and PKCVIII [GTHEYLAPEC] coupled to PPD (Lachmann *et al* .,(1986)) using Sulfo-SMCC(Pierce).

161

RESULTS AND DISCUSSION

    ATP induced fluorescence measurements enable non-invasive
estimations of PSII light harvesting efficiency: under
conditions where LHCII phosphorylation can occur then ATP-
dependant quenching provides a measurement of LHCII kinase
activity (Millner *et al* .,(1982)). At a peptide concentration
of 150μM, approximately comparable to the endogenous substrate
concentration, quenching observed was inversely correlated
with N-terminal peptide charge (Table 2) as expected.

Peptide	Sequence	Charge	Quenching (% Control)
LHCIIA	MRDSATTKKVAC	+2	67
LHCIIN	MRNle#SATTKKVAC	+3	40
LHCIIS	MRKSATTKKVAC	+4	9
LHCIIB	MRKKSATTKKVAC	+5	4

Table 2. ATP induced fluorescence quenching measurements
carried out in the presence of 150μM peptide.(Nle#~Norleucine)

    In isolation studies we have obtained an active fraction
following the isolation method of Coughlan and Hind (1986).
However, we believe the specific activity of our preparation
is too low to justify enzymological studies. Conversely, lack
of substantial activity does not preclude binding of kinase to
immobilised substrate. We therefore constructed an affinity
matrix using peptide LHCIIN as ligand. Other peptides were
significantly better substrates (LHCIIS, LHCIIB) in
specificity studies, but also showed extensive ion-exchange
properties (not shown). Affinity chromatography of ammonium
sulphate precipitated thylakoid membrane detergent extracts on
an LHCIIN peptide column (Fig. 1) yielded a small population
of candidate protein kinases. Following CHAPS extraction, five
polypeptides were isolated from a 30-45% ammonium sulphate
precipitated material, of 57,50,40,38, and 24 kDa(Fig.1A).
Elution of proteins was obtained using a stepped ATP gradient,
followed by a 4M NaBr wash. Of particular interest were
tightly bound 57,50, and 38 kDa polypeptides, eluted by the
chaotrope. These three proteins have all been electroeluted
from polyacrylamide gels and subjected to microsequencing.
Whilst, the 57 kDa protein has been successfully identified as
the α subunit of the thylakoid ATP synthetase, the 38kDa and
50kDa proteins were refractory to sequencing. Further
sequencing attempts will be made after limited proteolytic
dissection of the latter proteins. Small quantities of two
polypeptides, of 34 and 40 kDa, were isolated in a similar
fashion by chromatography of a 45-60% ammonium sulphate
precipitated fraction on the same column(Fig.1B).
    Finally, to circumvent problems caused by the kinase's
lability to detergent extraction, we have pursued

immunological identification of thylakoid kinases. Thylakoid
extracts have been probed with antibodies to conserved kinase
domains VI (Hanks *et al* ., (1988)) and VIII (Lawton *et al* .,
(1989)) which are implicated in ATP binding and substrate
recognition by Ser/Thr protein kinases. Figure 2A shows

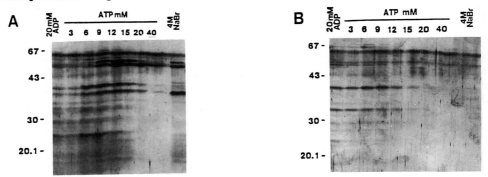

Figure 1. SDS-PAGE analysis of ammonium sulphate precipitated
fractions following affinity chromatography on LHCIIN-agarose.
A, 30-45% fraction; B, 45-60% fraction.

specific identification of cAMP dependant protein kinase
catalytic subunit by immunoblotting with anti-PKCVIII which
could be prevented by preabsorption of antiserum with peptide
antigen. This control has proved a critical feature of our
experiments, as identification of some polypeptides by
immunoblotting, using anti-peptide antibodies, can not always
be "blocked" in this fashion. Affinity purified anti-PKCVI
antibodies, cross-reacted with a polypeptide of 67kDa in a 30%
ammonium sulphate precipitated fraction of CHAPS solubilised
thylakoids(Fig.2B). Although we have yet to confirm the
specific immunoreactivity of this polypeptide, it may prove to
be the 64 kDa protein isolated by Coughlan and Hind (1986). We

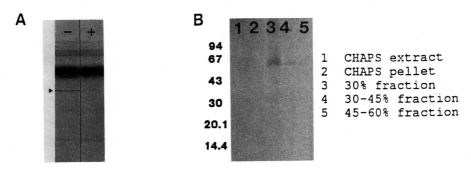

Figure 2. Immunological identification of A, cAMP dependant
protein kinase (arrowed), with anti-PKCVIII, [+/- preincubation with antigen] and B, putative thylakoid kinase with
affinity purified anti-PKCVI.

have also blocked immunoreactivity of a 38kDa protein in thylakoid membrane extracts using anti-PKCVIII serum(not shown), although identity with the affinity isolated 38kDa polypeptide or the 38kDa protein kinase isolated by Lin *et al* ., (1982) has not been shown.

In this article we have presented evidence supporting the requirement of the LHCII kinase for basic substrates. Purification of the kinase has been hampered by very low specific activities, necessitating alternative means of identification. To achieve this end we have raised antibodies specific for conserved domains of Ser/Thr kinases, and are able to show immunoreactivity with thylakoid polypeptides of 38 kDa and 67 kDa. Use of antibodies for identification of thylakoid protein kinases should afford easier developement of purification protocols. This should in turn, *via* protein chemical studies, allow elucidation of the regulatory processes involved.

REFERENCES

Allen,J.F., Bennett,J., Steinback,K.E., and Arntzen,C.J. (1981) Nature **291**, 21-25.

Bennett,J., Shaw,E.K., and Michel,H. (1988) Eur.J.Biochem. **171**, 95-100.

Cashmore,A.R., (1984) Proc.Natl.Acad.Sci.USA. **81**, 2960-2964.

Coughlan,S.J., and Hind,G. (1986) J.Biol.Chem. **261**, 11378-11385.

Doolittle, R.F. (1986) in: Of URFs and ORFs - A Primer on How to Analyze Derived Amino Acid Sequences. Oxford University Press. 85.

Gal,A., Schuster,G., Frid,D., Canaani,O., Schwieger,H.-G., and Ohad,I. (1988) J.Biol.Chem. **263**, 7785-7791.

Hanks,S.K., Quinn,A.M., and Hunter,T. (1988) Science **241**, 42-52.

Lachmann,P.J., Strangeways,L., Vyakarnam,A., and Evan.G. (1986) in: Synthetic Peptides as Antigens (Eds. Porter,R. and Whelan,J.) Ciba Foundation Symposium **119**, 25-57.

Lawton,M.A., Yamamoto,R.T., Hanks,S.K., and Lamb,C.J. (1989) Proc.Natl.Acad.Sci.USA. **86**, 3140-3144.

Lin,Z.K., Lucero,H.A., and Racker,E. (1982) J.Biol.Chem. **257**, 12153-12156.

Michel,H., Hunt,D.F., Shabanowitz,J., and Bennett,J. (1988) J.Biol.Chem. **263**, 1123-1130.

Millner,P.A., Widger,W.R., Abbott,M.A., Cramer,W.A., and Dilley,R.A. (1982) J.Biol.Chem. **257**, 1736-1742.

Westhoff,P., Farchaus,J.W., and Herrmann,R.G. (1986) Curr. Genet. **11**, 165-169.

White,I.R., O'Donnell,P.J., Keen,J.N., Findlay,J.B.C., and Millner,P.A. (1990) FEBS Lett. **269**, 49-52.

IV. CA^{2+} AND CYCLIC NUCLEOTIDE-INDEPENDENT PHOSPHORYLATION

A DIVERSITY OF ELEMENTS IN THE PROTEIN KINASE C SIGNAL TRANSDUCTION PATHWAY

P. J. Parker
Protein Phosphorylation Laboratory
Imperial Cancer Research Fund
44 Lincoln's Inn Fields
London WC2A 3PX
England

Introduction

Historically the discovery of this signalling system draws upon the convergence of two areas of research. The first was directed at the phenomenon of agonist induced inositol lipid turnover (reviewed in (Hokin, 1985)); the second was directed at the discovery and characterization of protein kinases with signalling potential, which lead to the identification of a calcium and phospholipid-dependent protein kinase termed protein kinase C (PKC) (see (Nishizuka, 1986; Nishizuka, 1988)). The critical link between these fields is the now established second messenger diacyglycerol (DAG) which is produced during agonist stimulated inositol lipid hydrolysis (by an inositol lipid-specific phospholipase C - PtdIns- PLC) and is itself responsible for the activation of PKC (figure 1). The inositol 1,4,5 trisphosphate head group which is concurrently produced on degradation of PtdIns 4,5 bisphosphate also serves a second messenger role in the release of Ca^{2+} from intracellular stores (Streb et. al., 1983). That DAG leads to PKC activation *in vivo* is evidenced by a variety of reports that short chain membrane permeable DAGs can activate PKC in intact cells as judged by the phosphorylation of particular proteins (e.g. (Kawahara et. al., 1980)). Furthermore such proteins can become phosphorylated in response to agonists that stimulate PtdIns breakdown providing evidence for a causal role in the events of agonist induced cellular responses.

This overview of what can be termed the PKC signal transduction pathway will focus on two key elements, namely the PtdIns-PLC activity responsible for lipid hydrolysis and PKC itself. In particular emphasis will be placed on the heterogeneity of the gene families represented by these two classes of enzymes.

NATO ASI Series, Vol. H 56
Cellular Regulation by Protein Phosphorylation
Edited by L. M. G. Heilmeyer, Jr.
© Springer-Verlag Berlin Heidelberg 1991

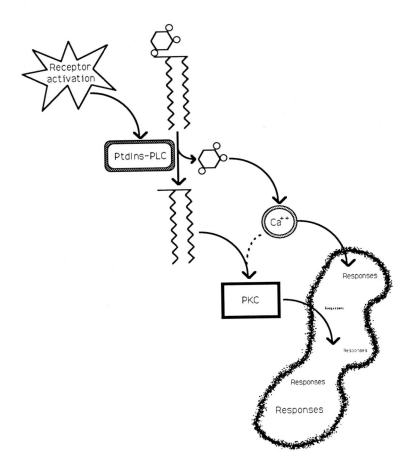

Figure 1. The Protein kinase C pathway. The diagram illustrates the primary role of the phosphatidylinositol-specific phospholipase C (PtdIns-PLC) in response to the activation of certain receptors at the cell surface. The consequent activation leads to the hydrolysis of the inositol 1,4,5 trisphosphate head group from the diacylglycerol backbone. The former is responsible for the release of Ca^{2+} from intracellular stores while the latter is responsible for the activation of protein kinase C (PKC). Both sets of signals are involved in eliciting cellular responses.

Protein kinase C

PKC was initially defined functionally on the basis of its ability to phosphorylate the lysine-rich histone h1 in a Ca^{2+} and phospholipid-dependent manner (Takai et. al., 1979). The activation by

DAG was shown to be effected through lowering the activation constants for Ca²⁺ and phospholipid (Takai et. al., 1979) although in a subsequently developed mixed micelle assay an absolute dependence upon DAG was observed (Hannun et. al., 1985). Based upon these properties, PKC was purified to apparent homogeneity to reveal a polypeptide of ~80kDa. The prior seminal observation that tumour promoters of the phorbol ester class were able to precisely mimic the effect of DAG *in vitro* also led to the identification of PKC as the major "receptor" for these compounds since the kinase and the phorbol ester binding activity copurified (reviewed in (Ashendel, 1985)).

A detailed structural analysis of PKC through protein sequence and the isolation of cDNAs, revealed the existence of multiple genes (see figure 2). To date there are six genes and at least seven distinct gene products that have been defined as part of this sub-family of protein kinases. The relationship between these proteins is evident both structurally as outlined in figure 2 and functionally as discussed further below.

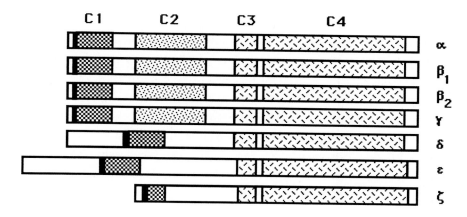

Figure 2. The domain structure of the PKC family. The figure shows the alignment of the various predicted gene products of the PKC family (see (Parker et. al., 1989) and references therein). Highlighted are the regions of greatest conservation, C_1-C_4. The chequered region (C_1) constitutes the phorbol ester/diacylglycerol binding domain. The speckled C_2 domain is thought to confer Ca²⁺-dependence (see text). The catalytic domains are essentially colinear and are made up of the C_3/C_4 regions. Immediately N-terminal to the C_1 domain in each predicted polypeptide are sequences (in black) that constitute the pseudo-substrate sites.

The basic domain structure of PKC consists of four highly conserved domains interspersed with more variable regions (see figure 2). Through mutagenesis and deletion analysis (Cazaubon et. al., 1990; Kaibuchi et. al., 1989; Ono et. al., 1989a) it is evident that the DAG/phorbol ester binding site is located in the C1 domain. This domain contains a tandemly repeated cysteine-rich motif $CX_2CX_{13}CX_2CX_7CX_{13}$ and it appears that either repeat is sufficient to confer phorbol ester binding. All the PKCs have at least one such cys-rich repeat and as such are expected to have some capacity to bind DAG/phorbol esters (ζ which has only one cys unit has been reported not to be activated by DAG (Ono et. al., 1989b); however if like PKC-ϵ (Schaap and Parker, 1990; Schaap et. al., 1989) and PKC-δ (Olivier and Parker manuscript in preparation) PKC-ζ is susceptible to proteolysis then this view may need to be modified).

The C_2 domain of PKC is only conserved in the PKC-α, $-\beta_1$, $-\beta_2$ and $-\gamma$ polypeptides. Based upon the behaviour of PKC-α and $-\beta$ in relation to that of PKC-ϵ it can be surmised that this C_2 domain confers the Ca^{2+}-dependence of these activities. This is most clearly illustrated by studies on the Ca^{2+}-dependence of membrane interactions. Thus (in the absence of agonists) PKC-ϵ distribution between the soluble and particulate fraction of cell extracts is independent of Ca^{2+} while the particulate association of PKC-α and $-\beta$ is driven by an excess of free Ca^{2+} (Kiley et. al., 1990).

The C_3 through C_4 domains make up a typical "protein kinase" catalytic domain. The intervening V_4 region represents a short insert that is present in the PKC-γ sequence only. Proteolysis of purified PKC with trypsin or calpain leads to activation of the holoenzyme through proteolysis of the V_3 region (defined for PKC -α, -β, -γ and $-\epsilon$) (Kishimoto et. al., 1989; Schaap et. al., 1989; Young et. al., 1988). Thus the catalytic domain fragment (C_3 to V_5) can express a constitutive activity independent of effectors; this implies that in the holoenzyme the regulatory domain serves to inhibit catalytic domain function. Evidence for the basis of this inhibition, has come from mutagenesis of the putative pseudosubstrate site domain. Thus mutation of PKC-α alanine-25 to glutamic acid (PKC-α E25) or deletion of the entire core of the pseudosubstrate site (PKC-Δ 22-28) leads to enzymatic activation (defined for purified PKC-α E25) and apparent activation *in vivo* (determined for both mutant forms) (Pears et. al., 1990).

Based upon the above considerations and the mapping of PKC-γ inhibitory monoclonal antibody epitopes to the V_2/C_2 region of PKC-γ (Cazaubon et. al., 1990), an activation model can be drawn up as shown in figure 3. This model indicates a number of features as described in the legend and it provides a framework for further structure-function investigation for these enzymes.

The functional analysis of these gene products has provided evidence that although all PKCs show phospholipid and DAG dependence, distinctions can be made between individual enzymes. Thus as noted above there are Ca^{2+}-dependent (-α, -β, -γ) (Huang et. al., 1988; Jaken and Kiley, 1987; Marais and Parker, 1989) and -independent (-δ, $-\epsilon$) PKCs (Schaap and Parker, 1990; Olivier and Parker manuscript in preparation). There are also differences in substrate specificity clearly distinguishing PKC-ϵ from PKC-α, -β, -γ (Schaap and Parker, 1990; Schaap et. al., 1989) and the

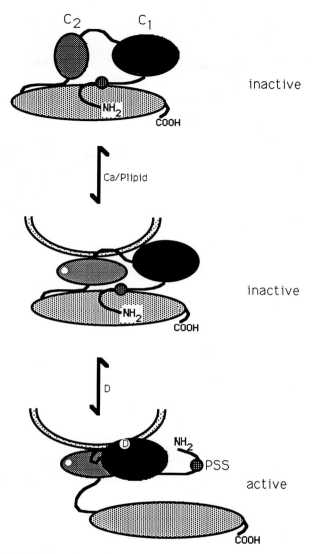

Figure 3. A model for PKC activation. The model illustrates the membrane association and subsequent activation of PKC (specifically the -α, -β and -γ enzymes which contain the C_2 domain). In the inactive state the enzyme activity is suppressed by the interaction of the pseudosubstrate site with the carboxyterminal catalytic domain. In this inactive state the protein can interact non-productively with membranes. This appears to be a Ca^{2+}-driven process (for -α, -β and -γ) and is likely to involve the C_2 domain (see text). Subsequent binding of diacylglycerol (D) leads to activation through conformational changes that release the pseudosubstrate site (PSS) from the catalytic domain.

structural basis of this specificity appears to reside at least in part in the regulatory domain (Pears, Schaap and Parker submitted). More recently we have investigated in detail the basic residue requirements for PKC-α, -β and -γ in the context of a synthetic peptide (based on residues 1-10 of glycogen synthetase). This study has demonstrated clear distinctions between PKC-α/β and PKC-γ, the latter showing a marked preference for synthetic peptides with a basic residue C-terminal to the phosphorylation site (Marais et. al., 1990).

In view of the specific functional differences in PKC isotypes it is anticipated that these enzymes will play distinct roles *in vivo*. Furthermore given the differential expression of the isotypes it is likely that the expression of a particular pattern of isotypes will confer on any given cell type a particular responsiveness. Thus the operation of these enzymes is not passive with respect to the quality or quantity of the signal transduction output and the expression of combinations of PKC isotypes will influence the cellular response.

Phospholipase C

While there is as yet little progress in defining the proteins and elucidating the structures of the enzymes responsible for diacylglycerol production from lipids other than inositol lipids, significant progress has been made in defining inositol lipid-specific phospholipases C (PtdIns-PLC). These studies (reviewed in (Meldrum et. al., 1990; Rhee et. al., 1989)) have revealed the existence of multiple genes that encode proteins capable of hydrolysing PtdIns, PtdIns-4 phosphate (PtdIns-4P) and PtdIns-4,5 bisphosphate (PtdIns-4,5P$_2$). These proteins are characterized by the presence of two conserved domains (figure 4) that would appear to confer catalytic function. The size of the proteins and the juxtaposition of the conserved domains defines three subfamilies of PtdIns-PLC (-β, -γ and -δ, as illustrated in figure 4).

The diversity of this enzyme family draws into question the function of each with respect to signal transduction. Formally it is possible that physiologically these proteins may show distinct specificities. However to date no evidence has been presented for the hydrolysis of lipids other than inositol lipids. There are however a number of inositol lipids that have been shown to occur physiologically, i.e. PtdIns, PtdIns-4P, PtdIns-4,5P$_2$, PtdIns-3P, PtdIns-3,4P$_2$ and PtdIns-3,4,5P$_3$. The first three of these inositol lipids are all substrates for the defined PtdIns-PLCs while the inositol-lipids with phosphate at the 3'-OH position do not appear to be(e.g. (Serunian et. al., 1989)). Of those inositol lipids that can be hydrolysed, different PtdIns-PLCs appear to show moderate differences in their relative rates of hydrolysis (see (Rhee et. al., 1989)). This is evidently of significance in the light of the above discussion of PKC isotypes. While PLC action on any of these lipids will produce DAG only the hydrolysis of PtdIns-4,5P$_2$ will lead to Ca^{2+} mobilisation through the action of inositol 1,4,5 trisphosphate (Streb et. al., 1983). The extent to which any of these proteins show differential hydrolysis of inositol lipids *in vivo* has yet to be determined.

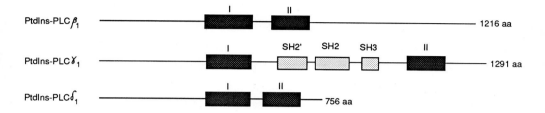

Figure 4. Alignment of members of the PtdIns-PLC family. One member of each of the subclasses of PtdIns-PLCs is shown. The predicted proteins each contain two highly conserved domains (I and II) that are considered to confer catalytic function. The proteins are essentially colinear over their aminoterminal stretches up to and including conserved domain I. Both PtdIns-PLC β$_1$ and δ$_1$ have non-homologous but highly charged sequences between domains I and II. By contrast PtdIns-PLC-γ$_1$ possesses a large domain between I and II containing the *src*-homology regions (SH2 and SH3). The related members within each subclass (e.g. -δ$_2$ and -δ$_3$; reviewed in (Meldrum et. al., 1990)) show the same conserved domains I and II but also show substantial similarity over their entire length.

Notwithstanding the above, it would appear that the critical distinction to be made with respect to these PtdIns-PLCs is in their mode of regulation. This may not be surprising since the agonists known to stimulate inositol lipid breakdown represent a diverse group of agents operating through distinct types of receptor.

An investigation into the mechanism by which EGF stimulates inositol lipid breakdown in A431 cells led to the identification of a PtdIns-PLC that became phosphorylated on tyrosine residues in response to an EGF challenge (Wahl et. al., 1988). It was subsequently shown that this PtdIns-PLC was the -γ$_1$ enzyme (Wahl et. al., 1989) and furthermore that this enzyme is tyrosine phosphorylated in a number of systems (e.g. (Meisenhelder et. al., 1989; Wahl et. al., 1989)). PtdIns-PLC-γ$_1$ and -γ$_2$ both contain between the two conserved domains regions showing homology to two domains within *src* i.e. the src homology 2 (SH2) and SH3 domains. There is evidence that SH2 domains are involved in complex formation with tyrosine phosphorylated receptors ((Moran et. al., 1990) and references therein) and PtdIns-PLC-γ$_1$ has been shown to become associated with activated (autophosphorylated) growth factor receptors (e.g. (Morrison et. al.,

1990)). Thus the association of PtdIns-PLC-γ with and/or phosphorylation by activated growth factor receptors would appear to be responsible for activation of this enzyme - an *in vitro* reconstituted response has yet to be described.

In contrast to PtdIns-PLC-γ, neither the -β nor -δ class of enzymes have been shown to be phosphorylated on tyrosine residues or to become associated in growth factor receptor complexes (reviewed in (Meldrum et. al., 1990; Rhee et. al., 1989)). However the PtdIns-PLC-β class of enzyme appears to be responsible for coupling inositol lipid metabolsim to G-protein linked receptors. An initial indication of this came from structural studies of a *drosophila* gene norpA which is required for photoreception. The cloning of this gene and comparison with known sequences identified the norpA product as a member of the PtdIns-PLC-β subfamily; since there is evidence for a G-protein linked PtdIns-PLC involved in *drosophila* photoreception it appeared likely that the -β enzymes were coupled in this manner (discussed in (Meldrum et. al., 1990)). More directly a cell permeabilisation system has been used to determine the effectiveness of G-protein coupling by a number of distinct PtdIns-PLCs. This has shown that PtdIns-PLC-β_1 but not -γ_1 or -δ_2 can reconstitute a G-protein response in permeabilised neutrophils (S. Cockcroft, personal communication).

From the foregoing discussion it appears that PtdIns-PLC-β and -γ are coupled to distinct classes of cell surface receptors. It would be convenient to classify the -δ subfamily in a similar manner however to date there is no clear indication of how the -δ enzymes may be regulated physiologically. Nevertheless, based upon the distinction between the -β and -γ enzymes it can be postulated that the -δ enzymes lie on a distinct receptor-driven path, feeding in to the same second messenger system.

Perspectives

The diversity of signalling components operating within the 'PKC pathway' is further exaggerated when one considers the multiplicity of receptor subtypes (e.g. PDGF a and b receptors, multiple muscarinic receptors) and the likelihood of multiple G-proteins also operating within this pathway. However in defining how these components operate, evidence is emerging for the selective operation of combinations of these components. Thus for example, the PtdIns-PLC-γ class appear to 'couple' with receptors of the tyrosine kinase class. Individual G-proteins may well be found to be responsible for coupling specific seven transmembrane receptors to particular members of the PtdIns-PLC-β subfamily.

A definitive rationalisation of the role of the PKC gene products downstream of second messenger production will also come no doubt from a detailed analysis of where and when they are activated and importantly what the (selective) targets are. Such future studies are likely to clarify the presently bemusing complexity of this pathway.

References

Ashendel, C. L. (1985). The phorbol ester receptor: a phospholipid-regulated kinase. Biochem. Biophys. Acta. 822: 219-242.

Cazaubon, S., Webster, C., Camoin, L., Strosberg, A. D. and Parker, P. J. (1990). Effector dependent conformational changes in protein kinase Cγ through epitope mapping with inhibitory monoclonal antibodies. Eur. J. Biochem. In Press:

Hannun, Y. A., Loomis, C. R. and Bell, R. M. (1985). Activation of protein kinase-C by Triton X-100 mixed micelles containing diacylglycerol and phosphatidyl serine. J. Biol. Chem. 260: 10039-10043.

Hokin, L. E. (1985). Receptors and phosphoinositide-generated second messengers. Ann. Rev. Biochem. 54: 205-235.

Huang, K.-P., Huang, F. L., Nakabayashi, H. and Yoshida, Y. (1988). Biochemical characterisation of rat brain protein kinase C isozymes. J. Biol.Chem. 263: 14839-14845.

Jaken, S. and Kiley, S. C. (1987). Purification and characterisation of three types of protein kinase C from rabbit brain cytosol. Proc. Natl. Acad. Sci. USA. 84: 4418-4422.

Kaibuchi, K., Fukumoto, Y., Oku, N., Takai, Y., Arai, K.-I. and Muramatsu, M.-A. (1989). Molecular genetic analysis of the regulatory and catalytic domain of protein kinase C. J. Biol. Chem. 264: 13489-13496.

Kawahara, Y., Takai, Y., Minakuchi, R., Sano, K. and Nishizuka, Y. (1980). Phospholipid turnover as a possible transmembrane signal for protein phosphorylation during human platelet activation by thrombin. Biochem. Biophys. Res. Commun. 97: 309-317.

Kiley, S., Schaap, D., Parker, P. J., Hsieh, L.-L. and Jaken, S. (1990). Protein kinase C heterogeneity in GH$_4$C$_1$ rat pituitary cells. J. Biol. Chem. 265: 15704-15712.

Kishimoto, A., Mikawa, K., Hashimoto, K., Yasuda, I., Tanaka, S.-I., Tommaga, M., Kuroda, T. and Nishizuka, Y. (1989). Limited proteolysis of protein kinase C subspecies by calcium-dependent neutral protease (calpain). J. Biol. Chem. 264: 4088-4092.
Marais, R. M., Nguyen, O., Woodgett, J. R. and Parker, P. J. (1990). Studies on the primary

sequence requirements for PKC-α, -β₁ and -γ peptide substrates. FEBS Letts. In Press:

Marais, R. M. and Parker, P. J. (1989). Purification and characterisation of bovine brain protein kinase C isotypes α, β and γ. Eur. J. Biochem. 182: 129-137.

Meisenhelder, J., Suh, P.-G., Rhee, S. G. and Hunter, T. (1989). Phospholipase C-γ is a substrate for the PDGF and EGF receptor protein-tyrosine kinases *in vivo* and *in vitro*. Cell. 57: 1109-1122.

Meldrum, E., Parker, P. J. and Carozzi, A. (1990). The PtdIns-PLC superfamily and signal transduction. Biochem. Biophys. Acta. In Press:

Moran, M. F., Koch, C. A., Anderson, D., Ellis, C., England, L., Martin, G. S. and Pawson, T. (1990). Src homology region 2 domains direct protein-protein interactions in signal transduction. Proc. Natl. Acad. Sci. USA. 87: 8622-8626.

Morrison, D. K., Kaplan, D. R., Rhee, S.-G. and Williams, L. T. (1990). Platelet-derived growth factor (PDGF)-dependent association of phospholipase C-γ with the PDGF receptor signalling complex. Mol. & Cell. Biol. 10: 2359-2366.

Nishizuka, Y. (1986). Studies and perspectives of protein kinase C. Science. 233: 305-312.

Nishizuka, Y. (1988). The molecular heterogeneity of protein kinase-C and its implications for cellular-regulation. Nature. 334: 661-665.

Ono, Y., Fujii, T., Igarashi, K., Kuno, T., Tanaka, C., Kikkawa, U. and Nishizuka, Y. (1989a). Phorbol ester binding protein kinase C requires a cysteine-rich zinc-finger-like sequence. Proc. Natl. Acad. Sci. USA. 86: 4868-4871.

Ono, Y., Fujii, T., Ogita, K., Kikkawa, U., Igarashi, K. and Nishizuka, Y. (1989b). Protein kinase C- ζ subspecies from rat brain: Its structure, expression and properties. Proc. Natl. Acad. Sci. USA. 86: 3099-3103.

Parker, P. J., Kour, G., Marais, R. M., Mitchell, F., Pears, C. J., Schaap, D., Stabel, S. and Webster, C. (1989). Protein kinase C - a family affair. Mol. Cell. Endocrinol. 65: 1-11.

Pears, C. J., Kour, G., House, C., Kemp, B. E. and Parker, P. J. (1990). Mutagenesis of the pseudosubstrate site of protein kinase C leads to activation. Eur. J. Biochem. In Press:

Rhee, S. G., Suh, P.-G., Ryu, S.-H. and Lee, S. Y. (1989). Studies of inositol phospholipid-specific phospholipase C. Science. 244: 546-550.

Schaap, D. and Parker, P. J. (1990). Expression, purification and characterization of protein kinase C-ε. J. Biol. Chem. 265: 7301-7307.

Schaap, D., Parker, P. J., Bristol, A., Kriz, R. and Knopf, J. (1989). Unique substrate specificity and regulatory properties of PKC-ε: a rationale for diversity. FEBS Lett. 243: 351-357.

Serunian, L. A., Haber, M. T., Fukui, T., Kim, T. W., Rhee, S. G., Lowenstein, J. M. and Cantley,

L. C. (1989). Polyphosphoinositides produced by phosphatidylinositol 3-kinase are poor substrates for phospholipases C from rat liver and bovine brain. J. Biol. Chem. 264: 17809-17815.

Streb, H., Irvine, R. F., Berridge, M. J. and Schulz, I. (1983). Release of Ca^{2+} from a non-mitochondrial intracellular store in pancreatic acinar cells by inositol-1,4,5-triphosphate. Nature. 306: 67-69.

Takai, Y., Kishimoto, A., Kikkawa, U., Mori, T. and Nishizuka, Y. (1979). Unsaturated diacylglycerol as a possible messenger for the activation of calcium-activated, phospholipid-dependent protein kinase system. Biochem. Biophys. Res. Commun. 91: 1218-1224.

Wahl, M. I., Daniel, T. O. and Carpernter, G. (1988). Antiphosphotyrosine recovery of phospholipase C activity after EGF treatment of A-431 cells. Science. 241: 968-971.

Wahl, M. I., Nishibe, S., Suh, P.-G., Rhee, S. G. and Carpenter, G. (1989). Epidermal growth factor stimulates tyrosine phosphorylation of phospholipase C-II independently of receptor internalization and extracellular calcium. Proc. Natl. Acad. Sci. USA. 86: 1568-1572.

Young, S., Rothbard, J. and Parker, P. (1988). A monoclonal antibody recognising the site of limited proteolysis of protein kinase C. Eur. J. Biochem. 173: 247-252.

"INDEPENDENT" PROTEIN KINASES : A CHALLENGE TO CANONS

Lorenzo A. Pinna
Dipartimento di Chimica Biologica
Università di Padova
Via Trieste, 75
35121 PADOVA

According to the canonical scheme of reversible protein phosphorylation (fig. 1) protein kinases are normally silent, converter enzymes whose
activation is triggered by a variety of impulses, mostly generated by
extracellular signals. Due to such a role of "regulatable regulators"
committed with the translation of signals into biochemical events, protein
kinases can be classified after either their responsiveness to various
stimuli or their capability to recognize definite structural features.

Classification of protein kinases according to their dependence on signals

The relationships between protein kinases and signals can be more or
less direct, as exemplified in fig. 2. The most direct linkage is provided
by those receptors whose intracellular domain is itself endowed with
protein kinase activity. The best known among these receptorial protein
kinases are tyrosine specific enzymes, like the EGF, the PDGF, the insulin
and the FGF receptors (Ullrich and Schlessinger, 1990).

A less direct connection with signals is provided by second
messengers, which are responsible for the activation of cAMP-dependent and
cGMP-dependent protein kinases (Beebe and Corbin,1986), of the various
isoforms of protein kinase-C (Nishisuka,1988) and of the whole family of
Ca^{2+} and calmodulin dependent protein kinases (Stull et al.,1986).

The connection with either primary signals or second messengers,
however, can be more complicated as it is exemplified by the phosphorylation cascades controlling kinases whose activity is finally triggered by
either phosphorylation (e.g. S6-kinase, Insulin stimulated protein kinase
(Dent et al., 1990)) or dephosphorylation (e.g. cdc-2 kinase (Draetta,
1990) and cellular tyrosine kinases of the c-src family (Cooper, 1990))
and by the theoretical possibility that a second messenger (e.g. Ca^{2+})
might turn on a converter enzyme which is not a kinase itself but can in
turn activate a protein kinase, e.g. by limited proteolysis. A mechanism
like that has been invoked for the so called "PAKs" (protease activated

NATO ASI Series, Vol. H 56
Cellular Regulation by Protein Phosphorylation
Edited by L. M. G. Heilmeyer, Jr.
© Springer-Verlag Berlin Heidelberg 1991

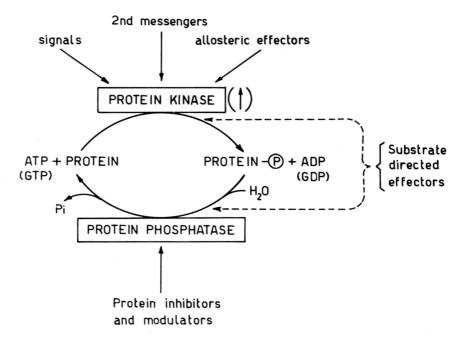

FIG. 1   Canonical model for the regulation of the biological properties of proteins by reversible phosphorylation.

kinases) and, under special circumstances, for the irreversible activation of PK-C.

A few protein kinases are allosteric enzymes whose positive effectors appear to be intracellular metabolites not commited with the function of second messengers. The AMP-activated protein kinase responsible for the phosphorylation of Acetyl CoA carboxylase and HMGCoA reductase (Hardie et al.,1989) and the mitochondrial pyruvate dehydrogenase kinase (Reed and Yeaman, 1986) can be ascribed to this group, although the former is also dependent on a phosphorylation cascade for its activity (Hardie et al., 1989).

A number of protein kinases however escape all the above classification criteria for being spontaneously active, or, as they often are less properly termed, "independent" enzymes. It is quite possible that some of them just represent irreversibly activated forms of "dependent" enzymes, generated e.g. by limited proteolysis that has removed their down-regulatory domains. It may be also conceivable that the "spontaneous" activity observed is, in some instances, physiologically irrelevant in comparison with the activity elicited by endogenous, as yet undetected,

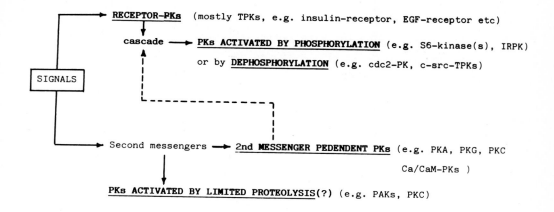

FIG. 2   Classification of protein kinases according to their correlation
with signals and effectors.
The abbreviations refer to the following protein kinases (PKs): TPKs,
tyrosine protein kinases (Hunter and Cooper, 1985); 6S-kinases, protein
kinases that phosphorylate the ribosomal protein S6; ISPK, insulin
stimulated protein kinase (Dent et al., 1990); cdc2 PK, protein kinase
expressed by the cell division cycle-2 gene (Draetta, 1990); c-src TPKs,
cellular tyrosine kinases of the src family (Cooper, 1990); PKA and PKG,
cAMP-and cGMP-dependent protein kinases (Beebe and Corbin 1986);
Ca/CaM-PKs, Ca^{2+} and calmodulin dependent protein kinases (Stull et al.,
1986); PKC, Ca^{2+}, phospholipid dependent protein kinase (Nishizuka, 1988);
PAK I and II, protease activated kinases I and II, the latter
corresponding to H4 kinase (Masaracchia et al., 1990); PDHK, pyruvate
dehydrogenase kinase (Reed and Yeaman, 1986); AMP-PK, protein kinase
activated by 5'-AMP, responsible for the phosphorylation of
acetyl-CoA-carboxylase and HMG CoA reductase (Hardie et al., 1989); CK1
and CK2, casein kinases 1 and 2 (Pinna, 1990) HSP and PRV-PKs protein
kinases expressed by Herpes simplex and Pseudorobies viruses (Zhang et
al., 1990); v-onc TPKs, tyrosine kinases expressed by viral oncogenes
(Hunter and Cooper, 1986).

effectors. In any event the existence of protein kinases which are
definitely active in their native form and apparently are not subjected to
any kind of "acute" regulation represents an intriguing puzzle and rises
the question as to how can their "independence" be compatible with the
ordered occurence of cellular functions. This question is especially
pertinent whenever the "uncontrolled" protein kinase is not the product of
an oncogene, but a normal component of the cell, involved in a variety of

physiological functions. This is the case, e.g., of the two protein kinases conventionally termed casein kinases-1 and -2 (Pinna, 1990) and of Glycogen synthase kinase-3, also known as FA kinase(¹Woodgett, 1990).

Overall classification of protein kinases according to their site specificity

Most if not all protein kinases are pleiotropic enzymes capable to phosphorylate a variety of proteins, often involved in quite different cellulary functions. Each of them must therefore be competent to recognize in all its substrates one or more exposed phosphorylatable residues speci-fied by a definite consensus sequence. Such a "recognition code" is often composed by either basic or acidic residues located at fixed positions relative to the target aminoacid. Consequently protein kinases can be roughly grouped into 5 categories according to their specificity for either serine/threonine or tyrosine, on one side, and to their preference (or lack of preference) for either basic or acidic residues, on the other (see table 1): 1) Ser/Thr-specific "basophilic" protein kinases affect serines and threonines whose recognition is determined by basic residues nearby. All known second messenger-dependent protein kinases belong to this category which also includes many other enzymes, either spontaneously active (HSV and PRV protein kinases) or controlled by cellular components (AMP-activated-PK, dsRNA activated PK, HCR etc.) 2) Ser/Thr-specific "acidophilic" protein kinases whose targeting of serine/threonine is dependent on crucial acidic residues. This group coincides with spontaneo-usly active casein kinases, either authentic, for being responsible for the phosphorylation of casein in vivo, or conventionally termed enzymes commited with the phosphorylation of a variety of regulatable proteins in nearly all organisms and tissues. 3) Ser/Thr specific protein kinases, which apparently don't exhibit any stringent requirement for either basic or acidic determinants, although they may well tolerate them. Cdc2 kinase (Draetta, 1990) can be ascribed to this category. It apparently shares with recently described "Pro-directed" protein kinases (Vulliet et al., 1989) the targeting of serines/threonines adjacent to proline(s) which are suspected to act as structural determinants in this case. GSK-3, considered a "phosphate directed" protein kinase (Fiol et al., 1990) might also belong to this group since it is not clear whether the phosphorylated side-chain acting as a positive determinant is actually recognized by virtue of its anionic nature. Some of the sites affected by GSK-3 or GSK-3 related protein kinases, moreover, seem to be substantially neutral and

Table 1 - Grouping of protein kinases according to their local structural
            determinants.
       Abbreviations not specified in fig. 2 are as follows: dsI,
dsRNA-activated inhibitor; HCR, haem controlled repressor (Proud et al.,
1991); GEF-CK, golgi enriched fraction casein kinase from lactating
mammary gland (Meggio et al., 1988); Pro-directed PK, protein kinase(s)
whose site specificity is determined by prolyl residues (Vulliet et al.,
1990); GSK-3, glycogen synthase kinase-3, also termed FA kinase (Woodgett,
1990; Picton et al., 1984) - Additional references on the specificity
determinants of protein kinases are reported elsewhere (Pinna et al. 1987,
Pinna, 1990).

Residue(s) affected	Specificity determinants	Protein Kinases
Ser/Thr	Basic residues	PK-A PK-G PK-C Ca/CaM-PKs AMP-PK PAKs PRV and HSV PKs dsI HCR
Ser/Thr	Acidic residues	CK-1 CK-2 GEF-CK
Ser/Thr	Neither basic nor acidic	cdc2 "Pro directed" PK GSK-3 ?
Tyr	Acidic residues	most TPKs(e.g.EGF-rec., insulin-rec.,abl-TPK, spleen TPKIIB)
Tyr	Neither basic nor acidic	Lyn-TPK(=TPKIIA=p40) (and other src-TPKs ?)

rich in proline. 4) Tyrosine specific acidophilic protein kinases. Most
autophosphorylation sites of tyrosine protein kinases are actually
characterized by acidic residues upstream of the target tyrosine. Studies
with model substrates however support the concept that a second category
of tyrosine kinases exists, mamely, 5) Tyr-specific protein kinases
exhibiting no special preference for either acidic or basic sites. The
whole family of src protein kinases may belong to this group (Donella-
Deana et al., 1990).

     Is it possible to predict the coarse site specificity of a given
protein kinase if its primary structure is known? An adfirmative answer
may come from the inspection of the residue homologue to PK-A Glu-170,
which has been shown to play a crucial role in the interaction with the

Table 2 - Correlation between the known specificity determinants of
          protein kinases and the nature of the residue homologue to PK-A
          Glu-170.
          Residues homologue to PK-A Glu-170 are deduced from Hanks et al.
(1988). For the features determining the site specificity see table 1.

Protein kinase	Residue homologue to PK-A Glu 170	Features determining the site specificity
PK-A	Glu	Basic
PK-G	Glu	Basic
PK-C	Asp	Basic
Myosin light chain kinase	Glu	Basic
Phosphorylase kinase	Glu	Basic
Ca/Calm.dependent PKII	Glu	Basic
PRV protein kinase	Glu	Basic
Casein kinase-2	His	Acidic
cdc-2 PK	Gln	Proline ?
GSK-3 (FA kinase)	Gln	Phosphate ?
EGF Receptor - TPK	Arg	Acidic
Insulin Receptor - TPK	Arg	Acidic
Lyn - TPK (=TPK-IIA)	Ala	Neutral ?
pp60 v-src	Ala	Neutral ?
p56 lck (=LSTRA-TPK)	Ala	Neutral ?

Table 3 - Grouping of protein kinases according to the nature of the
          residue homologue to PK-A Glu-170.
          Deduced from the compilation of protein kinases catalytic domain
sequences arranged according to the Intelligenetics format (S.K. Hanks and
A.M. Quinn, the Salk Institute for Biological Studies, March 25, 1990).

| | Number of PKs whose residue homologue to PK-A Glu-170 is: | | |
	ACIDIC	NEUTRAL	BASIC
Ser/Thr-PKs	38	23	3
Tyr-PKs	0	11	33

basic aminoacids representing the specificity determinants for this enzyme
(Taylor, 1989). As shown in table 2 a tight correlation indeed exists
between the nature of the residue homologue to PK A Glu-170 in a variety
of protein kinases, and the nature of the specificity determinants
recognized by each of them. Whenever the Glu-170 homolog is an acidic
residue the site specificity is invariably determined by basic aminoacids.

In CK-2 however, which is an acidophilic protein kinase, Glu-170 is replaced by histidine, while it is replaced by a glutamine in cdc-2 kinase, tentatively ascribed to the class of protein kinases requiring neither basic nor acidic residues. This rule applies also to tyrosine protein kinases, in which Glu-170 is replaced by either an arginine, in enzymes which are clearly "acidophilic", or by an alanine in kinases of the src family which don't display any stringent requirement for acidic residues, at least in peptide substrates.

Assuming the general validity of this concept one could predict that 49 out of the 78 Ser/Thr protein kinases whose sequence is known so far are suited to interact with basic sites for having an acidic residue at the position of PKA Glu-170 (Table 3). Only three of them are expected to be acidophilic, while the remaining are all predicted to be regardless of either basic or acidic groups. Acidophilic enzymes appear conversely to be predominant among tyrosine protein kinases (33 out of 44) all the remaining being expected to ground their site specificity on neither basic nor acidic determinants. Basophilic tyrosine protein kinases are neither predictable nor have been described so far.

Compartmentation as a regulatory device for mammary gland casein kinase

Among the spontaneously active protein kinases mentioned in fig. 2, the enzyme(s) responsible for the biosynthetic phosphorylation of casein in the lactating mammary gland represent a very special case since they are required to carry out a constitutive, non reversible, and exhaustive modification of a definite set of abundant protein molecules under particular metabolic conditions in a specific compartment of an individual type of cell. Detailed studies are available for the casein kinase termed GEF-CK. It is expressed only in secretorial cells of lactating mammary gland, being strictly associated with the golgi apparatus (Moore et al., 1985). Its specificity, deduced from the structure of the phosphorylated sites of casein fractions (Mercier, 1981) and later confirmed with the aid of peptide substrates (Meggio et al., 1988) enables it to recognize exclusively the triplet: Ser(Thr)-X-Glu(Asp). Thus, while the peptides SEEAAA and SAEAAA are phosphorylated, many other acidic peptides devoid of the Ser-X-Glu motif, like SAAEEE and SEAEEE are totally unaffected. Due to the profusion of casein molecules that have to be processed, moreover, not even all the suitable sites in them are becoming entirely phosphorylated, the less appropriate (e.g. for having Thr instead of Ser) remaining partially unoccupied (West, 1986). This presumably keeps "overbusy" GEF-CK which is consequently prevented from "mis-phosphorylating" other proteins despite its lack of short-term control mechanisms.

Presumably "independent" protein kinases with features and specificity similar to GEF-CK are responsible for the constitutive phosphorylation of other proteins, like pepsin, ovalbumin, fibrinogen, at phospho-sites sharing the motif SerP-X-Glu(Asp).

Are CK-1 and GSK-3 "secondary" protein kinases ?

The concept of "hierarchical" phosphorylation may provide a rationale for understanding the regulation of otherwise "uncontrollable" protein kinases. This could apply to spontaneously active casein kinase-1 (CK-1) and GSK-3 (see figure 2) both of which have been shown to depend on previously phosphorylated residues as specificity determinants. In early studies with casein fractions we firstly showed that the residues affected by CK-1 were distinct from those constitutively phosphorylated in native casein, for being located just downstream of them (Meggio et al., 1979). Invariably these CK-1 sites included very acidic clusters, composed by both phosphoserine and glutamic acids, adjacent to the N terminal side of the target residue. The additional finding that previous dephosphorylation of phosphoserines abolishes the subsequent activity of CK-1 prompted us to propose that CK-1 might be not a "primary" kinase, but rather a "secondary" one, depending on another protein kinase for its site recognition (Meggio et al., 1979). Recently this prediction has been elegantly confirmed by Flotow et al. (1990) showing that the phosphorylation of the glycogen synthase peptide PLSRTLSVASLPGL by CK-1 at Ser-10 occurred only after Ser-7 had been phosphorylated by PKA.

Likewise also the phosphorylation of a variety of protein and peptide substrates by GSK-3 is potentiated by the previous phosphorylation of residues nearby which apparently create the suitable sites for GSK-3 (e.g. Picton et al., 1982, Fiol et al., 1987). It is conceivable therefore that the regulation of the activity of "independent" protein kinases like CK-1 and GSK-3 might be at least partially committed to other protein kinases. It sounds somewhat disconcerting in this connection that the enzyme apparently in charge of controlling GSK-3 is casein kinase-2, i.e. the most typical specimen of a spontaneously active protein kinase. Quis custodiet custodes ? Next paragraph will try to give and answer to this question.

CK-2, the prototype of an "independent" protein kinase

A paradox inherent to the protein kinase generally termed casein kinase-2 is that this enzyme is spontaneously active despite two features that would argue in favour of its susceptibility to regulation, mamely: 1) A heterotetrameric structure with two catalytic and two non catalytic subunits, and, 2) its involvement in a miriad of cellular functions, with special reference to gene expression and cell proliferation (reviewed by Pinna, 1990).

A variety of observations on the other hand support the notion that CK-2 is indeed spontaneously active, at least toward many of its targets, a feature, incidentally, that favoured its early discovery, sooner than most of the other protein kinases.

Unlike the holoenzyme of PKA, heterotetrameric CK-2 is fully active and it doesn't dissociate unless under drastic conditions that compromise enzyme activity. The kinetic parameters of CK-2, moreover, are quite comparable to those of pure, typically dependent protein kinases under optimal conditions of activation: it seems hardly conceivable therefore that they might reflect just a basal activity, susceptible to acute enhancement in the presence of as yet unknown effectors. All known second messengers are ineffective on CK-2 and spontaneously active CK-2 cannot be generated by limited proteolysis from a larger regulatable kinase, since its structure corresponds to that deduced from its cDNAs. Although the β-subunit of CK-2 undergoes autophosphorylation there is no incontrovertible evidence that this non catalytic component might play a regulatory role. Actually the properties of CK-IIB, a maize monomeric CK-2 composed by a single catalytic subunit are very similar to those of the canonical CK-2 with tetrameric structure (G. Dobrowolska et al., 1987, and unpublished data). Additional support to the concept that CK-2 is spontaneously active also in situ is provided by the drastic effect of heparin, a specific inhibitor of CK-2, on the phosphorylation of proteins in crude extracts under basal conditions, i.e. whenever the dependent protein kinases are silent, and by the striking hyper-phosphorylation of CK-2 targets upon addition of okadaic acid (Issinger et al., 1988) a powerful inhibitor of protein phosphatases 2A and 1. Apparently therefore CK-2 cannot be turned off under conditions rendering its activity unnecessary and presumably undesired.

On the other hand it is noteworthy that a set of CK-2 targets is totally dependent on polybasic effectors, like polylysine, protamine, etc. for phosphorylation. Actually the influence of these cationic compounds appears to be strikingly conditioned by the protein substrate: while they are ineffective or even inhibitors with a few targets (including the β-subunit of CK-2 itself), and more or less stimulatory with many others, they are necessary for the appreciable phosphorylation of a number of substrates, including, e.g., calmodulin, clathrin β-chain, ornithine decarboxylase etc. Somewhat disappointing this property is not shared by polyamines which also variably stimulate CK-2 in a substrate directed manner but are unable to trigger an all-or-nothing response of CK-2 toward certain substrates, as observed with polybasic peptides.

Apparently therefore the substrates of CK-2 can be divided into two categories including those proteins whose phosphorylation is made possible by polybasic peptides, and those which are phosphorylated even in the absence of these effectors, respectively (fig. 3).

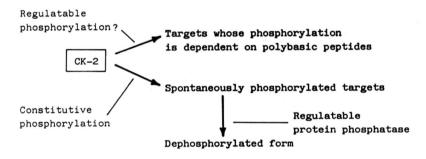

FIG. 3   Grouping of CK2 substrates according to their relying on polybasic effectors for phosphorylation.

The phosphorylation level of the latter set of substrates might be entirely controlled by the phosphatases committed with their dephosphorylation.

This is quite conceivable if the canonical phosphorylation-dephosphorylation cycle of fig. 1 is considered from an opposite angle (fig. 4A) i.e. assuming a preliminary, constitutive phosphorylation of all the molecules of the regulatable protein, which are subsequently variably converted into the dephosphorylated form by a phosphatase responsive to intra- and/or extra-cellular signals.

In this instance the regulatable protein phosphatase would antagonize the effect of the protein kinase, as it also does according to the canonical scheme of fig. 1. Alternatively however a mechanism can be envisaged where the two converter enzymes play a synergistic rather than

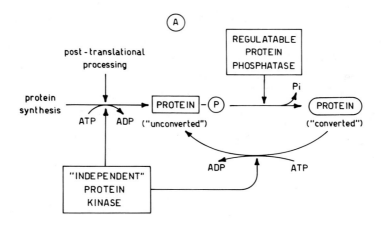

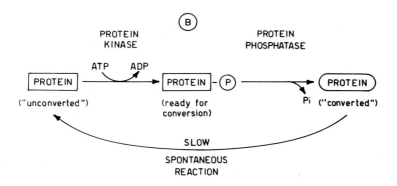

FIG. 4   Two models accounting for the controlled occurrence of
interconversions involving "independent" protein kinases.
    In A the kinase and the phosphatase still play their typical
antagonistic roles outlined in the canonical scheme of fig. 1. The
interconvertible substrate protein however is here assumed to normally be
in a constitutively phosphorylated form, the "converted" form being
therefore the dephosphorylated protein and the primary converter enzyme a
regulatable phosphatase, rather than the spontaneously active kinase.
    According to model B conversely the kinase and the phosphatase are
sequentially involved in a two-steps combined reaction implying a
synergistic role for them, suited for a "two-keys" safeguard mechanism.
Either both or just one of the converter enzymes are subjected to
short-term regulation, while the "converted" protein is reverted to its
"unconverted" form through a slow spontaneous reaction not implying
covalent modifications. Preliminary phosphorylation in this case is a
pre-requisite but not a sufficient condition for confering the required
alteration of biological properties. Depending on which enzymatic step is
rate-limiting, the phosphorylated form either will accumulate or it is
kept at negligible value, thus escaping detection. This latter case
appears to occur in the activation of the cytosolic form of protein
phosphatase-1 promoted by a phosphorylation-dephosphorylation reaction of
the modulator protein I-2 (see Ballou and Fischer, 1986).

antagonistic role, namely phosphorylation is merely a pre-requisite for dephosphorylation, the latter representing the very crucial step conferring novel properties to the phosphorylatable protein. This would imply that the conformation of the dephosphorylated form is different from that of the non-phosphorylated one, the latter being re-generated from the former through a slow, spontaneous process (fig. 4B). A mechanism like this would be not a mere fiction as it has been shown to operate, e.g., in the activation of the cytosolic form of protein phosphatase-1, occurring through the autocatalytic dephosphorylation of a threonyl residue of the regulatory subunit of the phosphatase (termed modulator protein or inhibitor-2) immediately after its phosphorylation by FA-kinase (=GSK-3) (Ballou and Fischer, 1986). Likewise it has been recently proposed that the potentiation of the estradiol receptor concomitantly requires both kinase and phosphatase activities in a combined reaction (Dayani et al., 1990). Conceivably such a mode of operation would require only one step to be subjected to control. This could be dephosphorylation rather than phosphorylation, thus providing a rationale for the function of "independent" protein kinases like CK-2, engaged with the supply of premodified molecules of the convertible protein. If however phosphorylation is the rate-limiting step, the phospho-form will never accumulate thus possibly accounting for the puzzling sub-stoichiometric phosphorylation of certain proteins, especially those affected at tyrosyl residues.

Focusing back on CK-2 it should be outlined that the mechanisms proposed in fig. 4, A and B, might operate also with those targets whose phosphorylation is dependent on polybasic effectors. Whichever might be in fact the natural compounds whose effect in vitro is mimicked by polylysine - e.g. certain histones and RAS proteins, but not polyamines, as discussed elsewhere (Pinna, 1990) - it is hard to assume for them a role like that of second messengers, capable to induce transient effects mediated by rapid fluctuations of their concentration. Reasonably polybasic peptides may ensure a persisting CK-2 activity toward selected targets in particular subcellular compartments, rather than a short-term regulation of CK-2 activity.

In any event the structure of the phosphoacceptor sites for CK-2 is expected to play a crucial role in determining their variable suscepti- bility both to polybasic enhancers of phosphorylation and to enzymatic dephosphorylation.

It should be recalled in this connection that the natural phosphorylation sites for CK-2, albeit all fulfilling the common Ser(Thr)-X-X-Glu(Asp) consensus sequence, are variable in several other respects (see Pinna, 1990). Namely some of them are extremely acidic, with stretches of five or more consecutive carboxylic residues downstream of serine, while others are moderately acidic, including just a couple of carboxylic residues. Acidic residues, furthermore, can be located either on the critical C terminal side alone or, sometimes, on both sides of the target serine. Studies with synthetic peptides have outlined the crucial relevance of the β-turn conformation for determining the susceptibility of the poorly acidic sites, but not of the highly acidic ones, to CK-2 mediated phosphorylation (see Pinna, 1990). They also support the concept that N terminal acidic residues may play a role in enhancing the responsiveness to polybasic effectors, since the deletion of the individual N terminal glutamic acid from the octapeptide ESVSSSEE fully abolishes the favourable effect of polylysine on its Km for CK-2 (Meggio et al.,1989). On the other hand dephosphorylation of phosphopeptides by protein phosphatase-2A can be dramatically improved by some features, like somewhat remote N terminal basic residues, the phosphopeptide RRREEESpEEE being rapidly dephosphorylated whereas ESpEEEEE is totally refractory to dephosphorylation (Agostinis et al., 1990).

Synthetic peptides also proved helpful for disclosing negative determinants that may compromise the phosphorylation of otherwise suitable substrates. Thus basic aminoacids are detrimental if they are adjacent to the N terminal edge of the target residue or located at variable positions on the C terminal side where neutral residues are accepted. Likewise the motif Ser-Pro is unacceptable, giving rise to peptides that may act as competitive inhibitors of CK-2. Such a narrow specificity may contribute to prevent "misphosphorylations" by an enzyme like CK-2 which could become noxious due to its lack of effective short-term control systems.

In this respect CK-2 could be considered as a potential oncogene whose triggering of proliferative events is normally controlled by the concomitant activity of protein phosphatase(s). Actually among the various growth related protein kinases (Ralph et al., 1990), CK-2 alone appears to be spontaneously active and devoid of any effective down regulation mechanism. Consequently it is a first choice candidate for mediating the effects of tumor promoters, like okadaic acid, whose primary biochemical effect is the inhibition of protein phosphatases. It should be also recalled in this connection that an abnormally high phosphorylation of

CK-2 targets will also bring about an increased activity of "secondary" protein kinases, like GSK-3, whose sites are specified by previous CK-2 mediated phosphorylation.

Acknowledgements: The secretarial aid of Miss Anna Maria Monaco is gratefully acknowledged. Studies carried out in our laboratory were supported by grants from italian C.N.R. (Target Project Biotechnology and Bioinstrumentation) and M.U.R.S.T.

References

Agostinis P, Goris J, Pinna LA, Marchiori F, Perich JW, Meyer HE and Merlevede W (1990) Synthetic peptides as model substrates for the study of the specificity of the polycation-stimulated protein phosphatase. Eur. J. Biochem. 189:235-241

Ballou LM and Fischer EH (1986) Phosphoprotein phosphatases. In: The Enzymes Vol. XVII (Boyer PD and Krebs EG eds) Academic Press Inc. pp. 311-361

Beebe SJ and Corbin JD (1986) Cyclic nucleotide-dependent protein kinases. In: The Enzymes Vol. XVII (Boyer PD and Krebs EG eds) Academic Press, Orlando pp. 44-111

Cooper JA (1990) The src family of protein tyrosine kinases. In: Peptides and protein phosphorylation (Kemp BE ed) CRC Press Inc. Boca Raton pp. 85-113

Dayani N, McNaught RW, Shenolikar S and Smith RG (1990) Receptor interconversion model of hormone action, 2. Requirement of both kinase and phosphatase activities for conferring estrogen binding activity to the estrogen receptor. Biochem. 29:2691-2698

Dent P, Lavoinne A, Nakielny S, Caudwell FB, Watt P and Cohen P (1990) The molecular mechanism by which insulin stimulates glycogen synthesis in mammalian skeletal muscle. Nature 348:302-308

Dobrowolska G, Meggio F and Pinna LA (1987) Characterization of multiple forms of maize seedling protein kinases reminiscent of animal casein kinases S (type 1) and TS (type 2) Biochim. Biophys. Acta 931:188-195

Donella-Deana A, Brunati AM, Marchiori F, Borin G and Pinna LA (1990) Different specificities of spleen tyrosine protein kinases as evidenced with synthetic peptide substrates Eur. J. Biochem. 194:773-777

Draetta G (1990) Cell cycle control in eukariotes: molecular mechanism of cdc2 activation Trends Biochem. Sci. 15:378-383

Fiol CJ, Mahrenholz AM, Roeske RW and Roach PJ (1987) Formation of protein kinase recognition sites by covalent modification of the substrate. Molecular mechanism for the synergistic action of casein kinase and glycogen synthase kinase 3. J. Biol. Chem. 262:14042-14048

Flotow H, Graves PR, Wang A, Fiol CJ, Roeske RW and Roach PJ (1990) Phosphate groups as substrate determinants for casein kinase I action. J. Biol. Chem. 265:14264-14269

Hardie DG, Carling D and Sim ATR (1989) The AMP-activated protein kinase: a multisubstrate regulator of lipid metabolism. Trends Biochem. Sci. 14:20-23

Hunter T, Cooper JA (1985) Protein tirosine kinases. Ann. Rev. Biochem. 54:897-930

Hunter T, Cooper JA (1986) Viral oncogenes and tyrosine phosphorylation. In: The Enzymes Vol. XVII (Boyer PD and Krebs EG eds) Academic Press, Orlando pp. 192-246

Issinger OG, Martin T, Richter WW, Olson M and Fujiki H (1988) Hyperphosphorylation of N60, a protein structurally and immunologically related to nucleolin, after tumor-promoter treatment EMBO J. 7:1621-1627

Masaracchia RA, Murdoch FE and Hassel TC (1990) Unique specificity determinants for an S6/H4 kinase and protein kinase-C.In: Peptides and Protein Phosphorylation (Kemp BE ed) CRC Press Inc. Boca Raton pp. 189-207

Meggio F, Donella-Deana A and Pinna LA (1979) Studies on the structural requirements of a microsomal cAMP-independent protein kinase FEBS Lett. 106:76-80

Meggio F, Perich JW, Meyer HE, Hoffmann-Posorske E, Lennon DPW, Johns RB and Pinna LA (1989) Synthetic fragments of -casein as model substrates for liver and mammary gland casein kinases. Eur. J. Biochem. 186:459-464

Mercier JC (1981) Phosphorylation of caseins, present evidence for an aminoacid triplet code posttranslationally recognized by specific kinases. Biochimie 63:1-17

Moore A, Boulton AP, Heid HW, Jarasch ED and Craig RK (1985) Purification and tissue specific expression of a casein kinase from the lactating guinea-pig mammary gland. Eur. J. Biochem. 152:729-737

Nishizuka Y (1988) The molecular heterogeneity of protein kinase-C and its implications for cellular regulation. Nature 334:661-665

Picton C, Woodgett J, Hemmings B and Cohen P (1984) Multisite phosphorylation of glycogen synthase from rabbit skeletal muscle. Phosphorylation of site 5 by glycogen synthase kinase-5 (casein kinase II) is a prerequisite for phosphorylation of site 3 by glycogen synthase kinase-3. FEBS Lett. 150:191-196

Pinna LA (1990) Casein kinase-2: An eminence grise in cellular regulation? Biochim. Biophys. Acta 1054:267-284

Pinna LA, Agostinis P and Ferrari S (1986) Selectivity of protein kinases and protein phosphatases: a comparative analysis. Adv. Prot. Phosphatases 3:327-368

Ralph RK, Darkin-Rattray S and Schofield P (1990) Growth-Related Protein kinases BioEssays 12:121-124

Reed LJ and Yeaman SJ (1987) Pyruvate dehydrogenase. In: The Enzymes Vol. XVIII (Boyer PD and Krebs EG eds) Academic Press, Orlando pp. 77-95

Stull JT, Nunnally MH and Michnoff CH (1986) Calmodulin dependent protein kinases. In: The Enzymes Vol. XVII (Boyer PD and Krebs EG eds) Academic Press, Orlando pp. 114-166

Taylor SS (1989) cAMP-dependent protein kinase. Model for an enzyme family J. Biol. Chem. 264:8443-8446

Ullrich A and Schlessinger J (1990) Signal transduction by receptors with tyrosine kinase activity. Cell 61:203-212

Vulliet PR, Hall FL, Mitchell JP and Hardie DG (1989) Identification of a novel proline-directed serine/threonine protein kinase in rat pheochromocytoma. J. Biol. Chem. 264:16292-16298

West DW (1986) Structure and function of the phosphorylated residues of casein. J. Dairy Res. 53:333-352

Woodgett JR (1990) Molecular cloning and expression of glycogen synthase kinase-3/factor A. EMBO J. 9:2431-2438

Zhang G, Stevens R and Leader DP (1990) The protein kinase encoded in the short unique region of pseudorobies virus: description of the gene and identification of its product in virions and in infected cells. J. Gen. Virol. 71:1757-1765

BIPHASIC ACTIVATION OF THE S6 KINASE: IDENTIFICATION OF SIGNALLING PATHWAYS

George Thomas
Friedrich Miescher Institute
P.O. Box 2543
4002 Basel, Switzerland

INTRODUCTION

In my first lecture I presented data which showed that following mitogenic stimulation of quiescent 3T3 cells, the M_r 70 kd S6 kinase becomes activated by serine/threonine phosphorylation. From this data we deduced two facts: first, that the S6 kinase lies on a kinase cascade initiated by the activation of the tyrosine kinase of the respective growth factor-receptor (Carpenter G and Cohen S, 1990), and second, that there must be at least one other serine/threonine kinase, activated by tyrosine phosphorylation, which couples the S6 kinase with the tyrosine kinase of the receptor, an S6 kinase kinase (Ballou LM et al, 1988a). Such a model would easily fit with earlier kinetic data showing that in quiescent cells stimulated with, for example, EGF, the S6 kinase is rapidly activated, reaching a maximum between five and ten minutes and then slowly decreasing back to basal level by about 60 minutes (Novak-Hofer I and Thomas, G, 1985).

Biphasic activation of S6 kinase

A more carefully conducted kinetic analysis revealed that activation of the S6 kinase is much more complex than the simple model outlined above. The results showed that the activation of the kinase was biphasic (Šuša M et al, 1989). The first phase of activation reached a maximum between ten and fifteen minutes and the second phase between 45 and 60 minutes. The most obvious explanation for this finding was that there were two different enzymes, one involved in the early phase of S6 kinase activation and a second responsible for the later phase. To test this possibility, cell extracts were prepared from quiescent cultures which had been stimulated with EGF for either ten or sixty minutes, and the extracts subjected to

NATO ASI Series, Vol. H 56
Cellular Regulation by Protein Phosphorylation
Edited by L. M. G. Heilmeyer, Jr.
© Springer-Verlag Berlin Heidelberg 1991

receptor kinase *in vitro* (Meyer T et al, 1989). A second analogue of the staurosporin CGP 42 700, has no effect on any of the three enzymes, PKC, EGF receptor or the S6 kinase. When we tested the ability of these two compounds to inhibit the S6 kinase activation induced by EGF we found that there was no effect of the inactive analogue. In the case of the active analogue, though, we found that the second phase of S6 kinase activation was totally abolished, strongly supporting our hypothesis that the second phase of S6 activation is under the control of PKC (M. Šuša and G. Thomas, unpublished).

Purification of S6 kinase from early and late phase

The results above indicate that the second phase of S6 kinase activation is under the control of PKC. However, the signalling mechanism controlling the early phase of S6 kinase activation has not been identified (Šuša M et al, 1989). From these results we inferred that the EGF triggers two distinct signalling pathways which converge at different times on the same kinase (Šuša M et al, 1989). However, homologues to a second S6 kinase of M_r 92 kd, which had previously only been identified in *Xenopus laevis*, have recently been detected in mouse and chicken cells stimulated with a number of mitogens (Chen RH and Blenis J, 1990; Sweet LJ et al, 1990). This last observation together with our earlier findings raised the question of the identity of the S6 kinase in the two phases of activation and the point at which the two signalling pathways converge to induce S6 kinase activation. In a previous study we purified and analyzed the phosphorylated amino acids in the M_r 70 kd kinase from cells stimulated for one hour with serum (Ballou LM et al, 1988a). In that study no precautions were taken against phosphotyrosine phosphatases. Therefore any effects on kinase activity due to tyrosine phosphorylation could have been partially or totally abrogated. To protect phosphotyrosine from dephosphorylation, sodium-orthovanadate and ammonium molybdate were added to the extraction buffer as potent phosphotyrosine phosphatase inhibitors (Šuša M and Thomas G, 1990). The kinetics of S6 kinase activation in response to EGF induction under these conditions was measured and compared with results obtained in the absence of phosphotyrosine phosphatase inhibitors. The amplitude and the kinetic appearance of each phase (Šuša M and Thomas G, 1990) of activity was very similar to those which we reported earlier, with the early phase peaking in ten to fifteen minutes and the late phase reaching a maximum between forty and sixty minutes following EGF induction (Šuša M and Thomas G, 1990). These results suggested that if phosphotyrosine were present in the S6 kinase it would play little role in activating the enzyme.

Identity of early and late phase S6 kinase

We next set out to purify both the early and late phase kinases labeled in vivo with ^{32}P-phosphate. Based on the data above, stimulation with EGF for 10 minutes was chosen for the early phase of S6 kinase activation, a time when the kinase activity was rising, and 60 minutes was used for the late phase of kinase activation, a time in which the activity in the early phase should have begun to return to basal levels. Following a four-step purification protocol, a final recovery of 30% was obtained (Lane HA and Thomas G, 1991). At each step of the purification the enzymes present in both phases eluted in the same position as a single peak of activity. To determine the identity of the kinase present in each phase the proteins present in the peak fraction of the last step of purification, a peptide affinity column, were separated on SDS-polyacrylamide gels and analyzed by autoradiography (Šuša M and Thomas G, 1990). In both cases, a single protein band with M_r 70 kd was observed, indicating that the kinase responsible in both the early and late phases of S6 kinase activation were identical.

To ensure that all of the radioactive ^{32}P-labeled protein eluting with S6 kinase activity from the peptide columns represented the M_r 70 kd enzyme, the active S6 kinase fractions from the peptide column were chromatographed on an analytical Mono Q column (Šuša M and Thomas G, 1990). All of the radioactivity from each column was found to co-elute with the S6 kinase activity. Next, a portion of each fraction was incubated with γ-^{32}P-ATP in vitro, to autophosphorylate the enzyme. The proteins present were then separated by SDS-PAGE and analyzed by autoradiography. The only protein band visible was again the M_r 70K protein. More importantly, this protein was found only in those fractions containing S6 kinase activity and the intensity of autophosphorylation directly paralleled S6 kinase activity (Šuša M and Thomas, G, 1990). Finally, the two M_r 70K autophosphorylated proteins were electrophoresed from the gel, digested with trypsin and the proteolytic products were analyzed by two-dimensional thin layer electrophoresis / chromatography. Although we found the relative intensities of the maps to be different, the identical phosphopeptides were observed, indicating that both kinases represented the same protein. Consistent with this finding, one-dimensional SDS-PAGE analysis of cyanogen bromide peptides of both autophosphorylated kinases were found to be identical. These results support our earlier hypothesis that the two phases of EGF-induced S6 kinase activation are mediated through the same enzyme.

Phosphorylation sites regulating S6 kinase activity

To determine whether the two EGF-induced pathways used the same or distinct phosphorylation sites to regulate S6 kinase activity we first analyzed the phosphorylated amino acids present in each protein. Phosphoserine, and to a much lesser extent, phosphothreonine, were detected in both enzyme preparations, and at approximately the same level (Šuša M and Thomas G, 1990). Hence, the different signalling pathways cannot be distinguished by the content of their phosphorylated amino acids. The absence of phosphotyrosine was further supported by the observations that alkaline treatment of gels containing the M_r 70K kinase from both the ten and the sixty-minute stimulated cells led to a complete removal of phosphate from either protein band and that antiphosphotyrosine antibodies failed to immunoprecipitate increased S6 kinase activity from EGF-stimulated cell extracts (M. Šuša and G. Thomas, unpublished). These results appear to rule out the phosphorylation and direct activation of the S6 kinase by the activated EGF receptor tyrosine kinase, again supporting the model that there has to be at least a single kinase coupling the EGF receptor with the S6 kinase during the early phase of activation.

Next, to determine whether the same phosphorylation sites were shared by the two pathways, tryptic peptides of in vivo ^{32}P-labeled S6 kinase from cells treated with EGF for 10 or 60 minutes were analyzed by two-dimensional thin layer electrophoresis / chromatography. The maps from the two enzymes showed very similar patterns, further indicating that the two kinases were identical. However, visual comparison of the maps revealed that the relative amount of phosphate incorporated did differ in some peptides (Šuša M and Thomas G, 1990). These differences could not be quantitated because of the low amount of ^{32}P incorporated in the protein. More striking was the appearance of a unique phosphopeptide in the late phase S6 kinase, which could not be detected in tryptic maps derived from the early phase of S6 kinase even after prolonged exposure (Šuša M and Thomas G, 1990). This was the only qualitative difference detected between these two kinase preparations. In the future it will be of great interest to determine whether this site is directly regulated by PKC.

First phase of activation: Is MAP2 kinase S6 kinase kinase?

In contrast to the second phase of S6 kinase activation, we as yet have no idea as to the identity of the kinase regulating the first phase of activation. Earlier it had been shown that the microtubule-associated protein 2 (MAP2) kinase (Ray LB and Sturgill TW, 1987; Ray LB and Sturgill TW, 1988), stimulated the activity of the M_r 92 kd S6 kinase from *Xenopus laevis* (Sturgill TW et al, 1988). More recently, it has also been claimed to have a similar effect on the M_r 70 kd kinase from rabbit liver (Gregory JS et al, 1989). In searching for insulin-activated kinases Ray and Sturgill (1987) used MAP2 as a substrate enabling the identification of a MAP2 kinase. Furthermore, they went on to show that the enzyme was phosphorylated at both serine and tyrosine residues (Ray LB and Sturgill TW, 1988) and was distinct from both the M_r 92 kd and the 70 kd S6 kinases (Ray LB and Sturgill TW, 1987). More interesting, they have recently shown that this enzyme requires both tyrosine and serine phosphorylation to maintain its activity (Anderson NG et al, 1990). This was demonstrated by complete inactivation of the enzyme with either phosphatase 2A or CD45, a tyrosine phosphatase from T cells (Tonks NK et al, 1988). To substantiate this finding they showed that the amino acids phosphatase 2A and CD45 specifically dephosphorylated were threonine and tyrosine, respectively. Thus the MAP2 kinase would seem to be a strong candidate for the activator of the S6 kinase.

MAP2 and S6 kinase in Swiss 3T3 cells

In quiescent 3T3 cells, serum or epidermal growth factor causes a transient increase in MAP2 kinase activity (Ballou LM et al, 1991). Maximum activation is 3-fold for serum and 3.4-fold for EGF at 2.5 minutes. S6 kinase under these conditions is activated to a higher level, although more slowly, increasing 6.8- and 9.3-fold after ten minutes with serum and EGF, respectively (Ballou LM et al, 1991). The faster response of MAP2 kinase would kinetically fit with this enzyme's putative S6 kinase kinase role *in vivo*. Furthermore, the EGF-stimulated MAP2 kinase appeared to be identical to the enzyme first described in insulin-treated adipocytes by Ray and Sturgill (Ray LB and Sturgill TW, 1987), determined by its chromatographic behavior, its ability to be immunoprecipitated with antibodies against phosphotyrosine and its inactivation by treatment with phosphatase 2A (Ballou LM et al, 1991).

The ability of the MAP2 kinase to reactivate the M_r 70K S6 kinase was tested using a control sample of S6 kinase and a preparation of the S6 kinase that was inactivated >90% by prior treatment with phosphatase 2A (Ballou LM et al, 1991). Both kinase preparations were then incubated with or without MAP2 kinase in the presence of ATP and a phosphatase inhibitor. The results of initial reactivation experiments were difficult to interpret because the S6 kinase lost as much as 80% of its activity during the incubation. This decrease was prevented by the addition of MAP2 kinase, resulting in the apparent activation of the control S6 kinase and reactivation of the phosphatase-treated enzyme. This effect was traced to stabilizing compounds in the MAP2 kinase buffer including ethylene glycol, Triton X-100 and bovine serum albumin. Another important factor in the reactivation buffer was the phosphatase inhibitor. Sturgill and colleagues (Sturgill TW et al, 1988) added sodium fluoride to inhibit phosphatase 2A after inactivating the S6 kinase. However, sodium fluoride does not inhibit phosphatases in the presence of magnesium chloride, a cation whose presence is required for kinase activity (S. Ferrari and G. Thomas, unpublished). In view of this finding, we instead used p-nitophenyl phosphate, which inhibited the phosphatase >90%. Under these conditions, the control and phosphatase-treated S6 kinases were stable over the incubation period in the absence of MAP2 kinase. When MAP2 kinase was added there was no effect on either S6 kinase preparation (Ballou LM et al, 1991). Similar results were obtained in experiments using five times more MAP2 kinase, S6 kinase purified from rat liver (Kozma SC et al, 1989; Lane HA and Thomas G, 1991), MAP2 kinase purified by immunoprecipitation with antiphosphotyrosine antibodies, and S6 kinase which had been only inactivated 50% by phosphatase 2A (LM Ballou and G Thomas, unpublished).

In agreement with the results above, the M_r 70 kd S6 kinase was not phosphorylated by MAP2 kinase *in vitro* (Ballou LM et al, 1991). These results conflict with a recent report claiming that a M_r 70 kd S6 kinase from rabbit liver, presumably equivalent to the M_r 70 kd enzyme from 3T3 cells or rat liver, was partially reactivated by MAP2 kinase from rat 1 cells (Gregory JS et al, 1989). However, the stability of the S6 kinase used in that study was not tested nor was phosphorylation of the enzyme by MAP2 kinase demonstrated. The failure of MAP2 kinase to reactivate the M_r 70K S6 kinase *in vitro* led us to search for *in vivo* evidence that would support this finding. If MAP2 kinase lies directly upstream of the S6 kinase then in a phosphorylation cascade the most potent activator of MAP2 kinase might be expected to be the most potent activator of the S6 kinase. However, such a correlation did not exist. For the ten agents tested the order of effectiveness in stimulating the MAP2 kinase was platelet-derived growth factor, > bombesin, > prostaglandin $F_{2\alpha}$, > EGF, > TPA, > serum, > orthovanadate, > prostaglandin E_1, > cAMP, > insulin. All these compounds except insulin induced maximum activation of the MAP2 kinase at 2.5 minutes. The order of effectiveness for activating S6 kinase was distinct; EGF > serum > TPA > platelet-derived growth factor > prostaglandin $F_{2\alpha}$, > orthovanadate, > bombesin, > insulin, > cAMP = prostaglandin E_1 (Ballou, L. and Thomas, G, unpublished). These results further support the argument that the MAP2 kinase is not the S6 kinase activator.

Effect of insulin on MAP2 kinase and S6 kinase

Surprisingly, of all the agents tried above, insulin had the least effect on MAP2 kinase although the enzyme was first detected in insulin-treated cells. A longer time course showed that insulin did not cause a substantial increase in MAP2 kinase activity, even after 25 minutes. Maximal stimulation of only 1.2-fold occurred at five minutes, whereas S6 kinase was activated 6.1-fold after 15 minutes with the hormone. An even more striking difference was seen when extracts were subjected to anion chromatography. MAP2 kinase activity from insulin-treated cells was barely above resting levels, while the activity in the MAP2 kinase peak after EGF treatment was 20 times higher. At insulin concentrations as high as 5×10^{-7}, MAP2 kinase activity in extracts never increased more than 1.9-fold, whereas S6 kinase was activated as much as 6.9-fold. Insulin-like growth factor 1 was also a poor activator of the MAP2 kinase, causing a maximal 2.1-fold stimulation after 2.5 minutes, but a 6.1-fold activation of the S6 kinase.

Together these data indicate that the MAP2 kinase is not the activator of the M_r 70K S6 kinase in fibroblasts. The results with insulin further imply that the activation of the two enzymes is accomplished through different signalling pathways. Indeed, it is unclear whether the cascade including MAP2 kinase and S6 kinase which was proposed earlier (Sturgill TW et al, 1988) operates *in vivo*. It has not been shown that insulin activates the MAP2 kinase in oocytes or that the S6 kinase activated by insulin in adipocytes is the M_r 92K species. The identity of the M_r 70K S6 kinase kinase remains obscure. Attention has focused on kinases that may be activated by tyrosine phoshorylation, but this is based on the assumption that there are only three components in the phosphorylation cascade. Furthermore, there may be more than one S6 kinase kinase, since S6 kinase is activated by a wide range of agents that act through different signalling pathways. The biphasic kinetics of S6 kinase activation induced by EGF could be explained by sequential activation of different S6 kinase kinases which act together to fully activate the enzyme. Future experiments designed to identify the S6 kinase activator should show reactivation and phosphorylation of the S6 kinase *in vitro* as well as data that the same sites which are dephosphorylated are the ones which become phosphorylated.

Immediate goals

One obvious approach to sorting out the problems listed above will be to map the *in vivo* sites of S6 phosphorylation. The aim will then be to reconstitute this map *in vitro*. Such an approach may be essential in view of the finding that activation of MAP2 kinase requires both threonine and tyrosine phosphorylation (Anderson NG et al, 1990) and that activation of the mammalian form of CDC2 kinase requires dephosphorylation of these same amino acids (see Draetta G, this volume). Thus, activation of the M_r 70K S6 kinase may require multiple phosphorylation events by different kinases at distinct sites.

Acknowledgements: I would like to thank Drs. S.C. Kozma and D. Reddy for their critical reading of the manuscript. I would also like to express my gratitude to Carol Wiedmer for her perseverance in transforming the tape of my talk into this written manuscript.

References

Anderson NG, Maller JL, Tonks NK and Sturgill TW (1990) Requirement for integration of signals from two distinct phosphorylation pathways for activation of MAP kinase. Nature 343:651-653

Ballou LM, Siegmann M and Thomas G (1988a) S6 kinase in quiescent Swiss mouse 3T3 cells is activated by phosphorylation in response to serum treatment. Proc. Natl. Acad. Sci USA 85:7154-7158

Ballou LM, Jenö P and Thomas G (1988b) Protein phosphatase 2A inactivates the mitogen-stimulated S6 kinase from Swiss mouse 3T3 cells. J Biol Chem 263:1188-1194

Ballou LM, Luther H and Thomas G (1991) MAP2 kinase and 70K S6 kinase lie on distinct signalling pathways. Nature 349:348-350

Blenis J and Erikson RL (1986) Stimulation of ribosomal protein S6 kinase activity by the Rous sarcoma virus transforming protein, serum, or phorbol ester. Proc Natl Acad Sci USA 83:1733-1737

Carpenter G and Cohen S (1990) Epidermal growth factor. J Biol Chem 265:7709-7712

Chen RH and Blenis J (1990) Identification of *Xenopus* S6 protein kinase homologs (pp90[rsk]) in somatic cells: phosphorylation and activation during initiation of cell proliferation. Mol Cell Biol 10:3204-3215

Gregory JS, Boulton TG, Sang B-C and Cobb MH (1989) An insulin-stimulated ribosomal protein S6 kinase from rabbit liver. J Biol Chem 264:18397-18401

Hannun YA, Loomis CR, Merill AH and Bell RM (1986) Sphingosine inhibition of protein kinase C activity and of phorbol dibutyrate binding *in vitro* and in human platelets. J Biol Chem 261:12604-12609

Kozma SC, Lane HA, Ferarri S, Luther H, Siegmann M and Thomas G (1989) A stimulated S6 kinase from rat liver: identity with the mitogen-activated S6 kinase of 3T3 cells. EMBO J 8:4125-4132

Lane HA and Thomas G (1991) Purification and properties of mitogen-activated S6 kinase from rat liver and 3T3 cells. In Methods of Enzymol 200, in press

Meyer T, Regenass U, Fabbro D, Alteri E, Rösel J, Müller M, Ceravatti G and Matter A (1989) A derivative of staurosporine (CGP 41 251) shows selectivity for protein kinase C inhibition and *in vitro* anti-proliferative as well as anti-tumor activity. Int J Cancer 43:851-856

Novak-Hofer I and Thomas G (1985) Epidermal growth factor-mediated activation of an S6 kinase in Swiss mouse 3T3 cells. J Biol Chem 260:10314-10319

Ray LB and Sturgill TW (1987) Rapid stimulation by insulin of a serine/threonine kinase in 3T3-LI adipocytes that phosphorylates microtubule-associated protein 2 *in vitro*. Proc Natl Acad Sci USA 84:1502-1506

Ray LB and Sturgill TW (1988) Insulin-stimulated microtubule-associated protein 2 kinase is phosphorylated on tyrosine and threonine *in vivo*. Proc Natl Acad Sci USA 85:3753-3757

Shibanuma M, Kuroki T and Nose K (1987) Inhibition of proto-oncogene c-*fos* transcription by inhibitors of protein kinase C and ion transport. Eur J Biochem 164:15-19

Sturgill TW, Ray LB, Erikson E and Maller JL (1988) Insulin-stimulated MAP-2 kinase phosphorylates and activates ribosomal protein S6 kinase II. Nature 334:715-718

Stabel S, Rodiguez-Pena A, Young S, Rozengurt E and Parker PJ (1987) Quantitation of protein kinase C by immunoblot-expression in different cell lines and response to phorbol esters. J Cell Physiol 130:111-117

Šuša M, Olivier AR, Fabbro D and Thomas G (1989) EGF induces biphasic S6 kinase activation: late phase is protein kinase C-dependent and contributes to mitogenicity. Cell 57:817-824

Šuša M and Thomas G (1990) Identical M_r 70,000 S6 kinase is activated biphasically by epidermal growth factor: a phosphopeptide which characterizes the late phase. Proc Natl Acad Sci USA 87:7040-7044

Sweet LJ, Alcorta DA, Jones SW, Erikson E and Erikson RL (1990) Identification of mitogen-responsive ribosomal protein S6 kinase PP90[rsk], a homologue of *Xenopus* S6 kinase II, in chicken embryo fibroblasts. Mol Cell Biol 10:2413-2417

Tabarini D, Heinrich J and Rosen OM (1985) Activation of S6 kinase activity in 3T3-L1 cells by insulin and phorbol ester. Proc Natl Acad Sci USA 82:4369-4373

Tamaoki T, Nomoto H, Takahashi I, Kato Y, Morimoto M and Tomita F (1986) Staurosporin, a potent inhibitor of phospholipid/Ca^{++} dependent protein kinase. Biochem Biophys Res Comm 135:397-402

Tonks NK, Charbonneau H, Diltz CD, Fischer EH and Walsh KA (1988) Demonstration that the leukocyte common antigen CD45 is a protein tyrosine phosphatase. Biochem 27:8695-8701

CHANGES ON THE ELECTROPHORETIC MOBILITY OF CD5 MOLECULES INDUCED BY PKC-MEDIATED PHOSPHORYLATION

Jose Alberola-Ila, Lourdes Places, Jordi Vives and Francisco Lozano.
Servei d'Immunologia. Hospital Clinic i Provincial de Barcelona.
Villarroel 170. Barcelona 08036.
Spain.

INTRODUCTION

Protein phosphorylation is the most common form of post-translational modification used to regulate cellular functions (Edelman et al., 1987). CD5 provides accessory signals in T lymphocyte activation and proliferation (Weiss and Imboden, 1987). CD5 has been shown to undergo hyperphosphorylation after treatment with phorbol esters (Chatila and Geha, 1988) which are tumor promoter agents (TPA) capable to translocate and activate the serine/threonine protein kinase C (PKC) (Nishizuka, 1984). Here we show the rapid and PKC-dependent induction of a hyperphosphorylated and more slowly migrating subset of CD5 molecules after stimulation with TPA on normal and lymphoblastoid T and B cells.

RESULTS AND DISCUSSION

Analysis on SDS-polyacrylamide gels of immunoprecipitates from ^{32}P-labeled peripheral blood mononuclear cells (PBMC) obtained with the Leu-1 mAb (CD5 specific) reveals that PMA treatment induces changes not only in phosphorylation but also in the electrophoretic mobility of CD5 molecules. CD5 is a constitutively weakly phosphorylated molecule and the treatment for 30 min with 100 ng/ml PMA induces the appearance of two hyperphosphorylated CD5 species with different electrophoretic mobility. The faster one has a M_r closely similar to the constitutively phosphorylated CD5 molecules, whilst the slower one has a significant higher M_r. These PMA-induced mobility changes seem to be characteristic for CD5 molecules since no similar changes were observed under these conditions for other constitutively phosphorylated lymphocyte surface molecules (HLA class I, CD7, CD18, CD43, CD44, CD45, CD45R) [Alberola-Ila et al, 1989]. These changes occur in all tested cell types (PBMC, thymocytes, tonsil B cells, B lymphocytes from chronic lymphocytic leukemia and the T cell lines CEM, Jurkat, 8402, Molt 4, HSB2 and HUT 78)

NATO ASI Series, Vol. H 56
Cellular Regulation by Protein Phosphorylation
Edited by L. M. G. Heilmeyer, Jr.
© Springer-Verlag Berlin Heidelberg 1991

(Fig. 1). The same shift in mobility could also be appreciated by the analysis of CD5 molecules from [^{35}S]cysteine-labeled PBMC (data not shown).

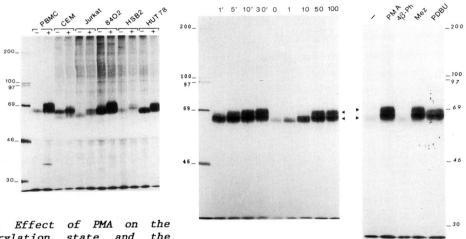

Fig. 1. Effect of PMA on the phosphorylation state and the apparent M_r of ^{32}P-labeled CD5 molecules from normal and lymphoblastoid T cells. Several cellular types were labeled with [^{32}P]orthophosphate, stimulated for 30 min in the presence (+) or absence (-) of 100 ng/ml PMA and immunoprecipitated with Leu-1 mAb. The SDS-PAGE analysis of the immunoprecipitates is shown.

Fig. 2. Dose and time response kinetics of TPA-induced CD5 mobility changes. Left. ^{32}P-labeled PBMC were treated either with PDBU at the indicated concentrations (ng/ml) or with different concentrations of PMA for 30 min. Right. Effect of different TPA on the phosphorylation state and electrophoretic mobility of CD5 molecules. ^{32}P-labeled PBMC were stimulated for 30 min with 100 ng/ml of different TPA. Cell lysates were immunoprecipitated with Leu-1 mAb and subjected to SDS-PAGE analysis.

The Western blot analysis under non-reducing conditions of PMA-treated PBMC lysates revealed also the appearance of Leu-1-reactive molecules with apparent higher M_r (Lozano et al, 1989). It becomes then clear that these higher M_r forms correspond to CD5 antigen and are not other associated molecules. Treatment of intact cells with neuraminidase showed that both CD5 forms were present at the cell surface (data not shown). These changes are induced by active (PMA, PDBU, mezerein), but not by inactive (4β-phorbol) tumor promoter agents (TPA) (Fig. 2). Additionally, they are dose-dependent and very rapid in time (are already evident at 5 min and picked at 30 min) (Fig 2). Preincubation of the cells with different PKC-inhibitors (staurosporine and H-7) was able to revert these TPA-induced changes (Fig 3). Phosphoaminoacid analysis showed that hyperphosphorylated TPA-induced CD5 forms are phosphorylated only at serine residues (data not shown). All these facts suggest that the effects of TPA on the electrophoretic mobility and the phosphorylation state of CD5 molecules are apparently mediated by PKC. Treatment of cells with Cycloheximide or Actinomycin D did not affect the mobility changes (Fig 3), discarding the need of de novo protein synthesis for the induction of the slow-mobility CD5 forms. As it is known that the addition

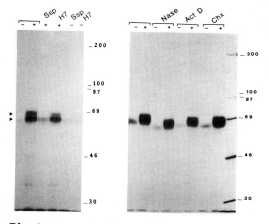

of sialic acid-containing sugar chains may affect the electrophoretic mobility of proteins in SDS-polyacrylamide gels [Krieger et al., 1989], we performed digestions of CD5 immunoprecipitates from PMA-treated and untreated ^{32}P-labeled PBMC with neuraminidase. This induced a slight drop in the apparent M_r of both slow and fast migrating hyperphosphorylated CD5 forms induced upon PMA-treatment (Fig 3). Therefore, differences in mobility are not due to alterations in their sialic acid content.

Fig.3. Dependence on PKC activation and independence on RNA and protein synthesis and glycosylation of the TPA-mediated CD5 changes. ^{32}P-labeled PBMC were incubated for 30 min in the presence (+) or the absence (-) of 10 ng/ml PMA. When indicated, prior PMA-treatment, cells were preincubated for 45 min with actinomicyn D (5μg/ml), cycloheximide (20μg/ml), staurosporine (100nM) or H-7 (250μM). Afterwards, cells were lysed and immunoprecipitated with Leu-1 mAb. Additionally, some immunoprecipitates were digested with neuraminidase.

As PKC seemed to be implicated in the changes of CD5 mobility induced by TPA, we decided to study whether they were due to phosphorylation. Digestion of CD5 immunoprecipitates from ^{35}S-labeled lymphoblastoid 8402 T cells with potato acid phosphatase (PAP) induced a decrease in the apparent M_r of the slow migrating form. PAP had no effect on the mobility of the CD5 molecules from unstimulated cells (Fig 4). Similar PAP-treatment of CD5 immunoprecipitates from ^{32}P-labeled, PMA-treated 8402 cells removed radioactivity from both the slow and fast migrating forms (Fig 4), although the slow migrating form underwent a bigger drop in its associated radioactivity. There was also a significative decrease in the radioactivity associated to CD5 in unstimulated cells. These experiments were also performed with PBMC with similar results. This indicates that phosphorylation could be the post-translational modification responsible for CD5 mobility changes.

A

B

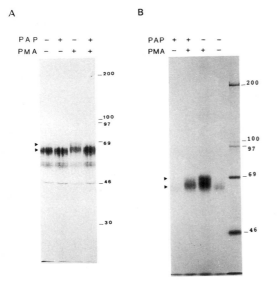

PAP − + − +
PMA − − + +

−200

−100
−97

−69

−46

−30

PAP + + − −
PMA − + + −

−200

−100
−97

−69

−46

Fig.4. Effect of phosphatase digestion on the TPA-mediated changes on the phosphorylation state and electrophoretic mobility of CD5 molecules. Lymphoblastoid 8402 T cells were metabolically labeled with [35S]cysteine (A) [32P]orthophosphate (B) and stimulated in the presence (+) or absence (−) of 100 ng/ml PMA for 30 min, followed by lysis and immunoprecipitation with the Leu-1 mAb. Afterwards, immunoprecipitates were digested (+) or not (−) for 1 h at 37°C with potato acid phosphatase (PAP) (30 μg/sample) followed by SDS-PAGE analysis.

Therefore, it could be suggested that the TPA-treatment cause phosphorylation of some critical sites responsible for the altered electrophoretic mobility of a subset of CD5 molecules on SDS-polyacrylamide gels.

The effects of phosphorylation in reducing the electrophoretic mobility on SDS-polyacrylamide gels for some cellular proteins have previously been referred [Sahyoun et al., 1984, Sprang et al., 1988]. Interestingly, TPA have been shown to induce similar changes on some membrane-associated (p56[lck]) [Marth et al. 1989], cytoplasmic (raf-1) [Morrison et al., 1988] and nuclear (L-myc) [Saksela et al., 1989] oncogene products deeply involved in the regulation of cellular activation and proliferation as it is also the case for the CD5 antigen. Therefore, a certain degree of structural similarity between CD5 molecules and these oncogene products could be investigated. In this respect, the analysis of the carboxyl-terminal region of CD5 has revealed a potential tyrosine phosphorylation site surrounded by acidic amino acids [13] similar to the tyrosine autophosphorylation region characteristic for protein-tyrosine kinases, as is the case for the products of the *src* gene family, namely p56[lck] [Hunter, 1987]. Confirming this suggestion, there are preliminary results showing detectable kinase activity associated with the CD5 antigen [Rudd et al. 1989].

In conclusion, the evidences here presented point out to the possibility that different post-translational modifications (namely phosphorylation) could exist for CD5 molecules, for mobility changes were observed only in a subset of CD5 molecules. Whether this different susceptibility to TPA reflects some structural and, subsequently, functional differences between both subsets remains to be elucidated.

REFERENCES

Alberola-Ila J., F. Lozano, L. Places, and J. Vives. 1989. Correlation between modulation and phosphorylation of lymphocyte surface antigens induced by phorbol esters. Immunología, 8:95.

Chatila, T.A. & Geha, R.S. (1988) Phosphorylation of T-cell membrane proteins by activators of protein kinase C. J. Immunol., 140: 4308-4314.

Edelman, A.M., Blumenthal, D.K. & Krebs, E.G. (1987) Protein serine/threonine kinases. Annu. Rev. Biochem. 56: 567-613.

Gahmberg, C.G. & Andersson, L.C. (1982) Role of sialic acid in the mobility of membrane proteins containing O-linked oligosaccharides on polyacrylamide gel electrophoresis in sodium dodecyl sulfate. Eur. J. Biochem., 122: 581-586.

Hunter, T. (1987) A thousand and one protein kinases. Cell, 50: 823-829.

Lozano, F., Alberola-Ila, J., Places, L. & Vives, J. (1989) Heterogeneity in the electrophoretic mobility of CD5 molecules after phorbol ester stimulation. Mol. Immunol. 26, 1187-1190.

Krieger, M., Reddy, P., Kozarsky, K., Kingsley, D., Hobbie, L. & Pennan, M. (1989) Analysis of the synthesis, intracellular sorting, and function of glycoproteins using a mammalian cell mutant with reversible glycosylation defects. Methods Cell Biol., 12: 57-84.

Marth, J.D., Lewis, D.B., Cooke, M.P., Mellins, E.D., Gearn, M.E., Samelson, L.E., Wilson, C.B., Miller, A.D. & Perlmutter, R.M. (1989) Lymphocyte activation provokes modification of a lymphocyte-specific protein tyrosine kinase (p56^{lck}). J. Immunol., 142: 2430-2437.

Morrison, D.K., Kaplan, D.R., Rapp, U. & Roberts, T.M. (1988) Signal transduction from membraane to cytoplasm: growth factors and membrane-bound oncogene products increase Raf-1 phosphorylation and associated protein kinase activity. Proc. Natl. Acad. Sci. USA, 85: 8855-8859.

Nishizuka, Y. (1984) The role of protein kinase C in cell surface signal transduction and tumor promotion. Nature, 308: 693-698.

Rudd, C.E., Odysseos, A.D. & Burgess, K.E. (1989) An assessment of protein kinase activity associated with human activation antigens. En Knapp, W., Dörken, B., Gilks, W.R., Rieber, E.P., Schmidt, R.E., Stein, H. & von dem Borne, A.E.G.Kr. (Eds.) Leukocyte Typing IV, Oxford University Press, Oxford, pp. 505-507.

Sahyoun, N., LeVine III, H. & Cuatrecasas, P. (1984) Ca^{2+}/calmodulin-dependent protein kinases from the neuronal nuclear matrix and post-synaptic density are structurally related. Proc. Natl. Acad. Sci. U.S.A., 81, 4311-4315.

Saksela, K., Mäkelä, T.P., Evan, G. & Alitalo, K. (1989) Rapid phosphorylation of the L-myc protein induced by phorbol ester tumor promoters and serum. EMBO. J., 8 :149-157.

Sprang, S.R., Acharya, K.R., Goldsmith, E.J., Stuart, D.I., Varvill, K., Fletterick, R.J., Madsen, N.B. & Johnson, L.N. (1988) Structural changes in glycogen phosphorylase induced by phosphorylation. Nature, 336: 215-221.

Weiss, A. and Imboden, J. B. (1987) Cell surface molecules and early events involved in human T lymphocyte activation. Advances in Immunology 41:1-38

OPPOSING EFFECTS OF PROTEIN KINASE C ON IgE-DEPENDENT EXOCYTOSIS AND InsP$_3$ FORMATION

Gat-Yablonski, G. and Sagi-Eisenberg, R.
Department of Chemical Immunology
The Weizmann Institute of Science
Rehovot 76100, Israel

INTRODUCTION

Aggregation of the receptors for Immunoglobulin E (IgE) by a multivalent antigen (Ag) leads to activation of phospholipase C (PLC) and secretion of serotonin from Rat Basophilic Leukemia (RBL) cells. The phorbol ester 12-0-tetradecanoyl-phorbol-13-acetate (TPA), exerts dual actions on the IgE-dependent responses in RBL cells: TPA potentiates IgE- induced secretion but it inhibits IgE-induced InsP$_3$ formation and the subsequent rise in internal Ca^{2+} concentrations. In addition, TPA synergizes with Ca^{2+} ionophores to trigger exocytosis. We have undertaken this study to examine whether the dual actions of TPA are both mediated by protein kinase C (PKC). It is relevant to evaluate the participation of PKC in TPA actions since some reports suggest that not all TPA actions are consequences of TPA binding and activation of PKC (Zick et al. (1985); Maraganore, J.M. (1987)). In addition, PKC consists of a family of enzymes that while closely related in structure (reviewed in Nishizuka, Y. (1988)), they differ in their cellular and intracellular disributions as well as in their responses to cofactors such as Ca^{2+}, diacylglycerol (DAG), phosphatidylserine (PS) and fatty acids (Nishizuka, Y. (1988)). We have therefore analyzed the possibility that different isozymes of PKC may be involved in mediating the opposing effects of TPA on the IgE-mediated responses in RBL cells.

RESULTS and DISCUSSION

To elucidate the molecular mechanism underlying the dual actions of TPA on IgE-mediated responses, we have established conditions where PKC was either activated or down regulated from the cells. This was achieved by subjecting the cells to TPA treatment for various incubation periods and following the subcellular location of the enzyme, its cellular quantity and catalytic activity. PKC levels and activity at each fraction were determined by immunoblotting using monoclonal antibodies against PKC and by following Ca^{2+}/PS-dependent phosphorylation of exogenous (histone) and endogenous substrates (Gat-Yablonski, G. and Sagi-Eisenberg, R. (1990)). We have shown that within 30 min of TPA treatment, PKC, initially found in the cytosolic fraction, translocates and is exclusively found in the membrane, where it is activated. Following longer incubation periods, PKC is degraded and is completely depleted from the cells. We next examined the effects of short and long-term incubation with TPA on receptor-induced responses. We could demonstrate that under conditions that PKC was activated, receptor-induced secretion was potentiated (Fig. 1 d,e). Depletion of PKC from the cells resulted in complete inhibition of secretion either in response to

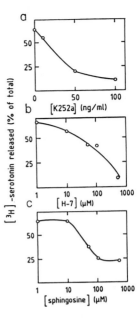

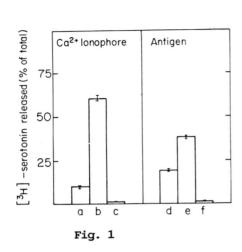

Fig. 1

Fig. 2

Fig 1: Effects of short and long-term treatment with TPA on serotonin secretion from RBL cells. IgE (DNP-specific)-bearing cells preloaded with ^{3}H-serotonin were incubated with TPA (100 ng/ml) for 30 min (b, e) or 18 h (c, f), or without it (a, d). The cells were subsequently triggered with either a Ca^{2+} ionophore (A23187, 100ng/ml) (a-c) or with antigen (DNP$_6$-BSA, 150 ng/ml) (d-f) for 20 min. Release is presented as a percentage of the total ^{3}H-serotonin taken up by the cells.

Fig 2: Effect of PKC inhibitors on antigen-induced serotonin secretion. IgE-bearing cells preloaded with ^{3}H-serotonin were preincubated for 30 min with the indicated concentrations of the PKC inhibitors K252a (a), H-7 (b), or sphingosine (c). Antigen was subsequently added and secretion was determined after a further 20 min. incubation.

IgE activation (Fig.1 f) or in response to TPA plus Ca^{2+} ionophore (Fig. 1 c). These results indicate that PKC plays an essential role in mediating secretion.

To further evaluate the involvement of PKC in exocytosis, the effects of several inhibitors of PKC, on IgE-dependent exocytosis were studied. The three inhibitors tested, K252a (Kase, H. et al. (1986)), H-7 (Kawamoto, S. and Holaka, S. (1984)), and sphingosine (Hannun et al. (1986)) were all found to inhibit IgE-induced serotonin secretion (Fig. 2). Inhibition was dose dependent with IC50 values of 25 ng/ml, 80 μM and 30 μM, respectively.

In contrast to its stimulatory effect on secretion, short incubation periods (30 min) with TPA completely inhibited IgE-induced InsP$_3$ formation (Fig. 3 lane c compared with lane b). In contrast, TPA had no inhibitory effect on InsP$_3$ formation following prolonged treatments that caused deple-tion of PKC from the cells (Fig. 3 lane d). Moreover, under these condi-tions the response to receptor activation was potentiated by 1.73+0.2 fold

217

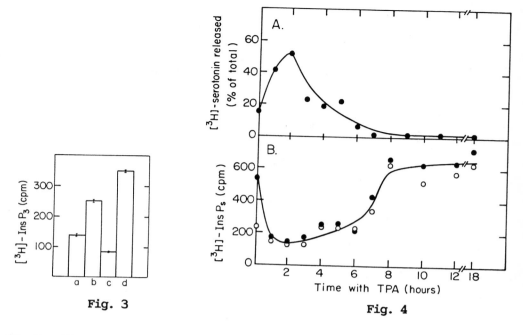

Fig. 3

Fig. 4

Fig 3: **Effect of short and long-term treatment with TPA on antigen- induced InsP₃ formation.** IgE-bearing cells preloaded with ³H-inositol were incubated with TPA for 30 min (c) or 18 h (d) or without it (a,b). The cells were subsequently incubated for 10 min with (b-d) or without antigen. Inositol phosphates formed were analysed on DOWEX-1 columns.

Fig 4: **Differential down-regulation of serotonin secretion and inhibition of InsP₃ formation by TPA.** IgE-bearing cells preloaded with either ³H-serotonin (a) or ³H-inositol (b) were incubated with TPA for the indicated time periods. The cells were subsequently triggered with antigen and serotonin secretion (a) and InsP₃ formation (b) were determined.

(n=4), indicating that also under physiological conditions, PKC inhibits PLC activity.

The effects of TPA were specific. They could not be mimicked by the vehicle (dimethyl sulphoxide) or by the inactive analogue of TPA (methyl-TPA). They could however be mimicked by phorbol dibutyrate, another active phorbol ester (results not shown).

RBL cells express both type II and type III isozymes of PKC (Huang, F.L. et al. (1989)). The subspecies of PKC have been shown to differ in their susceptibility to proteolysis (Huang, F.L. et al. (1989)). Moreover, they have been shown to undergo differential down-regulation in response to TPA (Huang, F.L. et al. (1989); Ase, K. et al. (1988)), type II is depleted faster than type III. These observations prompted us to investigate in detail the kinetics of inhibition of IgE-dependent exocytosis following TPA treatment and compare it with the kinetics of stimulation of InsP₃ production. As shown in Fig. 4, these processes clearly differ in their kinetics. At time periods that secretion was already completely blocked (5 hr of TPA treatment), IgE-induced InsP₃ production was still inhibited and 8 hr of TPA treatment were required in order to relieve this inhibition and potentiate this response. Based on these observations, we propose that PKC exerts its dual action by its different isozymes. Type II

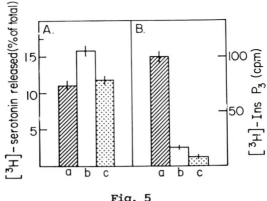

Fig. 5

Fig 5: Effect of sphingosine on TPA actions on antigen-induced serotonin secretion and InsP₃ formation. IgE-bearing RBL cells were preloaded with ³H-serotonin (A) or ³H-inositol (B). Antigen was subsequently added and serotonin secretion or InsP₃ formation were determined. a,Control cells; b, cells that were preincubated for 30 min with TPA; c, cells that were preincubated with sphingosine (100 μM) followed by 30 min incubation with TPA.

is essential for exocytosis while type III negatively regulates PIP₂ breakdown and the Ca²⁺ signal. This conclusion is further supported by our findings that the opposing effects of TPA also show different sensitivities to inhibitors of PKC. Whereas preincubation with sphingosine (100 μM, 30 min.) eliminated the potentiating effect of TPA on IgE-induced secretion (Fig. 5A), it failed to abolish the inhibition exerted by TPA on IgE-induced InsP₃ formation (Fig. 5B). In fact, none of the PKC inhibitors tested in this study exerted any effect on InsP₃ formation measured in the absence or presence of TPA (data not shown).

References

Ase K, Berry N, Kikkawa U, Kishimoto A, Nishizuka Y (1988) Differential down-regulation of protein kinase C subspecies in KM3 cells. FEBS Lett 236:396-400

Gat-Yablonski G, Sagi-Eisenberg R (1990) Differential down-regulation of protein kinase C selectively affects IgE-dependent exocytosis and inositol trisphosphate formation. Biochem J 270:679-684.

Hannun YA, Loomis CR, Merrill AJ Jr, Bell RM (1986) Sphingosine inhibition of protein kinase C activity and of phorbol dibutyrate binding in vitro and in human platelets. J Biol Chem 261:12604-12609

Huang FL, Yoshida Y, Cunha-Melo JR, Beaven MA, Huang KP (1989) Differential down-regulation of protein kinase C isozymes. J Biol Chem 264:4243

Kase H, Iwahasi K, Matsuda Y (1986) K-252a, a potent inhibitor of protein kinase C from microbial origin. J Antibiot 39:1054-1965

Kawamoto S, Holaka H (1984) 1-(5-Isoquinolinesulfonyl)-2-methylpiperazine (H-7) is a selective inhibitor of protein kinase C in rabbit platelets. Biochem Biophys Res Commun 125:258-261

Maraganore JM (1987) Structural elements for protein-phospholipid interactions may be shared in protein kinase C and phospholipases A_2. Trends Biochem Sci 12:176-177.

Nishizuka Y (1988) The heterogenicity and differential expression of multiple species of the protein kinase C family. Biofactors 1:17-20

Zick, Y, Grinberger, G, Rees-Jones, RW, Comi R (1985) Use of tyrosine containing polymers to characterize the substrate specificity of insulin and other hormone-stimulated tyrosine kinases. Eur J Biochem 148:177-182

PROTEINPHOSPHORYLATION IN PLATELETS-EVIDENCE FOR INCREASED PROTEIN KINASE C ACTIVITY IN ESSENTIAL HYPERTENSION

C.Lindsschau, H.Haller, P.Quass, A.Distler
Med. Klinik
Klinikum Steglitz
FU Berlin
Hindenburgdamm 30
D-1000 Berlin 45
FRG

Introduction

Several groups have reported that cytosolic free calcium $[Ca^{2+}]_i$ in the platelets of patients with essential hypertension is elevated (Bruschi et al. 1985, Erne et al. 1984, Cooper et al. 1987). It is generally believed that $[Ca^{2+}]_i$ is the primary messenger of smooth muscle contraction (Kuriyama et al. 1982). The increase in $[Ca^{2+}]_i$ in platelets of patients with essential hypertension has therefore been directly linked to the increased peripheral vascular tone in essential hypertension (Erne et al. 1984). While the $[Ca^{2+}]_i$ is important in the initiation of vascular smooth muscle contraction, the maintenance of the contractile response seems to be dependent on the activation of a calcium-, phospholipid-dependent kinase, the protein kinase C (PKC; Rasmussen et al. 1987). It has been suggested that PKC may have a role in hypertension, since phorbolester-induced contraction was shown to be increased in vascular smooth muscles strips from spontaneously hypertensive rats (Bruschi et al. 1988, Turla and Webb 1987).

After binding of an agonist to its specific receptor, two intracellular messenger systems are activated. One is the release of Ca^{2+} by inositoltrisphosphate (IP3) from intracellular stores. The increase in $[Ca^{2+}]_i$ leads, via activation of a calmodulin-dependent protein kinase, to the phosphorylation of a 20 kD protein, the myosin light chain (MLC). The other second messenger that is generated is diacylglycerol (DAG). DAG, together with the increase in $[Ca^{2+}]_i$, activates the PKC. In platelets, one of the substrates that are specifically phosphorylated by PKC is a protein with a molecular weight of 47 kD.

We investigated the activity of PKC and of the calcium/calmodulin pathway in thrombin- and TPA-stimulated platelets from patients with essential hypertension by measuring the phosphorylation of the 47 kD protein and of MLC. In parallel we measured intracellular free calcium $[Ca^{2+}]_i$ in these platelets. We sought to identify a possible relationship between the increased intracellular free calcium concentration and changes in protein phosphorylation in platelets.

NATO ASI Series, Vol. H 56
Cellular Regulation by Protein Phosphorylation
Edited by L. M. G. Heilmeyer, Jr.
© Springer-Verlag Berlin Heidelberg 1991

Patients and methods

We investigated 17 patients with mild to moderate essential hypertension (WHO I-II, blood pressure 159±4/104±2 mmHg) and 20 normotensive subjects (blood pressure 116±4/76±3 mmHg). Essential hypertension was diagnosed after secondary forms of hypertension had been ruled out by standard diagnostic procedures. The statistical analysis was conducted using a commercially available statistics software program.

Platelets were prepared by centrifugation of citrate-treated blood samples followed by gelfiltration of the Platelet-rich plasma. For phosphorylation studies, the platelet suspension was adjusted to a concentration of 10^9 cells/ml and incubated with carrier-free [^{32}P]orthophosphate (1 mCi/ml) for 1 h. Extracellular calcium was restored to 1 mM 20 min. prior to stimulation. 60 sec. prior to the end of incubation thrombin to a final concentration of 0.1 or 0.25 U/ml or TPA (12-myristate, 13-acetat phorbolester) to 100nM was added. Incubation was stopped by boiling the samples in SDS. All experiments were carried out in duplicates. The proteins were separated by SDS polyacrylamide gelelectrophoresis. Gels were stained and dried onto filter paper, followed by autoradiography. Results were analyzed by counting excised gel bands with a scintillation counter. The amount of phosphorylation of the 47kD and the 20kD protein (MLC) after stimulation with thrombin is expressed as percentage increase, setting basal phosphorylation to 100 %.

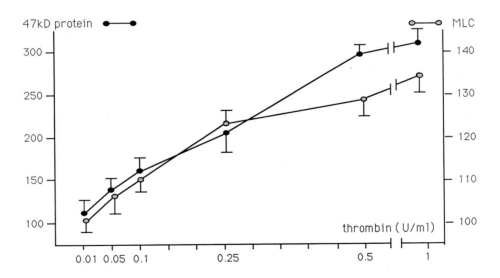

Fig. 1: Dose-response curve of protein phosphorylation in human platelets. Data represent results from five independent experiments carried out in duplicates.

For intracellular free calcium measurements PRP was incubated with Fura-2AM. Extracellular dye was removed by gelfiltration and platelets were suspended at 4-6 x 10^5 cells/ml. Fluorescence was measured in a spectrofluorometer at 505 nm emission wave length and excitation at 340 nm. Maximal fluorescence was determined by lysing the cells in Triton X-100, minimal fluorescence was obtained by adding $MnCl_2$. Shape changes of the platelets during the preparation were excluded by checking platelet form and size before and after chromatography with a platelet analyzer.

Results

Incubation of the platelets with thrombin led to a dose-dependent increase in phosphorylation of the MLC and the 47kD protein (Fig. 1).

The comparison of the phosphorylation of the MLC and the 47 kD protein in hypertensive patients and normotensive subjects after exposure to 0.1, 0.25 U/ml thrombin or 10^{-7} M TPA for 60 sec is shown on the following page (Fig. 2). Except for the TPA-stimulated phosphorylation of the MLC all other data show a significant increase in hypertensives compared to normotensives. There was no difference in basal activity between normo- and hypertensives.

To see whether the enhanced protein phosphorylation was related to blood pressure we performed statistical analysis. When both groups were analyzed together we observed a weak, but significant correlation between diastolic blood pressure and the phosphorylation of the MLC ($r=0.413$, $p<0.04$) and the 47kD protein ($r= 0.433$, $p< 0.03$). When both groups were analyzed separately no significant correlation between blood pressure and any changes in protein phosphorylation were identified.

Basal $[Ca^{2+}]_i$ in the normotensive subjects was 104 ± 5 nM. The concentration in the hypertensive patients was significantly increased to 124 ± 7 nM ($p<0.05$). There was a wide overlap between the two groups. In contrast to previous studies (Erne et al. 1984), we did not observe a significant correlation between the $[Ca^{2+}]_i$ and blood pressure. Comparison of the thrombin-stimulated $[Ca^{2+}]_i$ showed no significant differences between normo- and hypertensives.

In order to investigate the relation of the basal $[Ca^{2+}]_i$ in platelets with the observed changes in protein phosphorylation we correlated $[Ca^{2+}]_i$ with changes in protein phosphorylation. When both groups were analyzed together we found a weak, but significant correlation between the increased $[^{32}P]$ incorporation into the 47kD protein and $[Ca^{2+}]_i$ ($r=0.343$, $p<0.04$ for 0.25 U/ml thrombin, $r = 0.375$, $p< 0.022$ for 0.1 U/ml). On the other hand, the correlation between the changes in MLC phosphorylation and $[Ca^{2+}]_i$ were statistically not significant. When both groups were analyzed independently, no significant correlation between $[Ca^{2+}]_i$ and changes in protein phosphorylation were found.

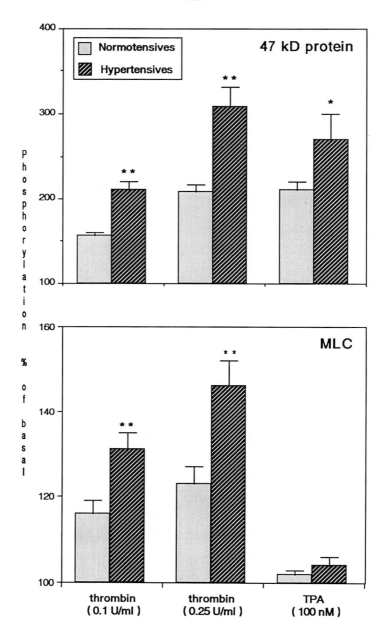

Fig. 2: Protein phosphorylation in human platelets after stimulation with thrombin or TPA for 1 Minute. Data represents results from 20 normotensive subjects and 17 age-matched patients with essential hypertension. Each experiment was carried out in duplicate.

Discussion

Our results indicate that both second messenger branches are more activated in the platelets of hypertensive patients. The higher phosphorylation of both proteins after stimulation in patients with essential hypertension could possibly result from defects in receptor function or the coupled G-protein, increased phospholipase C activity, an altered phospholipid content of the cell membrane, an increased intracellular free calcium concentration and/or an enhanced protein kinase C activity.

That at least the PKC activity contributes to the observed effects is supported by our results that exposure of the platelets to the phorbolester TPA which specifically activates PKC elicites a significantly higher phosphorylation in platelets from patients with essential hypertension.

We observed in our study a weak correlation between the $[Ca^{2+}]_i$ and the phosphorylation of the 47kD protein but not with the changes in the MLC. An increase in $[Ca^{2+}]_i$ could contribute to the activation of PKC by altering the sensitivtiy of the enzyme to DAG or the lipids of the plasma membrane. However, the rise in cytosolic free calcium which is necessary to bring about the translocation of PKC to the plasma membrane is considerably higher (in the range of 1 M) than the $[Ca^{2+}]_i$ observed in platelets. According to a hypothesis forwarded by Rasmussen (1989) the activation of PKC is linked to a simultaneous influx and efflux of calcium via the plasma membrane. This "cycling" of calcium across the plasma membrane leads to an increase in $[Ca^{2+}]_i$ within a confined space immediately beneath the plasma membrane. The small, but persistent increase in $[Ca^{2+}]_i$ of platelets from hypertensive patients could therefore be a reflection of the increased $[Ca^{2+}]_i$ in a submembraneous domain leading to a higher fraction of "preactivated" PKC. This hypothesis would also serve to explain the functional significance of the small rise in $[Ca^{2+}]_i$ observed in blood cells of hypertensive patients.

References

Bruschi G, Bruschi ME, Caroppo M, Spaggiari P, Orlandini G, Pavarani C (1985) Cytoplasmic free [Ca2+] is increased in the platelets of spontaneously hypertensive rats and essential hypertensive patients. Clin Sci:68-74

Bruschi G, Bruschi ME, Capelli P, Regolisti G, Borghetti A (1988) Increased sensitivity to proteinkinase C activation in aortas of spontaneously hypertensive rats. J Hypertension 6 (suppl 4):S248-S251.

Cooper RS, Shamsi N, Katz S (1987) Intracellular calcium and sodium in hypertensive patients. Hypertension 9:224-230.

Erne P, Bolli P, Burgisser E, Bühler FR (1984) Correlation of platelet calcium with blood pressure. N Engl J Med 310(17):1084-1089.

Kuriyama H, Ito Y, Suzuki H, Kitamura K, Itoh T (1982) Factors modifying contraction-relaxation cycle in vascular smooth muscles. Am J Physiol 243:H641-H676.

Rasmussen H, Takuwa Y, Park S (1987) Protein kinase C in the regulation of smooth muscle contraction. FASEB J. 1:177-185.

Rasmussen H (1989) The messenger function of Ca^{2+}: from PTH action to smooth muscle contraction. Bone and Mineral 5:233-248.

PROTEIN PHOSPHORYLATION IN LUTEAL MEMBRANE FRACTION

L.T.Budnik and A.K.Mukhopadhyay,
Institute for Hormone and Fertility Research,
D-2000 Hamburg 54, Grandweg 64, F.R.G.

Introduction

In corpus luteum, LH plays a key role in regulating various cellular
functions. It is well established that interaction of LH with its receptor
leads to the activation of adenylate cyclase and the formation of cyclic
AMP which then activates the protein kinase A (Hunzicker-Dunn et al
1985). Several substrate proteins in soluble fractions prepared from rat
(Richards & Kirschick 1984) and ovine (Hoyer & Kong 1989) luteal cells are
known to be phosphorylated by a cAMP dependent mechanism. In addition to
LH, the luteal steroidogenesis can also be stimulated by the phorbol ester
PMA or a diacylglycerol (Brunswig et al.,1986), via putative activation of
a Ca^{2+}/phospholipid- dependent protein kinase (PKC). Furthermore the
presence of substrates in luteal cell cytosolic compartments for
Ca^{2+}/phospholipid- or calmodulin- dependent protein kinases have been
reported (Maizels & Jungman 1983; Hoyer & Kong 1989). However, most of the
studies reported so far on this subject have dealt with the substrates
present in the cytosolic compartments. Since protein and peptide hormones
act on cells via interactions with the cell-surface receptors, we
considered it to be of interest to examine whether the particulate
fractions contain protein kinase activities amenable to hormonal
manipulation and whether the presence of endogenous substrates in the
particulate fraction for these kinases can be demonstrated.

NATO ASI Series, Vol. H 56
Cellular Regulation by Protein Phosphorylation
Edited by L. M. G. Heilmeyer, Jr.
© Springer-Verlag Berlin Heidelberg 1991

Methods

Preparation of membrane fractions from bovine luteal cells

Bovine luteal cells were purified as published elsewere (Budnik & Mukhopadhyay 1987), centrifuged at 90xg and the cell pellet was quickly frozen at -80°C. The membrane fraction prepared (Budnik & Mukhopadhyay 1990) from the cells, was washed once in 10 mM Tris/HCl containing 0.1 mM PMSF, 12 μM Leupeptin, 5 mM DTT, 5 mM EGTA 2 mM EDTA and centrifuged again at 105,000 g for 1 h at 4°C.

Phosphorylation of particular proteins

Endogenous phosphorylation was carried out in 100 μl of 50 mM Pipes buffer pH 7.0, 10 mM $MgCl_2$, 0.1 mM DTT, 5 pmol ^{35}S-thio ATP (65 Ci/mmol) 100 μg protein equivalent of the membrane fraction, 5 μM $CaCl_2$. 20 μg phosphatidylserine (PS), 10 nM PMA or 2.5 μg bovine luteal gonadotropin, bLH. Following incubation at 30°C for 5 min, the reaction was terminated by the addition of 50 μl stop solution (9 % SDS, 2.27 % Tris/HCl pH 6.9, 15 % glycerol, 0.015 % bromophenol blue). The samples were boiled for 1 min, 50 μl 10 % mercaptoethanol was added to each tube and the samples were kept frozen overnight. The phosphorylated proteins were analysed by SDS/PAGE (10% acrylamide) by the method of Laemmli (1970). The gels were stained, dried and exposed to autoradiography. The films were scanned with GS 300 Hoefer Densitometer using G360 Software.

Immunoblotting

After transfer of the proteins, the nitrocellulose membrane was saturated for 5h at 40°C in 50 mM Tris/ 150 mM NaCl (TBS), 4% BSA and then incubated with affinity purified anti-PKC antibodies (in TBS buffer, 1% BSA for 15h at 4°C). The nitrocellulose membrane was washed 4 times in TBS, 0.4% BSA, 0.05% Tween 20 and incubated for 2 h with alkaline phosphatase conjugated rabbit anti-mouse antibody. The immunocomplexes were detected using alkaline phosphatase colour reaction using nitroblue tetrazolium and 5-bromo-4-chloro-3-indolyl phosphate.

Results

Phospholipid sensitive, calcium-dependent protein kinase has been shown to utilize adenosine 5 'thiotriphosphate to thiophosphorylate histone H1 (Weise et al, 1986). Therefore, we have used this ATP form as a phosphate donor. At least six different protein bands were endogenously phosphorylated in particulate fraction of bovine luteal cells in a Ca^{2+}/ phospholipid-dependent manner (Fig 1A). It is evident that little phosphorylation occurs in the absence of Ca^{2+}. Calcium, phospholipid and PMA when added together caused a 3-5 fold increase of all the protein bands over controls. The strength of the phosphorylation signal obtained could be related to the amounts of membrane protein used. The extend of phosphorylation of each of the proteins depended upon duration of incubation and the maximum level was achieved by 20 min of incubation. The rate of phosphoprotein formation appears to be linear up to 10 min for all examined proteins (Budnik & Mukhopadhyay 1990). If the phosphorylation reaction was carried out in the presence of known PKC inhibitor, H7 the phosphorylation of 68-75 and 45 kDa proteins was completely abolished and the incorporation of label in 80 and 35 kDa proteins was drastically reduced. The observed phosphorylation reaction presumably could only be catalysed by endogenous kinases associated with particulate fraction since the soluble fraction was not added to the incubation cocktail in this experiments. The western blot analysis conducted with anti-PKC antibodies confirmed the presence of immunoreactive PKC in the particulate fraction of luteal cells (Fig 2). Interestingly, LH was able to mimic Ca^{2+}, PS and PMA induced phosphorylation pattern. All the major protein bands which were phosphorylated in the presence of Ca^{2+}, phosphatidyl serine and PMA were also phosphorylated when bLH alone or a combination of bLH and Ca^{2+} were added. However, in the presence of bLH a few additional protein bands with molecular weights of <20 kDa were found to be phosphorylated. The proteins corresponding to 80, 44 and 34 kDa show additive phosphorylation (Fig 1B) when LH was added together with Ca^{2+}, PS and PMA pointing to a possible link between the gonadotropin-induced receptor activation and cellular phosphorylation.

Fig 1A   **Fig 1B**

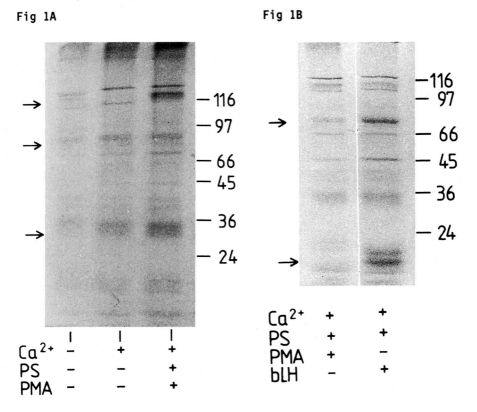

Ca²⁺	−	+	+
PS	−	−	+
PMA	−	−	+

Ca^{2+} + +
PS + +
PMA + −
bLH − +

Fig 1A, 1B : Comparison of the LH-induced and Ca²⁺- and phospholipid -dependent protein phosphorylation pattern in luteal cell membrane fraction (modified from Budnik and Mukhopadhyay 1990)

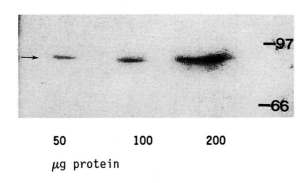

50   100   200

μg protein

Fig 2 : Western blot showing the presence of PKC in luteal cell membrane

Conclusions

In the present study, we have been able to demonstrate phosphorylation of several endogenous proteins in the particulate fraction prepared from bovine luteal cells. This in vitro phosphorylation was absolutely dependent upon the presence of calcium. Addition of phosphatidyl serine and PMA enhanced the degree of phosphorylation suggesting that the observed protein phosphorylation was being catalysed by endogenous calcium and phospholipid-dependent protein kinase, PKC. In addition we have shown that a similar protein phosphorylation pattern was observed when the membrane fraction was incubated either with LH or with a combination of calcium, phospholipid and PMA. The data obtained suggests that the interaction of LH with its receptor results in an activation of endogenous calcium and phospholipid dependent protein kinase which then results in the phosphorylation of its substrates present in the membrane fraction. Alternatively, it could be argued that LH possibly activates a protein kinase activity other than PKC, which then phosphorylates a set of common substrates. In any case, we appear to have a model of a differentiated cell where the primary signal LH is in the position to modulate its action through different signal transduction pathways, which perhaps can cross-talk.

Literature

Brunswig B., Mukhopadhyay A.K., Budnik L.T., Bohnet H.G. and Leidenberger F.A. (1986) Endocrinology 118:743-749
Budnik L.T. and Mukhopadhyay A.K. (1987) Mol. Cell. Endocrinol. 54:51-61
Budnik L.T. and Mukhopadhyay A.K. (1990) Mol. Cell. Endocrinol. 69:245-253
Hoyer P.B. and Kong W. (1989) Mol. Cell. Endocrinol. 62:203-215
Hunzicker-Dunn M., Lorenzini N.A., Lynch L.L. and West D.E. (1985) J. biol. Chem. 260:13360-13369
Laemmli U.K. (1970) Nature 227:680-685
Maizels E.T. and Jungman R.A. (1983) Endocrinology 112:1895-1902
Richards J.S. and Kirschick H.J. (1984) Biol. Reprod. 30:737-751
Wise BC., Glass D.B., Chou J.C.-H., Raynor R.L., Katoh N., Schatzman R.C., Turner R.S., Kibler R.F. and Kuo J.F. (1982) J. biol. Chem. 257: 8489-8495

Acknowledgements

We would like to thank Prof F.A. Leidenberger and Dr Ch. Weise for their support and encouragement throughout the work.

PURIFICATION OF BOVINE BRAIN PROTEIN KINASE C EMPLOYING METAL ION DEPENDENT PROPERTIES

Christoph Block and Detmar Beyersmann
Institut für Zellbiologie, Biochemie und Biotechnologie
Universität Bremen
Leobener Str.
2800 Bremen 33
F.R.G.

Within the sequence of protein kinase C (PKC), a Ca^{2+} and phospholipid dependent enzyme (reviewed in Nishizuka 1989), no conserved Ca^{2+} binding site could be identified and the enzyme does not bind Ca^{2+} as an isolated enzyme at physiological Ca^{2+} concentrations (Bazzi & Nelsestuen 1990). The primary structure of PKC reveals within the phorbol ester binding domain two "zinc finger like" consensus sequences as well as a number of highly conserved histidine and cysteine residues (Parker et al. 1986). Ono et al. (1989) demonstrated the necessity of putative Zn^{2+} ligands within the $CX_2CX_{13}CX_2C$ sequence for phorbol ester binding. Yet the functional consequences of metal binding to these structures or to other conserved potential ligands are poorly understood. Studies investigating the effects of divalent metal ions on PKC led to contradictory results, especially reporting activatory or inhibitory effects of zinc on the enzyme (Csermely et al. 1988, Speizer et al. 1989). Hence we developed a purification protocol which is suitable to characterize metal binding dependent properties of protein kinase C. Immobilized metal ion chromotography is suitable for preparative purification and subserves an analytical tool to evaluate surface accessible metal ion binding sites (Hemdan et al. 1989). The elucidation of metal binding properties of PKC may enhance our understanding of yet another mechanism regulating enzyme activation.

Experimental procedures

Materials All chemicals were of highest quality commercially available. Phenyl-Sepharose, chelating Sepharose and the Q-Sepharose HP column were from Pharmacia, Freiburg. The TSK-Gel Phenyl 5-PW column was obtained from Bio-Rad, Munich.

NATO ASI Series, Vol. H 56
Cellular Regulation by Protein Phosphorylation
Edited by L. M. G. Heilmeyer, Jr.
© Springer-Verlag Berlin Heidelberg 1991

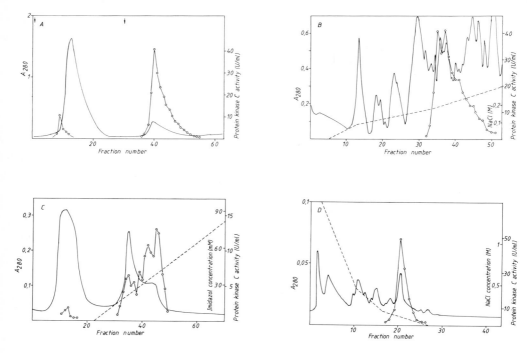

Fig. 1: Chromotographic steps of the PKC purification procedure, — absorption at 280 nm, o—o enzyme activity, ----- gradient applied; Ca^{2+} dependent hydrophobic chromatography (A), Q-Superose HP (B), Immobilised metal affinity chromatography (C), TSK-Gel Phenyl 5 PW (D)

Table 1: Purification of PKC

Step	Volume (ml)	Protein concen- tration (mg/ml)	Specific act. (U/mg)	Total act. (U)	Yield
Phenyl-S.	140	1.7	8.7	2080	100%*
Q-Seph.	48	0.8	30.3	1160	56%
Zn^{2+}-Seph.	50	0.066	134	441	21%
TSK-Phenyl	6	0.028	1100	190	9%

*Due to phosphatase activity in the homogenate the eluate of the first step was defined as 100% activity.

columns our results suggest the existence of a special zinc site on the surface of PKC. This site may be involved in membrane translocation of the enzyme as indicated in the work of Csermely et al. (1988). These authors have found an absolute Zn^{2+} requirement for activation of PKC induced by binding of anti CD3 antibody to mouse thymocytes. We do not assume the putative zinc-finger cysteines to be involved in binding to the immobilized zinc ions, since this type of metal binding site is not likely to be metal depleted by the purification procedure. In the case of the $CX_2CX_{13}CX_2C$ steroid hormone receptor zinc finger, which is identical to the spacing of conserved cysteines in PKC, it has been demonstrated that these sites are occupied even after prolonged dialysis against millimolar concentrations of EDTA at pH 7.8 (Pan et al. 1990). In a coordination sphere of four S^- donors, Zn^{2+} is not likely to expand its coordination sphere or exchange its ligands readily (Giedroc et al. 1987) and can thus only be removed by dialysis at low pH. Hence immobilized metal ion chromatography will provide a valuable tool to gain insight in possible modulatory interactions of PKC with divalent metal ions.

TSK Gel Phenyl-5PW chromatography (Fig. 1D) Final purification is performed on a high resolution hydrophobic column in the presence of chelator (Kikkawa et al 1986).

In summary this procedure using immobilized metal ion chromatography provides a valuable tool to gain insight in regulatory interactions of PKC with divalent metal ions.

References

Bazzi M D, Nelsestuen G L (1990) Protein kinase C interaction with calcium: A phospholipid-dependent process. Biochemistry 29, 7624-7630

Csermely P, Szamel M, Resch K, Somogyi J (1988) Zinc can increase the ativity of protein kinase C and contributes to its binding to plasma membranes in T lymphocytes. J Biol Chem 263, 6487-6490

Giedroc D P, Keating K M, Williams K R, Coleman J E (1987) The function of zinc in gene 32 protein from T4. Biochemistry 26, 5251-5259

Hemdan E S, Zhao Y-J, Sulowski E, Porath J (1989) Surface topography of histidine residues: A facile probe by immobilized metal ion affinity chromatography. Proc Natl Acad Sci USA 86, 1811-1815

Kikkawa U, Masayoshi G, Kuomoto J, Nishizuka Y (1986) Rapid purification of protein kinase C by high performance liquid chromatography. Biochem Biophys Res Commun 135, 636-643

Newton A C, Koshland D E (1989) High cooperativity, specificity, and multiplicity in the protein kinase C-lipid interaction. J Biol Chem 264, 14909-14915

Nishizuka Y (1989) Studies and prospectives of the protein kinase C family for cellular regulation. Cancer 63, 1892-1903

Ono Y, Fujii T, Igarashi K, Kuno T, Tanaka C, Kikkawa U, Nishizuka Y (1989) Phorbol ester binding to protein kinase C requires a cysteine-rich zinc-finger-like sequence. Proc Natl Acad Sci USA 86, 4868-4871

Pan T, Freedman L P, Coleman J E (1990) Cadmium-113 NMR studies of the DNA binding domain of the mammalian glucocorticoid receptor. Biochemistry 29, 9218-9225

Parker P J, Coussens L, Totty N, Rhee L, Young S, Chen E, Stabel S, Waterfield M D, Ullrich A (1986) The complete primary structure of protein kinase C-the major phorbolester receptor. Science 233, 853-859

Speizer L A, Watson M J, Kanter J R, Brunton L L (1989) Inhibition of phorbol ester binding and protein kinase C activity by heavy metals. J Biol Chem 264, 5581-5585

Walsh M P, Valentine K A, Ngai P K, Carruthers C A, Hollenberg M D (1984) Ca^{2+} dependent hydrophobic-interaction chromatography. Biochem J 224, 117-127

CARDIAC PROTEIN KINASE C ISOZYMES: PHOSPHORYLATION OF PHOSPHOLAMBAN IN LONGITUDINAL AND JUNCTIONAL SARCOPLASMIC RETICULUM

Bruce G. Allen and Sidney Katz
Division of Pharmacology and Toxicology
Faculty of Pharmaceutical Sciences
University of British Columbia
2146 East Mall, Vancouver, BC
Canada, V6T 1W5

Introduction

Cardiac contractility is regulated by hormones, neurotransmitters and drugs which interact with specific receptors located on the sarcolemmal membrane. These receptors are coupled to various intracellular second messenger systems. Cyclic AMP (cAMP) and calcium ions (Ca^{2+}) are two very important second messengers in the modulation of cardiac function. Substrates of the cAMP-dependent protein kinase in the heart include the sarcolemmal L-type Ca^{2+} channel, the sarcoplasmic reticulum (SR) proteolipid "phospholamban" and a high molecular weight SR protein thought to be the Ca^{2+} release channel. Phosphorylation of sarcolemmal Ca^{2+} channels increases Ca^{2+} influx thereby increasing the force of contraction. Phosphorylation of phospholamban stimulates Ca^{2+} sequestration by the SR to increase the rate of relaxation of the heart and provide more Ca^{2+} for release during subsequent contractions. A Ca^{2+}- and calmodulin-dependent protein kinase activity which is endogenous to the SR has also been shown to phosphorylate phospholamban and stimulate Ca^{2+}-uptake into SR vesicles. Since both Ca^{2+}- and cAMP- activated protein kinases phosphorylate phospholamban and since the phosphorylation state of this protein is correlated with the rate of Ca^{2+} transport into the SR phospholamban appears to play a key role in the regulation of cardiac function by integrating information from both intracellular and extracellular sources and modulating the activity of the SR Ca^{2+}-pump accordingly.

Protein kinase C (PKC) is a serine and threonine protein kinase which can be activated by diacylglycerol or phorbol esters in the presence of Ca^{2+} and acidic phospholipids. This kinase activity is ubiquitous in mammalian tissues (Kuo et al., 1980). In vitro PKC, purified from rat brain, has been shown to phosphorylate phospholamban (Iwasa and Hosey, 1984; Movesian et al., 1984) and to stimulate both Ca^{2+}-ATPase activity (Limas, 1980) and Ca^{2+}-uptake into the SR (Movesian et al., 1984). PKC catalytic activity has been reported in cytosol, sarcolemma and SR (Liu et al., 1989). Cloning and sequence analysis suggests as many as seven forms of PKC are present in brain and that while

NATO ASI Series, Vol. H 56
Cellular Regulation by Protein Phosphorylation
Edited by L. M. G. Heilmeyer, Jr.
© Springer-Verlag Berlin Heidelberg 1991

certain PKC isozymes are selectively expressed in brain, others are expressed in both brain and peripheral tissues (Ohno *et al.*, 1987). Variations in primary structure are found in both regulatory and catalytic domains (Ono *et al.*, 1987), suggesting these isozymes may differ both in activation and in target protein recognition. In the present study we have isolated and identified two PKC isozymes from canine heart. The ability of these kinases to phosphorylate phospholamban was examined in both the junctional and the longitudinal fractions of canine cardiac SR.

Methodology

Isolation of cardiac protein kinase C subtypes. Canine ventricular muscle was obtained fresh, trimmed of fat and minced in a grinder. The tissue was then homogenized in a Waring blender in 10 volumes of ice cold 20 mM Tris/HCL buffer (pH 7.5 at 20°C) which contained 0.3 M sucrose, 10 mM EGTA, 2 mM EDTA, 2 mM PMSF, 10 μg/ml leupeptin, 1.4 μg/ml pepstatin A, 5 μg/ml aprotinin, 2 mM dithiothreitol (DTT) and 0.5% (v/v) Triton X-100. The homogenate was centrifuged for 60 minutes at 30,000 g and 4°C. The resulting supernatant adjusted to 1% (v/v) Triton X-100 (final concentration). Further purification involved chromatography on DEAE-Sephacel (2.5 x 40 cm, eluted with 0.16 M NaCl), phenyl-Sepharose (2.5 x 10 cm, 1.0-0 M ammonium sulfate), poly-l-lysine agarose (1.5 x 10 cm, 0-0.8 M KCl) and hydroxylapatite (Bio-Gel HTP).
Preparation of Canine Cardiac Sarcoplasmic Reticulum. Crude SR was prepared according to the procedure described by Chamberlain *et al.* (1983) up to the "membrane vesicle" stage with the following modification: fresh canine ventricles were cut into 1x1 cm pieces and homogenized using a Waring blender on high for 2 x 30 sec in five volumes of 20 mM Imidazole/HCl buffer (pH 6.8 at 4°C) containing 0.3 M sucrose, 2 mM EGTA, 2 mM EDTA, 5 mM DTT, 5 mM sodium azide, 2 mM PMSF, 10 μg/ml leupeptin, 1.4 μg/ml pepstatin A and 5 μg/ml aprotinin. Each centrifugation step was adapted to available centrifuges according to pelleting efficiency and duration. Crude SR was then separated into longitudinal and junctional fractions by loading with calcium-phosphate followed by centrifugation into a discontinuous sucrose density gradient (Inui *et al.*, 1988).

Results and Discussion

Protein kinase C (PKC) was isolated from canine heart. This preparation eluted from hydroxylapatite as two peaks of Calcium and phospholipid dependent kinase activity (Fig. 1). In all preparations, peak 2 activity was 10 - 20 fold greater than peak 1. Using

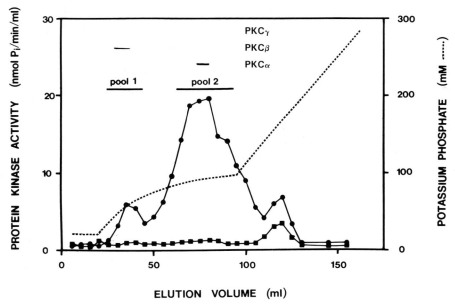

Figure 1. Hydroxylapatite chromatography of canine heart ventricular protein kinase C activity. PKC pooled from the poly-l-lysine column was dialysed against 20 mM potassium phosphate buffer (pH 7.5 at 4°C) containing 0.5 mM EGTA, 0.5 mM EDTA, 10% glycerol and 2 mM DTT (Buffer A) and then loaded onto a Bio-Gel HTP column (1.0 x 4 cm). The column was eluted (10 ml/hr, 2.5 ml/fraction) with an exponential gradient of 0-80 mM potassium phosphate in buffer A followed by a linear gradient of 80-280 mM potassium phosphate in Buffer A. Phospholipid- and calcium-dependent protein kinase activity (●) was measured for 3 minutes at 30°C in 20 mM HEPES/KOH buffer (pH 7.4 at 30°C) containing 1 mg/ml histones III-S, 10 mM $MgCl_2$, 10 mM DTT, 1 mM EGTA, 0.875 mM $CaCl_2$ (1 μM free), 80 μg/ml phosphatidyl serine (PS), 8 μg/ml 1-stearoyl-2-arachidonylglycerol (SAG) and protein kinase. The Ca^{2+}- and phospholipid-independent protein kinase activity (■) was measured under the same conditions without Ca^{2+} and phospholipid. **Inset.** Immunoblot of canine cardiac protein kinase C. Pools 1 and 2 from the hydroxylapatite column were probed with antisera to PKCγ, PKCβ_2 and PKCα.

isozyme specific antisera, peaks 1 and 2 were found to contain PKCβ_2 and PKCα, respectively (Fig. 1, inset). No cross-contamination was detected by immunoblotting. As antisera for PKCβ_1 was not employed in this study we cannot comment on the presence of this subtype in cardiac tissue. Previously PKCβ immunoreactivity was detected in PKC preparations isolated from bovine heart (Allen and Katz, submitted) but not from rat heart (Kosaka *et al.*, 1988).

Canine cardiac SR has been separated into junctional and longitudinal fractions by employing a Ca^{2+}-phosphate loading step followed by discontinuous sucrose density gradient centrifugation. Phosphorylation of a 28 kDa protein was increased by either the catalytic subunit of the cAMP-dependent protein kinase (PKA), the endogenous Ca^{2+}/calmodulin-dependent protein kinase activity or PKCα (Fig. 2). Upon boiling, the 28 kDa phosphoprotein was replaced by bands of lower molecular weight on the

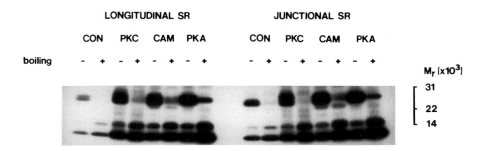

Figure 2. Autoradiogram showing phospholamban in either longitudinal or junctional SR. SR fractions (25 μg) were phosphorylated for 5 minutes at 30°C in 40 mM histidine/HCl buffer (pH 6.8) containing 10 mM MgCl$_2$, 100 μM [γ^{32}P]ATP, 5 mM DTT, 10 mM NaF, 1 mM Na$_3$VO$_4$. Additions to reaction medium were as follows: 1 mM EGTA (CON); 1 mM EGTA plus 0.1 μM PKA catalytic subunit (PKA); 0.2 mM CaCl$_2$, 80 μg/ml PS, 8 μg/ml SAG plus PKCα (PKC); or 0.2 mM CaCl$_2$ plus 0.1 μM calmodulin (CAM). Reactions were terminated with SDS-PAGE sample buffer and then incubated at either 100°C for 90 sec (+) or 37°C for 15 minutes (-) prior to SDS polyacrylamide gradient (5-20%) gel electrophoresis.

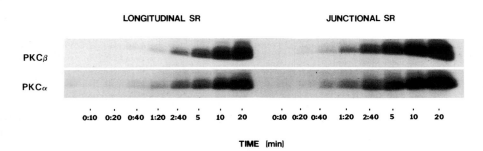

Figure 3. Autoradiogram showing the time course of phospholamban phosphorylation in longitudinal and junctional SR by 100 pmol/min/ml of either PKCβ or PKCα. The reaction medium was as described in Figure 2. Reactions were terminated at the time indicated by addition of sample buffer, heated to 37°C for 15 min and separated by SDS polyacrylamide gradient (5-20%) gel electrophoresis.

autoradiogram (Fig. 2). This reduction in apparent molecular weight upon boiling is characteristic of phospholamban (Lamers and Stinis, 1980). The intensity of phospholamban phosphorylation was similar in the two SR subfractions. A phosphoprotein of approximatly 15 kDa was also observed: however as this protein was phosphorylated by the Ca^{2+}/calmodulin-dependent protein kinase in addition to PKA and PKC, was present in both SR fractions and was sensitive to boiling, this band likely represents a complex of phospholamban monomers rather than the 15 kDa substrate for PKC and PKA reportedly present in sarcolemma (Presti, Scott and Jones, 1985). In the presence of equal amounts (100 pmol/min/ml) of either PKCβ or PKCα, a similar time course for the phosphorylation of phospholamban was observed in both junctional and longitudinal SR (Fig. 3). These results suggest that although PKCβ and PKCα differ structurally in a region proximal to the catalytic site (Ono *et al.*, 1987), they display similar affinity towards phospholamban as a target protein *in vitro*.

Acknowledgement

These investigations were funded by the British Columbia and Yukon Heart and Stroke Foundation. BGA was a recipient of a FEBS/NATO travel award.

References

Chamberlain BK, Levitsky DO, Fleischer S (1983) Isolation and characterization of canine cardiac sarcoplasmic reticulum with improved Ca^{2+} transport properties. J Biol Chem **258**:6602-6609

Inui M, Wang S, Saito S, Fleischer S (1988) Characterization of junctional and longitudinal sarcoplasmic reticulum from heart muscle. J Biol Chem **263**:10843-19850

Iwasa Y, Hosey, MM (1984) Phosphorylation of cardiac sarcolemma proteins by the calcium-activated phospholipid-dependent protein kinase. J Biol Chem **259**:534-540

Kosaka Y, Ogita K, Ase K, Nomura H, Kikkawa U, Nishizuka Y (1988) The heterogeneity of protein kinase C in various rat tissues. Biochem Biophys Res Commun **151**:973-981

Kuo JF, Andersson RGG, Wise BC, Mackerlova L, Salomonsson I, Bracket NL, Katoh N, Shoji M, Wrenn RW (1980) Calcium-dependent protein kinase:Widespread occurrence in various tissues and phyla of the animal kingdom and comparison of effects of phospholipid, calmodulin, and trifluoperazine. Proc Natl Acad Sci USA **77**:7039-7043

Lamers JMJ, Stinis JT (1980) Phosphorylation of low molecular weight proteins in purified preparations of rat heart sarcolemma and sarcoplasmic reticulum. Biochim Biophys Acta **624**:443-459

Limas CJ (1980) Phosphorylation of cardiac sarcoplasmic reticulum by a calcium-activated, phospholipid-dependent protein kinase. Biochem Biophys Res Commun **96**:1378-1383

Liu JD, Wood JG, Raynor RL, Wang Y-C, Noland TAJr, Ansari AA, Kuo JF (1989) Subcellular distribution and immunocytochemical localization of protein kinase C in myocardium, and phosphorylation of troponin in isolated myocytes stimulated by isoproterinol or phorbol esters. Biochem Biophys Res Commun **162**:1105-1110

Movesian MA, Nishikawa M, Adelstein RS (1984) Phosphorylation of phospholamban by calcium-activated, phospholipid-dependent protein kinase. Stimulation of cardiac sarcoplasmic reticulum calcium uptake. J Biol Chem **259**:8029-8032

Ohno S, Kawasaki H, Imajoh S, Suzuki K, Inagaki M, Yokokura H, Sakoh T, Hidaka H (1987) Tissue expression of three distinct types of rabbit protein kinase C. Nature (Lond) **325**:161-166

Ono Y, Fujii T, Ogita K, Kikkawa U, Igarashi K, Nishizuka Y (1987) Identification of three additional members of rat protein kinase C family: δ-, ε- and ζ-subspecies. FEBS Lett **226**:125-128

Presti CF, Scott BT, Jones LR (1985) Identification of an endogenous protein kinase C activity and its intrinsic 15-kilodalton substrate in purified canine cardiac sarcolemmal vesicles. J Biol Chem **260**:13879-13889

CHARACTERIZATION OF THE PHOSPHORYLATION SITES OF 40S RIBOSOMAL PROTEIN S6

H.R. Bandi, S. Ferrari, H.E. Meyer[*] and G. Thomas
Friedrich Miescher Institute
P.O. Box 2543
4002 Basel, Switzerland

Ruhr Universität Bochum[*]
4630 Bochum 1, W. Germany

Mitogenic stimulation of quiescent cells triggers a cascade of events which lead to cell division. One of the early obligatory steps in this pathway is a two- to three-fold increase in the rate of protein synthesis (Brooks RF, 1977; Rudland PS and Jiminez de Asua L, 1979; Thomas G et al., 1979). This increase is controlled at the level of initiation of protein synthesis (Stanners CP and Becker H, 1971; Rudland PS et al., 1975; Pain, OM 1986) and involves a large number of translational components. One of these components is 40S ribosomal protein S6 which becomes phosphorylated at multiple sites following mitogenic stimulation. This phosphorylation event takes place within minutes after the addition of the mitogen and is mediated by a highly specific oncogene/mitogen-activated S6 kinase. This kinase is a M_r 70K enzyme which has been recently purified, characterized and cloned in our laboratory (Jenö P et al., 1988; Kozma SC et al., 1989; Kozma SC et al, 1990). Whether S6 phosphorylation is an obligatory step for the increase in protein synthesis still remains to be investigated. A number of early studies carried out in vivo and in vitro favour this hypothesis. For example, 40S subunits containing the most highly phosphorylated derivatives of S6 have a selective advantage in entering polysomes in vivo (Thomas G et al, 1982; Duncan R and McConkey E, 1982), and such ribosomes are three to four times more efficient in carrying out protein synthesis in an in vitro reconstituted system than non-phosphorylated ribosomes (Traugh JA and Pendergast AM, 1986; Palen E and Traugh JA, 1987).

Experiments carried out in our laboratory using two-dimensional tryptic phospho peptide maps showed that the sites are phosphorylated in a specific order. This suggests that the more highly phosphorylated derivatives contain unique sites (Martin-Pérez J and Thomas G, 1983). To determine the role of each of the increasingly phosphorylated derivatives in protein synthesis it is prerequisite to identify the order of phosphorylation. To approach this problem we took advantage of the fact that intraperitoneal injection of cycloheximide leads to the activation of a number of early mitogenic reponses in the liver, including S6 phosphorylation (Krieg et al., 1988). The liver provided sufficient material to establish a strategy for sequencing the phosphorylation sites. The protein was purified by cation and reverse phase HPLC chromatography from 40S ribosomal subunits and then cleaved with cyanogen bromide. This treatment generated a single phosphopeptide containing all of the phosphoserines. The phosphoserines were then converted to S-ethycysteine (Meyer HE et al., 1986) and the peptide was sequenced (Krieg et al., 1988). The results showed that the phosphorylation sites are clustered in a stretch of 15 amino acids at the C-terminus of the S6 molecule (Figure 1).

At this point it was important to establish whether the phosphorylation sites in cycloheximide-treated rat liver S6 were the same as those in mitogen-stimulated Swiss 3T3 cells, our model system for cell growth. To carry out such studies required the development of a new protocol to handle small amounts of ribosomes. The details for this procedure will be published elsewhere. In summary, S6 was isolated from crude 80S ribosomes rather than purified 40S subunits, the protein was cleaved with Endoproteinase Lys C (Endo Lys C) rather than cyanogen bromide and the peptides were separated on an inert HPLC system employing a narrow bore column. Use of 80S ribosomes reduced losses traditionally encountered in the purification of 40S ribosomes. Endo Lys-C instead of cyanogen bromide cleavage generated a 19 as opposed to a 32 amino acid peptide which still contained all of the phosphorylation sites. This improvement alone increased yields >2-fold during Edman degradation. Finally, employing inert HPLC prevented large losses of phosphorylated peptide, making it possible to use significantly smaller amounts of initial starting material. Preliminary amino acid sequencing experiments indicate that the sites of phosphorylation are the same in both systems (see Figure 1). This finding is consistent with parallel studies which show that the same apparent M_r 70K S6 kinase is activated in both mitogen-stimulated 3T3 cells and the liver of cycloheximide-injected rats (Kozma et al., 1989).

Having once established the sites of phosphorylation it will be important to determine the order of phosphorylation *in vivo*. Earlier we showed that the phosphate which is incorporated into S6 is metabolically stable during the first 60 minutes following serum stimulation (Thomas G et al., 1979). Since S6 becomes fully phosphorylated during this time it should be possible to harvest ribosomes at different times following the addition of serum and to selectively enrich for each of the increasingly phosphorylated derivatives. Each derivative containing an additional mole of phosphate should then represent the product of the "on" or phosphorylation reaction rather than a mixed reaction due to phosphorylation and dephosphorylation.

To establish conditions for identifying each of the sites two approaches are presently being exploited. Both approaches utilize 40S ribosomes phosphorylated *in vitro* using the M_r 70K kinase in the presence of $^{32}P\gamma$ATP. In the first case each of the increasingly phosphorylated derivatives were separated on 2-D gels. Each phospho derivative was excised from the gel, electroeluted and digested with Endo Lys C. The phosphopeptides were then separated on either IEF-PHAST gels (Pharmacia) or tricine gels and the products visualized by autoradiography. In the second approach, S6 was phosphorylated to different extents and then purified by cation and reverse phase chromatography, cleaved with Endo Lys C and the fragments separated on a C18 HPLC reverse phase column. The radioactive peaks were manually collected and counted for Cerenkov cpm. Each fraction was analyzed on a tricine gel, a system established for separating small peptides in the molecular weight range of 800 to 2000 daltons (Schägger H and Jagow G, 1987). Taken together the results show that the endoproteinase cannot cleave the more highly phosphorylated derivatives of S6, suggesting that Endo Lys C does not recognize Lys243 as a proteolytic site if serine Ser244 and possibly 240 are phosphorylated (Figure 1). The final interpretation of these results awaits sequencing of each of the digestion products which were separated by either reverse phase HPLC or tricine gels. When the sequencing procedure is optimized it will be applied to the *in vivo*-^{32}P-labeled S6.

Perspectives

The sequence information outlined has already been used for the design of synthetic peptides which correspond to the C-terminus of S6 and which have been modified to search for the minimal recognition sequence. If these peptides serve as good substrates for phosphorylation *in vitro* they will be injected into 3T3 cells as competitive inhibitors. In addition, polyclonal antibodies have been prepared against the C-terminus of S6 and are being employed in immune-electronmicroscopy studies for the localization of the S6 phosphorylation sites within the 40S subunit. If located in the anticodon-codon interaction site we would like to further test the effect of S6 phosphorylation on the conformation of the ribosome. Finally, in the long term, site-directed mutagenesis studies will be carried out on the C-terminal serine residues using a previously isolated S6 cDNA clone (Lalanne JL et al, 1987). By expressing mutagenized S6 cDNAs under an inducible promoter we hope to directly evaluate the effect of specific sites of S6 phosphorylation on protein synthesis during the mitogenic response.

S6 Phosphorylation sites

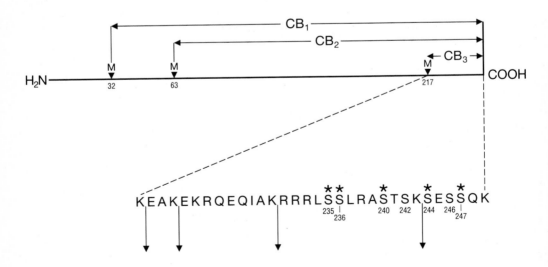

CB = cyanogen bromide fragment

∗ = *in vivo* Ser-P site

↓ = Endo LysC site

Acknowledgements: We would like to thank Drs. R. Layden, D. Reddy and R. Mayer for critical reading of the manuscript, and Carol Wiedmer for typing the manuscript.

References

Brooks RF (1977) Continuous protein synthesis is required to maintain the probability of entry into S phase. Cell 12:311-317

Duncan R and McConkey E (1982) Preferential utilization of phosphorylated 40S ribosomal subunits during initiation complex formation. Eur J Biochem 123:539-544

Jenö P (1988) Identification and characterization of a mitogen-activated S6 kinase.

Kozma SC et al. (1989) A stimulated S6 kinase from rat liver: identity with the mitogen-activated S6 kinase of 3T3 cells. EMBO J 8:4125-4132

Kozma SC et al. (1990) Cloning of the mitogen-activated S6 kinase from rat liver reveals an enzyme of the second messenger subfamily. Proc Natl Acad Sci USA 87:7365-7369

Krieg J et al. (1988) Identification of the 40S ribosomal protein S6 phosphorylation sites induced by cycloheximide. J Biol Chem 263:11473-11477

Lalanne J-L et al (1988) Complete sequence of mouse ribosomal protein S6. Nuc Ac Res 15:4990

Martin-Pérez J and Thomas G (1983) Ordered phosphorylation of 40S ribosomal protein S6 after serum stimulation of quiescent 3T3 cells. Proc Natl Acad Sci USA 80:926-930

Meyer HE (1986) Sequence analysis of phospho-serine-containing peptides. Modification for picomolar sensitivity. FEBS Lett 204:61-66

Pain OM (1986) Initiation of protein synthesis in mammalian cells. Biochem J 235:625-637

Palen E and Traugh JA (1987) Phosphorylation of ribosomal protein S6 by cAMP-dependent protein kinase and mitogen-stimulated S6 kinase differentially alters translation of globin mRNA. J Biol Chem 262:1293-1297

Rudland PS and Jimenez de Asua L (1979) Action of growth factors in the cell cycle. Biochem Biophys Acta 560:91-133

Rudland PS et al. (1975) Changes in RNA metabolism and accumulation of presumptive messenger RNA during transition from the growing to the quiescent state of cultured mouse fibroblasts. J Mol Biol 96:745-766

Schägger H and Jagow G (1987) Tricine-sodium dodecyl sulfate-polyacrylamide gel electrophoresis for separation of proteins in the range from 1 to 100 kDa. Anal Biochem 166:368-379

Stanners CP and Becker H (1971) Control of macromolecular synthesis in proliferating and resting Syrian hamster cells in monolayer culture. J Cell Phys 77:31-42

Thomas G, Siegmann M, Gordon J (1979) Multiple phosphorylation of ribosomal protein S6 during transition of quiescent 3T3 cells into early G, and cellular compartmentalization of the phosphate donor. Proc Natl Acad Sci USA 76:3952-3956

Thomas G et al (1982) The effect of serum, EGF, $PGF_{2-\alpha}$ and insulin on S6 phosphorylation and the initiation of protein and DNA synthesis. Cell 30:235-242

Traugh JA and Pendergast AM (1986) Regulation of protein synthesis by phosphorylation of ribosomal protein S6 and aminoacyl-tRNA synthetases. Prog Nuc Ac Mol Biol 33:195-230

V. CYCLIC NUCLEOTIDE-DEPENDENT SIGNALLING

G Protein Oncogenes

Henry R. Bourne
Departments of Pharmacology and Medicine
University of California, San Francisco, CA 94143
USA

This review will describe experiments aimed at answering two questions raised by the recent convergence of research on signal-transducing G proteins and oncogenes: What roles do individual G proteins play in normal regulation of cell proliferation? How do dominantly acting G protein oncogenes contribute to tumorigenesis, in experimental systems and in human cancer?

G proteins transduce extracellular signals, detected by cell surface receptors for hormones or sensory stimuli, into cellular responses mediated by altered function of effector enzymes or ion channels (Bourne, 1990; Freissmuth, 1989). Each G protein is composed of an α chain that binds and hydrolyzes GTP and β and γ chains that form a tightly bound βγ subunit. βγ presents the GDP-bound form of the α chain to ligand-activated receptors, which promote exchange of GDP for GTP by increasing the rate constant for dissociation of GDP from αβγ. The distinctive α chains interact specifically with receptors and effectors.

Thirteen α chain genes have been identified. These encode α chains of: G_s, which mediates hormonal stimulation of adenylyl cyclase; three distinct "G_i" proteins (G_{i1}, G_{i2}, and G_{i3}), named for the ability of one or more of them to inhibit adenylyl cyclase; G_o, which is abundant in brain and whose physiological role is not known; two retinal transducins (G_t), which mediate retinal phototransduction; G_{olf}, thought to mediate olfactory transduction; and five additional G proteins of unknown function, identified by virtue of cloned α chain genes or cDNAs (Fong, 1988; Matsuoka, 1988; Strathmann, 1989). Signaling through G_t, G_o, and the three G_i proteins is blocked by pertussis toxin (PTX), which uncouples these proteins from receptors by catalyzing ADP-ribosylation of a conserved cysteine residue in their α subunits.

G proteins amplify and sort signals. Amplification results from a slow rate of GTP hydrolysis ($k_{cat\cdot GTP}$ = 3-5 min^{-1}) (Bourne, 1990; Freissmuth, 1989),

NATO ASI Series, Vol. H 56
Cellular Regulation by Protein Phosphorylation
Edited by L. M. G. Heilmeyer, Jr.
© Springer-Verlag Berlin Heidelberg 1991

which is slowed even further by cholera toxin (CTX)-catalyzed ADP-ribosylation of α_s and α_t (Abood, 1982; Cassel, 1978) and by mutations that convert α_s and α_{i2} into oncogenic proteins (see below). The signal-sorting function is mediated by the G protein's ability to discriminate among subsets of receptors and effectors.

G proteins and mitogenesis

Before the recent discovery in human tumors of oncogenic mutations in genes for G protein α chains, two kinds of evidence indicated involvement of G proteins in signaling pathways that stimulate proliferation: 1. Mitogenic effects of many ligands that activate G protein-coupled receptors, including many neuroendocrine peptides, lysophosphatidic acid (LPA) (van Corven, 1989), acetylcholine (Ashkenazi, 1989), serotonin (5HT) (Julius, 1989; Kavanaugh, 1988; Nemecek, 1986; Seuwen, 1988), and trophic endocrine hormones that work via G_s and cAMP — e.g., growth hormone releasing hormone (GHRH) and thyrotropin (TSH). 3. Ability of PTX to abrogate mitogenic effects of some growth-stimulating ligands, including bombesin (Letterio, 1986; Zachary, 1987), thrombin (Paris, 1986), LPA (van Corven, 1989), and serotonin (5HT) (Seuwen, 1988).

Except in the case of G_s and cAMP, neither the specific molecular species of α chain nor the downstream mitogenic pathway has been identified in any specific case. Stimulation of PI-PLC may mediate mitogenic effects of acetylcholine in astrocytes (Ashkenazi, 1989) and of 5HT in NIH-3T3 cells expressing a recombinant 5HT1c receptor (Julius, 1989). In other cases, PTX or ligands specific for pharmacologically distinct receptors can often uncouple the mitogenic effect of the extracellular ligand from activation of the phosphoinositide/Ca^{2+} pathway (Kavanaugh, 1988; Letterio, 1986; Seuwen, 1988; van Corven, 1989; Zachary, 1987).

The *gsp* oncogene

GHRH stimulates proliferation of GH-secreting pituitary somatotrophs *via* intracellular cAMP (Billestrup, 1986). A subset of human GH-secreting tumors exhibit constitutively elevated GH secretion, cellular cAMP, adenylyl cyclase activity, and G_s activity (Vallar, 1987) . Because such tumors are excellent candidates for harboring oncogenic α_s (Bourne, 1987), we reverse-transcribed mRNA from the tumors, amplified α_s cDNAs using the polymerase chain reaction (PCR), and sequenced the α_s coding regions

(Landis, 1989). All GH-secreting tumors with constitutively elevated adenylyl cyclase and G_s activity exhibited α_s mutations, while α_s mutations were not found in GH-secreting tumors with normal adenylyl cyclase and G_s activities. The mutations in these and subsequently analyzed α_s DNAs (Lyons, 1990) encoded amino acid substitutions at either of two codons — Q227, cognate to Q61 of $p21^{ras}$, and R201, the residue whose ADP-ribosylation by CTX inhibits the GTPase of α_s (Cassel, 1978; Van Dop, 1984). When expressed in cyc^- cells, mutant α_s with either kind of substitution exhibited low GTPase activity and caused constitutive activation of adenylyl cyclase.

Using a PCR/allele-specific oligonucleotide screening procedure, we screened 25 GH-secreting tumors for the presence of gsp mutations (Landis, 1990). The 10 patients whose tumors contained gsp mutations came to surgery with significantly smaller tumors and lower pre-operative GH levels. Their tumors were also more susceptible to glucose suppression of GH release than those without mutations. The activating mutations identify a subgroup of GH-secreting tumors with a distinctive clinical course that probably results from a shared oncogenic mechanism. More extensive clinical studies will determine whether presence or absence of gsp mutations is a clinically useful indicator of prognosis.

To explore the mitogenic effect of gsp in tissue culture cells, we expressed α_s-Q227L in Swiss 3T3 cells (Zachary, 1990) — cells in which cAMP stimulates mitogenesis in synergistic combinations with other growth-promoting agents, such as insulin (Rozengurt, 1988). Expression of α_s-Q227L constitutively elevated adenylyl cyclase activity measured in particulate extracts. In the presence of a phosphodiesterase (PDE) inhibitor (but not in its absence), basal and forskolin-stimulated cAMP accumulation were considerably elevated, and insulin alone stimulated incorporation of radioactive thymidine into DNA in these cells. We suspect that the gsp-expressing cells counteracted the gsp-induced elevation of adenylyl cyclase activity by inducing compensatory mechanisms (e.g., elevated PDE). Just as carcinogenesis requires more than one oncogenic mutation, Swiss 3T3 cells require a (pharmacologic) "second hit" — in this case a PDE inhibitor — to exhibit the mitogenic effect of constitutively active α_s and adenylyl cyclase.

Because TSH stimulates secretion of thyroid hormones and proliferation of thyroid cells, we used an allele-specific oligonucleotide technique to look for gsp mutations in thyroid tumors. So far, we have found a high frequency of gsp mutations in multinodular goiters and in papillary

256
and follicular carcinomas of the thyroid (P. Goretzki, et al., submitted for publication).

The *gip2* oncogene

Our belief in the functional significance of conserved amino acid sequence prompted us to look for oncogenic mutations in genes for other G protein α chains. We postulated that the two conserved amino acids whose mutational replacements activate α_s play similar functional roles in α subunits of other G proteins and therefore that mutations replacing cognate amino acids in other α chain genes would activate the corresponding G proteins. We screened 268 human tumors for mutations in the R179 and Q205 codons of the α_{i2} gene, using a strategy in which appropriate PCR-amplified regions of genomic DNA are hybridized with allele-specific oligonucleotides at high stringency (Lyons, 1990).

Mutations substituting cysteine or histidine for R179 of α_{i2} (corresponding to R201 of α_s) were found in 3 of 11 tumors of the adrenal cortex and in 3 of 10 ovarian sex cord stromal tumors (Lyons, 1990). Because allele-specific hybridization can detect point mutations only in samples in which substantial expansion of a mutant clone (a hallmark of neoplasia) has occurred, and because these mutations were found at high frequency in a very specific subset of tumors, we conclude that the R179 mutations convert the α_{i2} gene into an oncogene, which we dubbed *gip2*. Fragmentary evidence (summarized in (Lyons, 1990)) is consistent with the possibility that G_{i2} mediates mitogenic effects of certain hormones and growth factors on adrenocortical and ovarian cells in tissue culture.

To explore mitogenic effects of the *gip2* oncogene, we expressed wild type α_{i2} or mutant α_{i2} with an R179C substitution (α_{i2}-R179C) in Rat-1 cells. This cell type seemed an appropriate target for oncogenic effects of mutated α_{i2}, because LPA had been reported to stimulate DNA synthesis and to inhibit adenylyl cyclase in Rat-1 cells; both effects were inhibited by PTX (van Corven, 1989). The α_{i2}-R179C cells grew to a 3-fold higher saturation density in 10% calf serum, and — like cells neoplastically transformed by *ras* oncogenes, formed colonies in soft agar. Moreover, subcutaneous injection of α_{i2}-R179C cells led to formation of tumors in nude mice. We concluded that *gip2* mediates neoplastic transformation of Rat-1 cells.

We similarly infected NIH-3T3 cells with wild type α_{i2}, α_{i2}-R179C, and also α_{i2}-Q205L. In this cell type expression of mutant α_{i2} did not alter growth in culture, nor did it cause formation of tumors in nude mice. NIH-3T3 cell populations carrying mutant α_{i2} did show decreased adenylyl cyclase activities, however, suggesting that mutant α_{i2} inhibits adenylyl cyclase.

Further Questions

The discovery of G protein oncogenes strongly confirms previous indications that extracellular stimuli can work through G proteins to regulate mitogenesis as well as differentiated cell functions. Discovery of *gsp* and *gip2*, unlike that of other oncogenes, occurred after the normal cellular functions of the corresponding proteins were known, either in specific detail (α_s) or in broad outline (α_{i2}). Now the challenge is to use this knowledge in tackling hard questions about the complex network of regulatory pathways that control proliferation of normal and cancerous cells.

Are there more α chain oncogenes? If so, in which human tumors? The *gsp* and *gip2* precedents make it very likely that other human tumors will exhibit activating mutations in genes encoding α chains other than α_s and α_{i2}. Using PCR and allele-specific oligonucleotides, we are currently searching for such mutations.

What is the "host range" of cells and tissues in which mutant α_s and α_{i2} are oncogenic? We suspect that this range will be narrow and probably specific for each G protein oncogene, despite ubiquity of the corresponding normal proteins, because the signaling pathway(s) triggered by an individual G protein can trigger proliferation in a limited number of cell types. Further experiments in tissue culture cells and in transgenic mice will be required to test this notion.

What signaling pathways are triggered by G protein oncogenes? Although cAMP probably mediates the effects of *gsp* in pituitary and thyroid cells, we do not yet know whether *gip2* is tumorigenic in Rat-1 cells (or in adrenal cortex and ovary) because it inhibits adenylyl cyclase and thereby relieve an inhibitory constraint on growth

Does *gip2* (or any other α chain oncogene) modify the biological behavior of tumor cells by altering their invasiveness, adhesiveness, or tendency to metastasize? Suggestive answers may come out of searches for mutant α chain genes in human tumors and analysis of the biological behavior of tumors in transgenic mice.

REFERENCES

Abood ME, Hurley JB, Pappone M-C, Bourne HR, Stryer L (1982) Functional homology between signal-coupling proteins: Cholera toxin inactivates the GTPase activity of transducin. J Biol Chem 257:10540-10543

Ashkenazi A, Ramachandran J, Capon DJ (1989) Acetylcholine analogue stimulates DNA synthesis in brain-derived cells via specific muscarinic receptor subtypes. Nature 340:146-150

Billestrup N, Swanson LW, Vale W (1986) Growth hormone-releasing factor stimulates proliferation of somatotrophs *in vitro*. Proc Natl Acad Sci USA 83:6854-6857

Bourne HR (1987) G proteins and cAMP. Discovery of a new oncogene in pituitary tumours? Nature 330:517-518

Bourne HR, Sanders DA, McCormick F (1990) The GTPase superfamily. A conserved switch for diverse cell functions. Nature 348:125-132

Cassel D, Pfeuffer T (1978) Mechanism of cholera toxin action: Covalent modification of the guanyl nucleotide-binding protein of the adenylate cyclase system. Proc Natl Acad Sci USA 75:2669-2673

Fong HKW, Yoshimoto KK, Eversole-Cire P, Simon MI (1988) Identification of a GTP-binding protein α subunit that lacks an apparent ADP-ribosylation site for pertussis toxin. Proc Natl Acad Sci USA 85:3066-3070

Freissmuth M, Casey PJ, Gilman AG (1989) G proteins control diverse pathways of transmembrane signaling. FASEB J 3:2125-2131

Julius D, Livelli TJ, Jessell TM, Axel R (1989) Ectopic expression of the serotonin 1c receptor and the triggering of malignant transformation. Science 244:1057-1062

Kavanaugh WM, Williams LT, Ives HE, Coughlin SR (1988) Serotonin-induced deoxyribnucleic acid synthesis in vascular smooth muscle cells involves a novel, pertussis toxin-sensitve pathway. Mol Endocrin 2:599-605

Landis CA, Harsh G, Lyons J, Davis RL, McCormick F, Bourne HR (1990) Clinical characteristics of acromegalic patients whose pituitary tumors contain mutant G_s protein. J Clin Endocrinol Metab in press

Landis CA, Masters SB, Spada A, Pace AM, Bourne HR, Vallar L (1989) GTPase inhibiting mutations activate the α chain of G_s and stimulate adenylyl cyclase in human pituitary tumours. Nature 340:692-696

Letterio JJ, Coughlin SR, Williams LT (1986) Pertussis toxin-sensitive pathway in the stimulation of c-myc expression and DNA synthesis by bombesin. Science 234:1117-1119

Lyons J, Landis CA, Harsh G, Vallar L, Grünewald K, Feichtinger H, Duh QY, Clark OH, Kawasaki E, Bourne HR, McCormick F (1990) Two G protein oncogenes in human endocrine tumors. Science 249:655-659

Matsuoka M, Itoh H, Kozasa T, Kaziro Y (1988) Sequence analysis of cDNA and genomic DNA for a putative pertussis toxin-insensitive guanine nucleotide-binding regulatory protein α subunit. Proc Natl Acad Sci USA 85:5384-5388

Nemecek GM, Coughlin SR, Handley DA, Moskowitz MA (1986) Stimulation of aortic smooth muscle cell mitogenesis by serotonin. Proc Natl Acad Sci USA 83:674-678

Paris S, Pouysségur J (1986) Pertussis toxin inhibits thrombin-induced activation of phosphoinositide hydrolysis and Na^+/H^+ exchange in hamster fibroblasts. EMBO J 5:55-60

Rozengurt E, Erusalimsky J, Mehmet H, Morris C, Nånberg E, Sinnett-Smith J (1988) Signal transduction in mitogenesis: Further evidence for multiple pathways, Cold Spring Harbor Symp. Quant. Biol. 53:945-954

Seuwen K, Magnaldo I, Pouysségur J (1988) Serotonin stimulates DNA synthesis in fibroblasts acting through $5-HT_{1B}$ receptors coupled to a G_i-protein. Nature 335:254-256

Strathmann M, Wilkie TM, Simon MI (1989) Diversity of the G-protein family: Sequences from five additional α subunits in the mouse. Proc Natl Acad Sci USA 86:7407-7409

Vallar L, Spada A, Giannattasio G (1987) Altered G_s and adenylate cyclase activity in human GH-secreting pituitary adenomas. Nature 330:566-568

van Corven EJ, Groenink A, Jalink K, Eichholtz T, Moolenaar WH (1989) Lysophosphatidate-induced cell proliferation: Identification and dissection of signaling pathways mediated by G proteins. Cell 59:45-54

Van Dop C, Tsubokawa M, Bourne HR, Ramachandran J (1984) Amino acid sequence of retinal transducin at the site ADP-ribosylated by cholera toxin. J Biol Chem 259:696-698

Zachary I, Masters SB, Bourne HR (1990) Increased mitogenic responsiveness of Swiss 3T3 cells expressing constitutively active $G_{s\alpha}$. Biochem Biophys Res Comm 168:1184-1193

Zachary I, Millar J, Nanberg E, Higgins T, Rozengurt E (1987) Inhibition of bombesin-induced mitogenesis by pertussis toxin-dissociation from phospholipase C pathway. Biochem Biophys Res Comm 146:456-463

A NEW ROLE FOR βγ-SUBUNITS OF G-PROTEINS?

E.J.M. Helmreich
Department of Physiological Chemistry
University of Würzburg
Koellikerstraße 2
D-8700 Würzburg, FRG

The prevailing view of the role of βγ-subunits in signal transmission of G-protein-linked hormone receptors ascribes to βγ-subunits a role in facilitating the binding of the α-subunits in heterotrimeric G-proteins in the inactive GDP-bound form to the receptor (cf: Bourne et al. 1990). In addition, there is good evidence that βγ-subunits serve as anchors in membrane attachment of G-proteins (Sternweis 1986). The evidence which is now presented supports a role of βγ-subunits in the attachment of α-subunits to the β-adrenoceptor and, moreover, indicates that separate βγ-subunits can also bind to the activated (Im et al. 1988) and to the non-activated β-receptor as well. These findings suggest that the β-receptor may form a long-lived complex with βγ-subunits alone. As that may be, this complex does not seem to dissociate with activation when the α-subunit charged with GTP[S] is discharged from the receptor. A detailed account of this work is in press (Kurstjens et al. 1990).

Experimental design

All these experiments were carried out with reconstituted proteoliposomes (Kurstjens et al. 1990). The β_1-receptor was purified from turkey erythrocyte membranes (see: Hekman et al. 1984), the α- and βγ-subunits were derived from G_o purified to

NATO ASI Series, Vol. H 56
Cellular Regulation by Protein Phosphorylation
Edited by L. M. G. Heilmeyer, Jr.
© Springer-Verlag Berlin Heidelberg 1991

homogeneity from bovine brain (Sternweis & Robishaw 1984). To simplify the assay, β-receptor and G-protein subunits were tagged with fluorescein isothiocyanate (FITC). Different from our previous reconstitution assay, special attention was directed to the preparation of a rather homogenous population of unilamellar tightly sealed liposomes with the β-adrenoceptor oriented inside-out. As a result of these efforts two preparations of liposomes were prepared: one was lipid-rich and larger with a diameter of about 200 nm, the other was lipid-poor with a diameter of about 50 nM. Since the latter vesicles were heavier than the former lipid-rich vesicles and since both vesicle preparations were tightly sealed, it became possible to separate them by isopycnic sucrose density gradient centrifugation. When reconstitution with G-proteins alone or with the separate, purified β-adrenoceptor was carried out, the heterotrimeric G-protein or separate α_0 - and βγ-subunits partitioned themselves exclusively in the lipid-rich fraction, whereas the β-receptor distributed itself in the heavier lipid-poor fraction. Moreover, with the lipid mixture used for the formation of the liposomes which consisted of dimyristoylphosphatidylcholine, dipalmitoylphosphatidylethanolamine, dipalmitoylphosphatidylserine and cholesterol in a 54:1:10:35 molar ratio, the β-adrenoceptor was oriented inside-out in the liposomes. Among the factors responsible for the orientation were probably charge interactions between the many basic (lysyl) residues of the receptor and the negatively charged phosphatidylserine in the outer leaflet of the liposome. In support of this conjecture is that the β-adrenoceptor assumed a more randomized orientation and was shifted to a lighter lipid-rich fraction when phosphatidylserine was omitted from the lipid mixture used for reconstitution. But there may be other factors which might be as important for the orientation of the receptor in liposomes as interactions of charged amino acid side chains with the phospholipid head groups. For example, the restricted area available in small vesicles with a small internal volume could force the more bulky portion of the receptor protein to the outside. The nearly exclusive inside-out orientation of the receptor incorporated into tightly sealed liposomes where the binding domains of the β_1-adrenoceptor for G-proteins were facing outside made it possible to put the hypothesis of M. Chabre (Chabre 1987) to test, namely that interactions of G-proteins with the β-receptor can occur at the aqueous membrane interface as was originally proposed for the transducin-rhodopsin interactions.

Summary of results

These properties of the tight liposome preparations where one population of vesicles (fraction I) contains only G-protein monomers or oligomers and the other (fraction II) the β_1-adrenoceptor oriented in a way which is favorable for interaction with G-proteins made these sealed vesicle preparations seem an attractive tool for studying the binding of the separate α- and $\beta\gamma$-subunits of G-proteins to the receptor. In Fig. 1A is shown that G_s reconstituted alone with lipids is recovered in the light vesicle fraction I, but when G_s and nonactivated β_1-adrenoceptor were reconstituted together (Fig. 1B), the G_s-preparation was carried by the receptor to the heavier vesicle fraction II which contains the receptor. This transfer is apparently driven by the mutual affinity of the

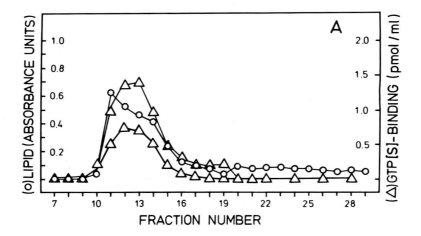

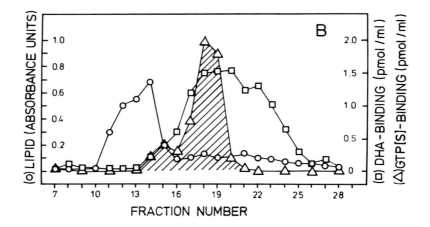

Fig. 1. *Distribution of G_s in lipid vesicles in the presence and absence of β_1-adrenoceptor.* A: G_s alone; B: G_s and β_1-adrenoceptor, (O) phospholipid phosphorus, (□) [³H]dihydro-alprenolol binding, (△) is [³⁵S]GTP[S] binding and the upper trace in A is [³⁵S]GTP[S] binding in the presence of 0.2% Lubrol PX. The shaded area in B shows the distribution of G_s in vesicles co-reconstituted with β_1-adrenoceptor.

receptor for G_s, since this shift does not occur in the absence of receptor by heterotrimeric G_s or G_0 or by α-subunits or by that matter by $\beta\gamma$-subunits alone. We have made use of this characteristic feature and have used this shift in order to study the role of $\beta\gamma$-subunits in the binding of the α-subunit to the receptor. These experiments are shown in Fig. 2. When

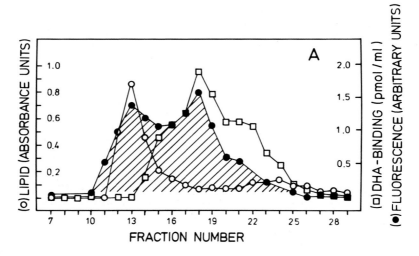

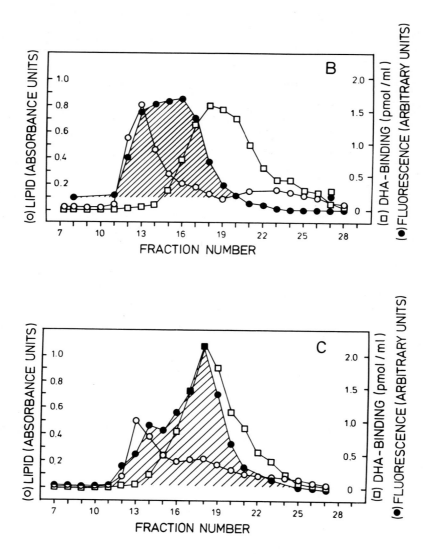

Fig. 2. *Distribution of bovine brain α_0- and $\beta\gamma$-subunits separate or combined in the presence of β_1-adrenoceptor.* A: $\beta\gamma$ labelled with FITC and β_1-adrenoceptor. B: α_0 labelled with FITC and β_1-adrenoceptor. C: α_0 labelled with FITC combined with unlabelled $\beta\gamma$ and β_1-adrenoceptor. (○) Phospholipid phosphorus, (□) [^{3}H]dihydroalprenolol binding, (●) fluorescence emission.

bovine brain $\beta\gamma$-subunits were co-reconstituted with the β-receptor at a 1:2 molar ratio, about 50% of the vesiculated $\beta\gamma$-fraction was found in the β-receptor-containing fraction II.

The efficiency of incorporation of the β-adrenoceptor in liposomes was nearly 100%. One estimates based on the size of the lipid vesicles the lipid and the protein concentrations on the average about one receptor per 100 vesicles. In the case of separate α_o- and βγ-subunits only between 40 to 70% of the added subunits were vesiculated. But even when the molar ratio of α_o to β-adrenoceptor was increased to 1:3 and 1:4 yielding up to one receptor per 25 vesicles, much less α_o was transferred to the receptor on co-reconstitution as compared to βγ-subunits. When, however, the same amount of α_o was combined with the same amount of β_1-adrenoceptor in the presence of βγ-subunits, the α_o-subunits were now almost completely transferred to the β-receptor containing vesicle fraction II. Binding interactions between all three components seem to be involved, but it is apparent that the βγ-subunits are mainly responsible for the transfer of α_o to the nonactivated β-adrenoceptor. It should be noted that nearly all (>90%) of α_o which was transferred to the receptor-containing fraction II was accessible from the outside, since the increase in binding of GTP[S] on solubilization with 0.02% lauroyl sucrose or 0.2% Lubrol PX was negligible. These data show therefore that β-receptor/G-protein interactions can take place at the aqueous membrane interface. They point to the primary importance of protein-protein interactions for the recognition of G-proteins by a receptor and are therefore compatible with M. Chabre's ideas (Chabre 1987).

In Fig. 3 is shown that the interaction between α_o and the receptor in the presence of βγ-subunits increased its GTPase activity. Replacement of the β_1-adrenoceptor by unrelated proteins, bovine serum albumin, ovalbumin, γ-globulins or addition of a heat-denatured β_1-adrenoceptor preparation abolished the receptor-promoted stimulation of the GTPase activity. Activation of the receptor by the agonist l(-)isoproterenol had little further activating effect, whereas addition of the antagonist d,l-propranolol failed to inhibit the GTPase. On the other hand with ADP-ribosylated G_o, the receptor-promoted increase in basal GTPase activity was about the same. This was

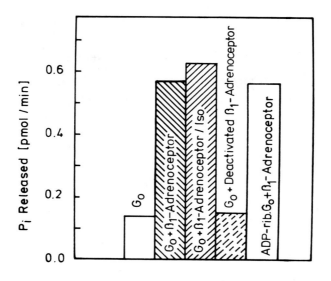

Fig. 3. *GTPase activity of G₀ in the presence of β₁-adrenocep-tor.* α₀- and βγ-subunits were reconstituted with or without β-receptor, as described in (Hekman et al. 1987) and the GTPase activity was determined. ADP-ribosylation of α₀ was carried out as described in (Kurstjens et al. 1990). The β₁-adrenoceptor was incubated with either l(-)isoproterenol or d,l-propranolol or in the absence of ligands. In control experiments, the β₁-adrenoceptor was heat-denatured or replaced by other un-related proteins.

to be expected, since the penultimate carboxyterminal part of the α₀- or α₁-subunit with the ADP-ribosylated cysteine residue is apparently not involved in basal GTP-GDP exchange and GTPase activity when the receptor is not activated and occupied by an agonist (Haga et al. 1985; Kurose et al. 1986; Sunyer et al. 1989).

The data as a whole show therefore that binding interactions between α₀- and βγ-subunits with a nonactivated β₁-adrenoceptor stimulate the GTPase activity of α₀, whereas activation of the β-receptor by l(-)isoproterenol had little further effect. These experiments support the notion that binding of α₀ to the β-receptor and the GTPase activity resulting from this interac-tion are dependent on βγ-subunits.

Conclusions

As we had reported previously when βγ-subunits purified from bovine brain are added to purified β-adrenoceptor and G$_s$ from turkey erythrocytes in lipid vesicles, they promote the inter-action of the activated β-adrenoceptor with G$_s$ and increase GTPase activity (Hekman et al. 1987). The finding that βγ-subunits are necessary to increase the GTPase activity of G$_{t\alpha}$ or G$_{i\alpha}$ in the presence of bleached rhodopsin (Fung 1983; Cerione et al. 1985) is in accordance with that. Moreover, βγ-subunits have been reported to interact with muscarinic ace-tylcholine receptors co-reconstituted with α$_o$ and α$_i$ into lipid vesicles (Florio & Sternweis 1985; Haga et al. 1988). We have now shown that βγ-subunits promote the binding of α$_o$ to the nonactivated β$_1$-adrenoceptor and that the relatively high basal GTPase activity of G$_o$ is stimulated even further when α- and βγ-subunits are bound to the β$_1$-adrenoceptor. These observa-tions are reminiscent of findings reported by Cerione and coworkers where stimulation of GTPase activity was observed on co-reconstitution of G$_i$ with β$_1$-adrenoceptor in lipid vesicles (Cerione et al. 1985). This suggests that the basal activity in the hormonally nonactivated state is due to a precoupled G-protein/β-receptor complex. Furthermore, one might question whether the βγ-subunit is released from the receptor together with the α-subunit on activation. In previous experiments we have shown that the β$_1$-adrenoceptor from turkey erythrocytes binds even more bovine brain βγ-subunits when activated with l(-)isoproterenol (Im et al. 1988). But as long as information on the steady state distribution of free and receptor-bound βγ-subunits in the hormonally stimulated and the resting state in cells and tissues is not available, the biological significance of the partitioning between receptor-bound and free βγ or βγ bound to other targets remains an open question (Gilman 1989; Boege et al. 1990).

Although, the information reported here is qualitative, it points to the importance of the binding of βγ-subunits to the

β_1-adrenoceptor for the association of the latter with α-subunits. Since α-subunits may obligatorily require $\beta\gamma$-subunits for binding to G-coupled receptors, one is tempted to suggest that the $\beta\gamma$-subunits may play a role equally important as that of the α-subunit for coupling of receptor and G-protein (Birnbaumer et al. 1989). In that respect, one would like to know how specific is the binding of $\beta\gamma$ to the receptor. This is an important issue which has not yet been considered. The question raised concerns a possible role of $\beta\gamma$-subunits in distinguishing among receptors. Such a sorting-out function might be important considering the multiplicity of coupling reactions of receptors and G-proteins. With that in mind, quantitative data of the binding of $\beta\gamma$-subunits from transducin and bovine brain to the β-receptor are presently being collected in this laboratory (Fröhlich M, Heithier H, Baumann M, Dees C, Häring M, Gierschik P, Schiltz E, Vaz W, Hekman M, Helmreich EJM (1991) Subunit interactions of GTP-binding proteins and with β-adrenoceptor. Eur J Biochem, submitted). A detailled account of this work is being published in (Kurstjens et al. 1990).

The work was made possible through grants from the Deutsche Forschungsgemeinschaft He 22/44-1, the Sonderforschungsbereich 176, Project A1, the Volkswagen-Stiftung and the Fonds der Chemischen Industrie e.V.

Literature

Birnbaumer L, Yatani A, Codina J, Van Dongen A, Graf R, Mattera R, Sanford J, Brown AM (1989) Multiple roles of G proteins in coupling of receptors to ionic channels and other effectors. In Molecular Mechanisms of Hormone Action (Gehring U, Helmreich E, Schultz G, eds) pp 147-177. Springer Berlin Heidelberg.

Boege F, Neumann E, Helmreich EJM (1990) Structural heterogeneity of membrane receptors and GTP-binding proteins and its functional consequences for signal transduction. Eur J Biochem, in press.

Bourne HR, Sanders DA, McCormick F (1990) The GTPase superfamily: a conserved switch for diverse cell functions. Nature 348:125-132.

Cerione RA, Staniszewski C, Benovic JL, Lefkowitz RJ, Caron MG, Gierschik P, Somers R, Spiegel AM, Codina J, Birnbaumer L (1985) Specificity of the functional interactions of the β-adrenergic receptor and rhodopsin with guanine nucleotide regulatory proteins reconstituted in phospholipid vesicles. J Biol Chem 260:1493-1500.

Chabre M (1987) The G protein connection: is it in the membrane or the cytoplasm? TIBS 12:213-215.

Florio VA, Sternweis PC (1985) Reconstitution of resolved muscarinic cholinergic receptors with purified GTP-binding proteins. J Biol Chem 260:3477-3483.

Fung BKK (1983) Characterization of transducin from bovine retinal rod outer segments. J Biol Chem 258:10495-10502.

Gilman AG (1989) G proteins and regulation of adenylyl cyclase. JAMA 262:1819-1825.

Haga K, Haga T, Ichiyama A, Katada T, Kurose H, Ui M (1985) Functional reconstitution of purified muscarinic receptors and inhibitory guanine nucleotide regulatory protein. Nature 316:731-733.

Haga T, Haga K, Berstein G, Nishiyama T, Uchiyama H, Ichiyama A (1988) Molecular properties of muscarinic receptors. TIPS February supplement, 12-18.

Hekman M, Feder D, Keenan AK, Gal A, Klein HW, Pfeuffer T, Levitzki A, Helmreich EJM (1984) Reconstitution of β-adrenergic receptor with components of adenylate cyclase. EMBO J 3:3339-3345.

Hekman M, Holzhöfer A, Gierschik P, Im MJ, Jakobs KH, Pfeuffer T, Helmreich EJM (1987) Regulation of signal transfer from $β_1$-adrenoceptor to adenylate cyclase by βγ-subunits in a reconstituted system. Eur J Biochem 169:431-439.

Im MJ, Holzhöfer A, Böttinger H, Pfeuffer T, Helmreich EJM (1988) Interactions of pure βγ-subunits of G-proteins with purified $β_1$-adrenoceptor. FEBS Lett 227:225-229.

Kurose H, Katada T, Haga T, Haga K, Ichiyama A, Ui M (1986) Functional interaction of purified muscarinic receptors with purified inhibitory guanine nucleotide regulatory proteins reconstituted in phospholipid vesicles. J Biol Chem 261:6423-6428.

Kurstjens NP, Fröhlich M, Dees C, Cantrill RC, Hekman M, Helmreich EJM (1990) Binding of α- and βγ-subunits of G_0 to $β_1$-adrenoceptor in sealed unilamellar lipid vesicles. Eur J Biochem, in press.

Sternweis PC (1986) The purified α subunits of G_0 and G_1 from bovine brain require βγ for association with phospholipid vesicles. J Biol Chem 261:631-637.

Sternweis PC, Robishaw JD (1984) Isolation of two proteins with high affinity for guanine nucleotides from membranes of bovine brain. J Biol Chem 259:13806-13813.

Sunyer T, Monastirsky B, Codina J, Birnbaumer L (1989) Studies on nucleotide and receptor regulation of G_1 proteins: effects of pertussis toxin. Mol Endo 3:1115-1124.

CONDITIONS FAVOURING PHOSPHORYLATION INHIBIT THE ACTIVATION OF ADENYLATE CYCLASE IN HUMAN PLATELET MEMBRANES

IA Wadman, RW Farndale, BR Martin
Department of Biochemistry
University of Cambridge
Tennis Court Rd
Cambridge CB2 1QW
United Kingdom

Summary

1. Incubation of human platelet membranes with p[NH]ppG causes a time dependent increase in the G_S mediated activation of adenylate cyclase. 2. Activation by p[NH]ppG follows first order kinetics in both the presence and absence of the assay components; cAMP, ATP, creatine phosphate and creatine kinase. 3. Inhibition by the assay components occurs throughout the activation process, reducing both the rate of activation and the maximum activity obtained. 4. Maximal activation of adenylate cyclase after prolonged incubation with p[NH]ppG slowly reverses in the presence of the assay components.

Introduction

Persistent activation of G_S can be achieved by the non-hydrolysable GTP analogue guanosine 5'[βγ-imido]triphosphate (p[NH]ppG). The degree of activation is dependent on the concentration of both p[NH]ppG and Mg^{2+} (Iyengar & Birnbaumer, 1981). Previous work in this laboratory found that preincubating human platelet plasma membranes with p[NH]ppG (10^{-5}M) at physiological concentrations of Mg^{2+} (1mM) for a period of about 20min doubled the adenylate cyclase activity measured on transferring membranes from the preincubation to an assay. The assay contained ATP, cyclic AMP (cAMP) and an ATP regenerating system of creatine phosphate and creatine kinase. The activation was markedly reduced when ATP, cAMP and the ATP regenerating system were also included in the preincubation. This effect was due to increased activity of G_S, and not a decrease in G_i activity (Farndale et al. 1987).

Recently it has become evident that phosphorylation provides a means of modulating the

NATO ASI Series, Vol. H 56
Cellular Regulation by Protein Phosphorylation
Edited by L. M. G. Heilmeyer, Jr.
© Springer-Verlag Berlin Heidelberg 1991

activity of the adenylate cyclase complex. This mechanism of regulation could occur at the level of the receptor, G proteins or catalytic unit. Receptor phosphorylation has been widely documented (reviewed by Sibley *et al.* 1987) and two possible phosphorylation sites have been identified from a cDNA sequence of the catalytic unit (Krupinski *et al.* 1989). Regulation of G proteins particularly G_i, by protein kinase C has been proposed (Katada *et al.* 1985; Pyne *et al.* 1989). There is little evidence for G_s phosphorylation.

The above inhibitory effect of cAMP, ATP, creatine phosphate and creatine kinase is consistent with the idea that a change in the phosphorylation state of G_s may modulate its activity. The activation by p[NH]ppG may involve phosphatase activity possibly on G_s, and the inhibition may be due to phosphorylation of G_s by kinase A. In support of this a 45kDa protein was found to be phosphorylated in platelet membranes in a cAMP dependent manner (Farndale *et al.* 1987). In the present study the nature of the inhibitory effects of cAMP, ATP, creatine phosphate and creatine kinase on p[NH]ppG stimulated adenylate cyclase activity were investigated further.

Methods

Membrane preparation:- A crude preparation of plasma membranes were made by a modification of the method of Stein & Martin (1983) using outdated human platelet concentrate (the kind gift of the East Anglian Regional Blood Transfusion Service, Cambridge).

Membrane preincubation:- Membranes diluted to about 1mg/ml into 25mM Tris/HCl pH 8.0, 1mM DTT were warmed to 30°C for 5min. Preincubations were composed of equal volumes of pre-warmed membranes and preincubation mix and contained 10μM p[NH]ppG and 1mM Mg^{2+}. At intervals, 50μl samples were removed for adenylate cyclase assay.

Adenylate cyclase assay:- 15-60μg of membrane protein was assayed for a period of 5min by the method of Salomon *et al.* (1974) in 25mM Tris/HCl pH 8.0, 1mM DTT, 1mM $MgCl_2$, 0.5mM cAMP, 0.1mM ATP, 1×10^6 c.p.m. of [α-^{32}P]ATP, 8mM creatine phosphate and 0.3mg/ml creatine kinase in a total volume of 100μl. The reaction was started by adding 50μl of preincubated membranes.

Results

Fig 1 shows the effect of including ATP, cAMP, creatine phosphate and creatine kinase in the preincubation with p[NH]ppG and Mg^{2+}. The data were analysed according to first order kinetics. Activation under both sets of conditions followed first order kinetics, the rate constants

Fig 1 Inhibition of p[NH]ppG activation of platelet adenylate cyclase on including ATP, cAMP, creatine phosphate and creatine kinase in the membrane preincubation

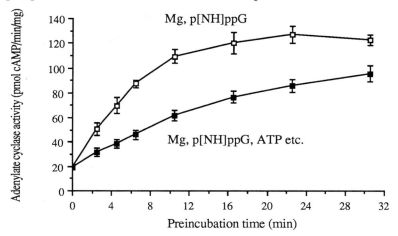

Two preincubations were set up at 30°C containing 10μM p[NH]ppG and 1mM Mg^{2+} either with (■) or without (□) 50μM cAMP, 100μM ATP, 8mM creatine phosphate and 0.3mg/ml creatine kinase (final concentrations). Triplicate samples containing 15μg of protein were removed to assay adenylate cyclase activity. Values shown are means of 12 data points ± S.D.

Fig 2 Activation by p[NH]ppG is inhibited on the delayed addition of the assay components to the preincubation

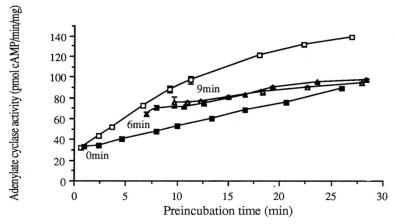

Platelet membranes were preincubated at 30°C with 10μM p[NH]ppG and 1mM Mg^{2+} in the presence or absence (□) of 50μM cAMP, 100μM ATP, 8mM creatine phosphate and 0.3mg/ml creatine kinase (final concentrations). These additions were present from the start of the preincubation (■) or added after 6min (▲) and 9min (△) in a volume that was 5% of the final preincubation volume (the necessary adjustments being made to the p[NH]ppG and Mg^{2+} concentrations). Triplicate samples containing 23μg of protein were withdrawn at 6s intervals to assay the adenylate cyclase activity and results are means ± S.E.M.

in the presence and absence of ATP, cAMP, creatine phosphate and creatine kinase were respectively 0.1690min^{-1} and 0.0926min^{-1}. These additions slowed the activation process and reduced the maximum adenylate cyclase activity observed during the time period of the experiment. Experiments were next carried out to investigate whether the activation of adenylate cyclase could still be inhibited on the delayed addition of the assay components to a preincubation. Addition of ATP, cAMP, creatine kinase and creatine phosphate blocked subsequent activation when added after 6 or 9 minutes of preincubation (Fig 2).

Finally the effect of the assay components on a maximully activated adenylate cyclase complex was investigated. Membranes preincubated with p[NH]ppG and Mg^{2+} until peak adenylate cyclase activity was observed were subsequently diluted into a solution containing ATP, cAMP, creatine kinase and creatine phosphate. The effect on the adenylate cyclase activity is shown in Fig 3. The assay components caused a slow decline in the adenylate cyclase activity, indicating that the process of activation could be reversed.

Fig 3 Activation of platelet adenylate cyclase by p[NH]ppG is reversed in the presence of the assay components

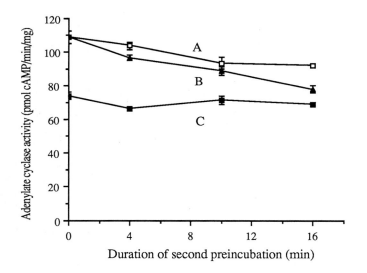

Two preincubations of platelet membranes (0.76mg/ml) were set up at 30°C in the presence of 10μM p[NH]ppG and 1mM Mg^{2+} with and without 100μM ATP, 50μM cAMP, 0.3mg/ml creatine phosphate and 8mM creatine kinase. After 25min membranes were diluted into a second preincubation (0.38mg/ml). Those initially preincubated with only p[NH]ppG and Mg^{2+} were diluted into p[NH]ppG and Mg^{2+} in the presence (B, ▲) and absence (A, □) of ATP, cAMP, creatine phosphate and creatine kinase (concentrations as above). Membranes from the initial preincubation in the presence of p[NH]ppG, Mg^{2+} and the assay components were diluted into a solution of the same (C, ■). Triplicate samples containing 19μg of membrane protein were withdrawn at 6s intervals for adenylate cyclase assay and values are means ± S.E.M.

Discussion

The activation kinetics of adenylate cyclase were quite different in the presence and absence of cAMP, ATP, creatine phosphate and creatine kinase. The first order rate constants were $0.1690min^{-1}$ and $0.0926min^{-1}$ respectively. The slowness of these reactions may explain why similar effects in other tissues have not been widely recognised. However gradual changes in G_s activity have been observed in rat liver plasma membranes (Wong & Martin, 1986).

The assay conditions support the activity of protein kinases present in the membrane preparation. These results are consistent with the idea that the activation of adenylate cyclase may be due to a de-phosphorylation event, and the inhibition observed with cAMP, ATP, creatine phosphate and creatine kinase is a result of phosphorylation, possibly of G_s by protein kinase A (Farndale et al. 1987). However the precise identity of the phosphorylation substrate is the subject of investigation in this laboratory.

References

Farndale RW, Wong SKF, Martin BR (1987) Activation of adenylate cyclase in human platelet membranes by guanosine 5'-[βγ-imido]triphosphate is inhibited by cyclic-AMP-dependent phosphorylation. Biochem J 242:637-643

Iyengar R, Birnbaumer L (1981) Hysteretic activation of adenylyl cyclases. J Biol Chem 256:11036-11041

Katada T, Gilman AG, Watanabe Y, Baller S, Jakobs KH, (1985) Protein kinase C phosphorylates the inhibitory guanine-nucleotide-binding regulatory component and apparently suppresses its function in hormonal inhibition of adenylate cyclase. Eur J Biochem 151:431-437

Krupinski J, Coussen F, Bakalyar HA, Tang WJ, Feinstein PG, Orth K, Slaughter C, Reed RR, Gilman AG (1989) Adenylyl cyclase amino acid sequence: possible channel- or transporter-like structure. Science 244:1558-1564

Pyne NJ, Murphy GJ, Milligan G, Houslay MD (1989) Treatment of intact hepatocytes with either the phorbol ester TPA or glucagon elicits the phosphorylation and functional inactivation of the inhibitory guanine nucleotide regulatory protein G_i. FEBS Lett 243:77-82

Salomon Y, Londos C, Rodbell M, (1974) A highly sensitive adenylate cyclase assay. Anal Biochem 58:541-548

Sibley DR, Benovic JL, Caron MG, Levkowitz RJ (1987) Regulation of transmembrane signalling by receptor phosphorylation. Cell 48:913-922

Stein JM, Martin BR (1983) The role of GTP in prostaglandin E_1 stimulation of adenylate cyclase in platelet membranes. Biochem J 214:231-234

Wong SKF, Martin BR (1986) Activation of rat liver adenylate cyclase by guanosine 5'-[βγ-imido]triphosphate and glucagon. Biochem J 233:845-851

VI.   PROTEIN TYROSINE PHOSPHORYLATION

REGULATION OF TYROSINE KINASES BY TYROSINE PHOSPHORYLATION

Jonathan A. Cooper, Alasdair MacAuley, and Andrius Kazlauskas
Fred Hutchinson Cancer Research Center
1124 Columbia St
Seattle, WA 98104
U. S. A.

Protein kinases can be classified according to whether they phosphorylate phenolic (tyrosine) or aliphatic (serine and threonine) hydroxyl groups (Hunter and Cooper, 1985). Although there are now a few examples of kinases that seemingly break the rule (Howell et al., 1990), most tyrosine kinases differ from the serine/threonine kinases in their primary amino acid sequences. Consequently, as open reading frames in cDNA clones have been sequenced it has been relatively easy to predict whether they might encode tyrosine kinases (Hanks et al., 1988). This has lead to a rapid expansion of the tyrosine kinase family.

The tyrosine kinases can be divided into two major groups, according to whether they straddle the cell membrane or are contained within the cell (Hunter and Cooper, 1985). Those that cross the membrane have their amino-termini outside and carboxy-terminal kinase domains inside. In many cases these proteins are receptors for extracellular ligands, and binding of ligand stimulates phosphorylation by the receptor kinase domain (Yarden and Ullrich, 1988). Probably, ligand binding induces dimerization, bringing the kinase domains of two receptors into proximity (Schlessinger, 1988). This permits intermolecular autophosphorylation and causes activation of the catalytic domains. In many cases, phosphorylation of the intercellular portion of the receptor is important for stimulation of catalytic activity towards other substrates, presumably via allosteric effects. Phosphorylation may also serve other functions, as described below.

The intracellular tyrosine kinases fall into several sequence families. Since these kinases do not have an extracellular ligand binding domain, it is not evident how they are regulated. One family, comprising Src and 7 close relatives in mammals (there are additional members of the family in invertebrates), appears to be regulated by phosphorylation (Cooper, 1990). What regulates phosphorylation is less clear. One of the Src family, Lck, forms a tight complex with the cytoplasmic tails of either of two cell surface proteins, CD4 and CD8 (Eiseman and Bolen, 1990). The CD4.Lck and CD8.Lck complexes are analogous to transmembrane receptors. Like authentic receptor tyrosine kinases, inducing dimerization, by cross-linking CD4, causes changes in Lck phosphorylation state and kinase activity. However, whether this is the normal mechanism of Lck regulation is unsure.

NATO ASI Series, Vol. H 56
Cellular Regulation by Protein Phosphorylation
Edited by L. M. G. Heilmeyer, Jr.
© Springer-Verlag Berlin Heidelberg 1991

In this review, we discuss two examples of tyrosine kinase regulation by phosphorylation, chicken Src and the human PDGF receptor (β subunit), in which phosphorylation of specific residues regulates either activity or signal transduction.

Regulation of Src Kinase Activity by Phosphorylation

Carboxyterminal to the kinase domain of Src lies a short stretch of residues that is highly conserved between members of the Src family (at least in vertebrates; invertebrate Src-like kinases show less conservation)(Cooper, 1990). Eleven residues from the kinase domain, and 4-6 residues from the extreme carboxy terminus of the protein, there is a conserved tyrosine residue (Figure 1A). This tyrosine is phosphorylated, by unidentified tyrosine kinases, when the Src protein is expressed in fibroblasts (Cooper et al., 1986). Lck is also phosphorylated on the corresponding tyrosine when expressed artificially in fibroblasts (not its usual environment) or naturally in T lymphocytes. Mutation of the codon for the carboxyterminal tyrosine to phenylalanine activates Src and all other Src-family kinases tested, as shown by increased tyrosine phosphorylation of cell proteins and malignant transformation (Hunter, 1987; Cooper, 1990). In vitro, dephosphorylation of the Src terminal tyrosine, Tyr527, stimulates the activity of the kinase domain. Therefore, it seems that phosphorylation of the terminal tyrosine inhibits Src, and probably other Src-family kinases.

Establishing the relevance of this phosphorylation to physiological regulation of Src activity requires finding conditions where Src is activated, and assessing whether the phosphorylation state is reduced. Other than mutations in Src, 3 conditions are known to activate Src. In polyoma virus-transformed cells, Src (and its relatives Fyn and Yes) bind to the medium T antigen encoded by polyoma. The bound Src molecules are activated and dephosphorylated at Tyr527 (Cartwright et al., 1986). At mitosis, Src becomes phosphorylated near the amino terminus at one serine and 2 threonine residues, and is more active (Chakalaparampil and Shalloway, 1988). Although the novel phosphorylations could explain the increased activity, experiments with mutant Src molecules suggest that a reduction in phosphorylation of Tyr527 could also be important. In PDGF-treated cells, a subpopulation of Src molecules associates with the activated PDGF receptor (Kypta et al., 1990). There is also a small increase in net Src kinase activity (Gould and Hunter, 1988). Evidence so far suggests that activation may result from novel serine and tyrosine phosphorylations near the amino terminus, but does not exclude the possibility that a small population of Src molecules may be dephosphorylated at Tyr527.

In the case of Lck, activation by cross-linking of CD4 or CD8 causes changes in phosphorylation state. Phosphorylation at the terminal tyrosine, Tyr505, may be incomplete in T lymphocytes (Luo and Sefton, 1990), and cross-linking of CD4 causes little change in Tyr505

A. Residues found in C-termini of vertebrate members of the Src family

516	517	518	519	520	521	522	523	524	525	526	527	528	529	530	531	532	533
Leu	Glu	Asp	Phe	Phe	Thr	Ala	Thr	Glu	Pro	Gln	Tyr	Gln	Pro	Gln	Pro	ter	ter
		Asp		Tyr	Tyr		Ser	Ala		Gly		Glu	Gln	Gly	Asp	Asn	Leu
								Ser				Leu			Glu	Gln	Thr

B. Residues found in non-transforming mutants of Src

518	519	520	521	522	523	524	525	526	527	528	529
Asp	Tyr	Phe	Thr	Ser	Thr	Glu	Pro	Gln	Tyr	Gln	Pro
Phe	Trp	Ser	Cys	Ser	Asp	Gly	Glu			**Trp**	**Arg**
Ser	**Ser**	**Ile**	Gly	**Ile**		**Ala**	**Gly**			**Lys**	**Leu**
	Cys	**Pro**	Ala			**Ser**	**Leu**			**His**	**Ser**
			Asn			**His**					
						Gln					
						Arg					

Figure 1. (A) Compilation of C-terminal sequences of vertebrate members of Src family (Src, Fyn, Yes, Fgr, Lck, Hck, Lyn, Blk). The numbering of residues is for chicken Src, which terminates at residue 533. Other family members terminate at the homologous residue or 2 residues upstream (ter). (B) Residues found in non-transforming mutants of Src. Codons for residues 518-526, 528 and 529 were subjected to random mutagenesis, mutant proteins were expressed in Rat2 fibroblasts, and non-transforming mutants sequenced (MacAuley and Cooper, 1990). Differences from the Src sequence (top line) are listed. Boldface type: residues not encountered in Src family members (see part A).

phosphorylation. Instead, there is increased phosphorylation of Tyr394, in the kinase domain (Luo and Sefton, 1990). This is probably important for activation (Abraham and Veillette, 1990). In Src, the corresponding tyrosine, Tyr416, is phosphorylated in all forms of Src that are activated by either mutation or by agents like polyoma virus, mitosis, or PDGF, but is not phosphorylated in inactive Src. *In vitro* studies on the importance of this phosphorylation are hard to do, because dephosphorylated molecules become phosphorylated at Tyr416 during *in vitro* assay. Results with phenylalanine mutants suggest this tyrosine is important for maximal activity (Kmiecik et al., 1988).

In summary, there are two tyrosine residues implicated in regulating Src-family kinases. The carboxyterminal tyrosine can inhibit activity when phosphorylated. The kinase domain tyrosine can enhance activity when phosphorylated. Currently, it seems unlikely that phosphorylation of either residue is due to intramolecular autophosphorylation. Instead, experiments with kinase-inactive mutants show that phosphorylation at either position can be catalyzed in trans, by other kinases (Cooper and MacAuley, 1988). Src catalyzes intermolecular phosphorylation (autophosphorylation) primarily at Tyr416, but also detectably at Tyr527. In fibroblasts, Tyr527 is extensively phosphorylated, and Tyr 416 hardly phosphorylated at all. Even so, the phosphorylation of Tyr527 in fibroblasts could be catalyzed by Src if cellular factors affect the specificity.

We and others have made unproductive attempts to detect kinase activities capable of phosphorylating Tyr527 in cell lysates. Recently one laboratory has succeeded in purifying a kinase with apparent specificity for Tyr527, from neonatal rat brain (Okada and Nakagawa, 1990). Presently it is unclear whether this kinase is expressed in fibroblasts, and it is possible that other kinases, or Src itself, maintains Tyr527 in a phosphorylated state in fibroblasts. Even inefficient phosphorylation could allow high levels of net phosphorylation if dephosphorylation is slow. Unfortunately, the slow turnover of the ATP pool prohibits the use of labeling of intact cells for accurate analysis of phosphate turnover rate .

Recently, we have tried to evaluate the sequence specificity of the kinases that phosphorylate Src Tyr527 in Rat2 fibroblasts, by expressing mutant molecules in which Tyr527 is presented in a different sequence context (MacAuley and Cooper, 1990). Mutants that do not cause morphological transformation exhibit low *in vitro* kinase activity and are highly phosphorylated at Tyr527. A summary of substitutions found in normal mutants shows greater diversity than found in the known Src family (Figure 1B). At most of the residues subjected to mutagenesis, non-conservative substitutions are compatible with normal phosphorylation and regulation. The exceptions are Asp518 and Glu524, suggesting that these acidic residues may be essential for phosphorylation or regulation. Interpretation of the sequences of transforming (i.e. deregulated) mutants is complicated because most of the mutants contain more than one substitution and it is not possible to ascribe the activated phenotype to a single changed residue. However, several of the transforming mutants contain alterations at 518 or 524 in conjunction with conservative substitutions elsewhere, so we are reasonably sure that 518 and 524 are key residues. One of the transforming mutants has Ser522 changed to Arg as the only mutation, suggesting that this residue, although replaceable by some non-conservative residues (Figure 1B), cannot be replaced by Arg without perturbing phosphorylation or regulation.

The phosphorylation states of some of the mutants have been assessed by ^{32}P labeling and peptide mapping (MacAuley and Cooper, 1990). Generally, non-acidic substitutions at Glu524 greatly reduce the phosphorylation of Tyr527 in Rat2 cells. Transforming mutations at Asp518 or Ser522 reduce phosphorylation to intermediate levels. These mutations begin to define the sequence requirements for phosphorylation in fibroblasts. Decreased phosphorylation could result from reduced recognition by kinases, or increased dephosphorylation by phosphatases.

We are starting to test whether the mutant Src molecules are phosphorylated to different extents by known kinases. Initially, we are testing autophosphorylation by Src. This can be done by expressing the mutants in yeast. Src is phosphorylated at Tyr527 in yeast, as it is in fibroblasts, although the stoichiometry is lower. In yeast, unlike fibroblasts, a kinase-inactive mutant is not tyrosine phosphorylated, so all the Tyr527 phosphorylation seems to be catalyzed by Src (Cooper and MacAuley, 1988). Therefore, the relative levels of phosphorylation of the mutants in yeast will indicate their relative autophosphorylation ability. So far, mutants with low phosphorylation in fibroblasts are scarcely phosphorylated at all in yeast, and high phosphorylation in fibroblasts corresponds to wild-type levels of phosphorylation in yeast (MacAuley, unpublished). The simplest interpretation is that the kinase that phosphorylates Tyr527 in Rat2 cells has the same substrate requirements that Src itself has. This does not prove that Tyr527 phosphorylation in fibroblasts is catalyzed by Src, since other kinases may have the same specificity. We will try to address this question in the future.

Regulation of the PDGF Receptor by Phosphorylation

PDGF stimulates phosphorylation of its receptor at 2 or more tyrosines (Kazlauskas and Cooper, 1989). The major site is Tyr857, which corresponds to Tyr416 of Src and Tyr394 of Lck. A minor site is at Tyr751, which lies in a sequence in the kinase domain that is not found in most tyrosine kinases and is called a "kinase insert". In vitro, immuno-precipitated PDGF receptors phosphorylate at Tyr751, so this is an autophosphorylation site. Whether Tyr751 phosphorylation in PDGF-treated cells is catalyzed by the receptor or by another kinase is not known. Tyr857 is a lesser autophosphorylation site in vitro. As with Src Tyr416 and 527, it is not clear whether the different ratios of phosphorylation of PDGF receptor Tyr751 and 857 in vivo and in vitro is due to other kinases, phosphatases, or environmental factors.

Phosphorylation of Tyr857 may be important for stimulation of kinase activity by PDGF (Kazlauskas, unpublished). Receptors immunoprecipitated from PDGF-treated cells, which have increased phosphorylation of Tyr857, have increased in vitro activity towards added substrates. Mutant molecules, with Phe at 857, do not show this increase. Phe751 mutants do. Also, cells expressing Phe857 mutant receptors show little increase in tyrosine phosphorylation of PDGF receptor substrates (GAP and PLCγ, see below) in response to PDGF (Figure 2). Thus, like Tyr416 in Src and Tyr394 in Lck, phosphorylation of the homologous Tyr857 in the PDGF receptor may increase activity.

Phosphorylation of Tyr751 seems to have quite different effects. Wild-type PDGF receptors contain PDGF-inducible binding sites for specific intracellular proteins. Thus PDGF causes the binding of phosphatidyl-inositol 3 kinase (PI3 kinase), phospholipase C (PLC) γ, Ras GTPase activating protein (GAP) and Src to the PDGF receptor (Kaplan et al., 1987; Kazlauskas and Cooper, 1989; Morrison et al., 1990; Kazlauskas et al., 1990; Kypta et al., 1990). The binding of PI3 kinase requires both a kinase-active receptor and Tyr at 751 (neither Gly nor Phe mutants can bind), and correlates with phosphorylation at Tyr751 (Figure 2; Kazlauskas and Cooper, 1989). Tyrosine is not required at residue 857. One model is

	DNA synthesis	PI kinase binding	GAP binding	GAP P.Tyr	PLCγ P.Tyr
WT	+ +	+ +	+ +	+ +	+ +
R635	-	-	-	-	-
F751	+	-	+	+ +	+ +
F857	+	+ +/+	+	-	-
F751/857	-	-	+	-	-

Figure 2. Properties of mutant forms of the human PDGF receptor β subunit, expressed in dog epithelial TRMP cells. WT: wild-type. R635: a kinase-inactive mutant. F751, F857 and F751/857: single and double mutants with Tyr to Phe mutations. For details see: Kazlauskas and Cooper, 1989; Kazlauskas et al., 1990; and Kazlauskas, unpublished.

that phosphorylation at Tyr751 creates a binding site for PI3 kinase. An alternative model is that phosphorylation of Tyr751 causes a conformation change that exposes a binding site elsewhere. The requirement for phosphorylation has been confirmed by *in vitro* studies using immunoprecipitated PDGF receptors and adding cell lysates as a source of PI3 kinase (Kazlauskas and Cooper, 1990). The receptors have to be phosphorylated to permit binding of PI3 kinase. Preincubation with ATP, but not with βγimido ATP, allows phosphorylation at Tyr751 and subsequent PI3 kinase binding. Phe751 mutant receptors do not work in this assay. The assay also suggests that PI3 kinase does not have to be phosphorylated by the receptor in order to bind, because PI3 kinase from lysates of cells not treated with PDGF can bind to receptors under conditions that do not allow phosphorylation.

Experiments with kinase-inactive receptors suggest that the binding of PLCγ, GAP and Src also requires phosphorylation. Mutation of either Tyr751 or 857, or both, reduces GAP binding, although not to the extent of a kinase-inactivating mutation (Figure 2; Kazlauskas et al., 1990). This is difficult to interpret in terms of the simple, single-site model proposed for PI3 kinase. One possibility is that Tyr751 is needed to form a binding site and that Tyr857 is needed for full kinase activity, to allow phosphorylation of GAP. In this model, phosphorylation of both Tyr751 and GAP would be needed for binding. If phosphorylation of GAP is needed, then a simple *in vitro* binding assay, described above for PI3 kinase, would not work for GAP. Very recently we have been able to detect GAP binding to *in vitro*-phosphorylated PDGF receptors (Kazlauskas, unpublished). It is not clear how well this *in vitro* system approximates the situation in the cell. The binding is weak and requires large amounts of GAP (synthesized using Baculovirus vectors) to drive the binding. In this assay, phosphorylation of the receptor is required but GAP phosphorylation is not. Given that *in vitro* phosphorylated PDGF receptors contain little phosphate at Tyr857, it is likely that phosphorylation of this residue is not needed. Possibly the Phe857 mutation inhibits binding *in vivo* by altering the conformation of the receptor. This will be addressed in future experiments.

GAP, PLCγ, PI3 kinase and Src have a shared sequence element, the SH2 domain (Pawson, 1988). Three laboratories have evidence that SH2 domains have an affinity for tyrosine phosphorylated sequences. For example, a bacterial fusion protein containing just the SH2 domains of GAP or PLCγ can bind to phosphorylated EGF receptors in a cell lysate (Anderson et al., 1990). Also, SDS-denatured tyrosine-phosphorylated medium T antigen binds to the 85 K component of PI3 kinase immobilized on a transfer membrane (Cohen et al., 1990). Our mutagenesis data suggest that the specific sequence context of the phosphotyrosine will be important, because PI3 kinase does not bind to PDGF receptors phosphorylated at Tyr857 and not Tyr751 (Kazlauskas and Cooper, 1989).

What is the significance of binding of PI3 kinase, PLCγ and GAP to activated receptors? The binding of PI3 kinase seems to be quantitative. PI3 kinase is a soluble enzyme, but when bound to the receptor it is membrane-associated. This would tend to bring the enzyme close to its substrate phosphoinositides. In fact, the products of PI3 kinase, $PI3,4P_2$ and $PI3,4,5P_3$, accumulate in the membranes of PDGF-treated cells but are undetectable in the membranes of control cells (Auger et al., 1989). Increased phosphorylation may result from the translocation of PI3 kinase or from an actual change in enzymatic activity. PLCγ is also a soluble enzyme, and after PDGF or EGF treatment becomes bound to the membrane via the cognate receptor. Growth factors stimulate $PI4,5P_2$ cleavage, evinced by decreases in $PI4,5P_2$ and increases in $1,4,5IP_3$ and diacylglycerol. Translocation of PLCγ into proximity with membrane $PI4,5P_2$ may be important for this, although very recent evidence suggests that tyrosine phosphorylation of PLCγ may be sufficient to stimulate its activity in $vitro$ (Nishibe et al., 1990).

In the case of GAP, interpretation is complicated by uncertainty regarding the true function of GAP (McCormick, 1990). GAP stimulates the hydrolysis of GTP bound to Ras, and GAP may also be an effector of Ras. PDGF-induced translocation to the membrane could bring GAP close to Ras.GTP, so increased hydrolysis of GTP bound to Ras might be expected. The situation is complicated, however, because some of the products of PI turnover, including diacylglycerol and phosphatidic acid, inhibit GAP in $vitro$ (Tsai et al., 1990; Yu et al., 1990). In fact, in the cell, PDGF causes an increase in the concentration of Ras.GTP relative to Ras.GDP (Satoh et al., 1990). Whether this is due to inhibition of GAP is not known. Any model to account for these observations has to allow for the apparently low stoichiometries of GAP tyrosine phosphorylation and of association with the membrane via the PDGF receptor.

<u>Summary</u>

Protein tyrosine kinases are themselves regulated by tyrosine phosphorylation. Src and its relatives are inhibited by carboxy-terminal phosphorylation by presently-unidentified tyrosine kinases. The specificity of these inhibitory kinases resembles that of Src itself, and one possibility is that Src undergoes an autoinhibitory self-phosphorylation. Src, the PDGF receptor, and many other tyrosine kinases are also subject to phosphorylation at a conserved tyrosine residue in the kinase domain. Phosphorylation of this residue correlates well with activity, and is probably important for full enzymatic activity. Phosphorylation in the kinase insert of the PDGF receptor creates a binding site for PI3 kinase. Binding of GAP, PLCγ and Src to activated PDGF receptors may similarly involve interactions with phosphotyrosine residues in the kinase insert or elsewhere in the PDGF receptor. These binding reactions may be important for translocating soluble signal transduction proteins to the membrane.

<u>References</u>

Abraham N, Veilette A. Activation of p56^{lck} through mutation of a regulatory carboxy-terminal tyrosine residue requires intact sites of autophosphorylation and myristylation. <u>Mol. Cell. Biol.</u> 1990; 10:5197-5206.

Anderson D, Koch CA, Grey L, Ellis C, Moran MF, Pawson T. Binding of SH2 domains of phospholipase Cγ1, GAP, and Src to activated growth factor receptors. <u>Science</u> 1990; 250:979-982.

Auger KR, Serunian LA, Soltoff SP, Libby P, Cantley LC. PDGF-dependent tyrosine phosphorylation stimulates production of novel polyphosphoinositides in intact cells. <u>Cell</u> 1989; 57:167-175.

Cartwright CA, Kaplan PL, Cooper JA, Hunter T, Eckhart W. Altered sites of tyrosine phosphorylation in pp60^{c-src} associated with polyoma middle tumor antigen. <u>Mol. Cell. Biol.</u> 1986; 6:1562-1570.

Chakalaparampil I, Shalloway D. Altered phosphorylation and activation of pp60^{c-src} during fibroblast mitosis. <u>Cell</u> 1988; 52:801-810.

Cohen B, Yoakim M, Piwnica-Worms H, Roberts TM, Schaffhausen BS. Tyrosine phosphorylation is a signal for the trafficking of pp85, an 85-kDa phosphorylated polypeptide associated with phosphatidylinositol kinase activity. <u>Proc. Natl. Acad. Sci. USA</u> 1990; 87:4458-4462.

Cooper JA, Gould KL, Cartwright CA, Hunter T. Tyr^{527} is phosphorylated in pp60^{c-src}: implications for regulation. <u>Science</u> 1986; 231:1431-1434.

Cooper JA, MacAuley A. Potential positive and negative autoregulation of p60^{c-src} by intermolecular autophosphorylation. <u>Proc. Natl. Acad. Sci. USA</u> 1988; 85:4232-4236.

Cooper JA. The src-family of protein-tyrosine kinases, in Kemp B (ed) Peptides and Protein Phosphorylation. CRC Press Inc, Boca Raton, FL 1990, pp85-113.

Eiseman E, Bolen JB. src-related tyrosine protein kinases as signaling components in hematopoietic cells. <u>Cancer Cells</u> 1990; 2:303-310.

Gould KL, Hunter T. Platelet-derived growth factor induced multisite phosphorylation of pp60^{c-src} and increases its protein-tyrosine kinase activity. <u>Mol. Cell. Biol.</u> 1988; 8:3345-3356.

Hanks SK, Quinn AM, Hunter T. The protein kinase family: conserved features and deduced phylogeny of the catalytic domains. <u>Science</u> 1988;

241:42-52.

Howell, BW, Lew J, Douville EMJ, Icely PLE, Gray DA, Bell JC. STY: a developmentally regulated kinase cloned from p19 embryonal carcinoma cells. Mol. Cell Biol. 1990; in press.

Hunter T, Cooper JA. Protein-tyrosine kinases. Annu. Rev. Biochem. 1985; 54:897-930.

Hunter T. A tail of two src's: mutatis mutandis. Cell 1987; 49:1-4.

Kaplan DR, Whitman M, Schaffhausen B, Pallas DC, White M, Cantley L, Roberts TM. Common elements in growth factor stimulation and oncogenic transformation: 85 kd phosphoprotein and phosphatidylinositol kinase activity. Cell 1987; 50:1021-1029.

Kazlauskas A, Cooper JA. Autophosphorylation of the PDGF receptor in the kinase insert region regulates interactions with cell proteins. Cell 1989; 58:1121-1133.

Kazlauskas A, Cooper JA. Phosphorylation of the PDGF receptor Ã subunit creates a tight binding site for phosphatidylinositol 3 kinase. The EMBO J. 1990; 9:3279-3286.

Kazlauskas A, Ellis C, Pawson T, Cooper JA. Binding of GAP to activated PDGF receptors. Science 1990; 247:1578-1581.

Kmiecik TE, Johnson PJ, Shalloway D. Regulation by the autophosphorylation site in overexpressed $pp60^{c-src}$. Mol. Cell. Biol. 1988; 8:4541

Kypta RM, Goldberg Y, Ulug ET, Courtneidge SA. Association between the PDGF receptor and members of the src family of tyrosine kinases. Cell 1990; 62:481-492.

Luo K, Sefton BM. Cross-linking of T-cell surface molecules CD4 and CD8 stimulates phosphorylation of the lck tyrosine protein kinase at the autophosphorylation site. Mol. Cell. Biol. 1990; 10:5305-5313.

MacAuley A, Cooper JA. Acidic residues at the carboxyl terminus of $p60^{c-src}$ are required for regulation of tyrosine kinase activity and transformation. The New Biologist 1990; 2:828-840.

McCormick F. The world according to GAP. Oncogene 1990; 5:1281-1283.

Morrison DK, Kaplan DR, Rhee SG, Williams LT. Platelet-derived growth factor (PDGF)-dependent association of phospholipase C-Ó with the PDGF receptor signaling complex. Mol. Cell. Biol. 1990; 10:2359-2366.

Nishibe S, Wahl MI, Hernandez-Sotomayor SMT, Tonks NK, Rhee SG, Carpenter G. Increase in catalytic activity of phospholipase C-γ1 by tyrosine phosphorylation. Science 1990; 250:1253-1256.

Okada M, Nakagawa H. A protein-tyrosine kinase involved in regulation of $pp60^{c-src}$ function. J. Biol. Chem. 1989; 264:20886-20893.

Pawson T. Non-catalytic domains of cytoplasmic protein-tyrosine kinases: regulatory elements in signal transduction. Oncogene 1988; 3:491-495.

Satoh T, Endo M, Nakafuku M, Akiyama T, Yamamoto T, Kaziro Y. Accumulation of $p21^{ras}$·GTP in response to stimulation with epidermal growth factor and oncogene products with tyrosine kinase activity. Proc. Natl. Acad. Sci. USA 1990; 87:7926-7929.

Schlessinger J. Signal transduction by allosteric receptor oligomerization. <u>TIBS</u> 1988; 13:443-447.

Tsai M-H, Yu C-L, Stacey DW. A cytoplasmic protein inhibits the GTPase activity of H-Ras in a phospholipid-dependent manner. <u>Science</u> 1990; 250:982-985.

Yarden Y, Ullrich A. Growth factor receptor tyrosine kinases. <u>Annu. Rev. Biochem.</u> 1988; 57:443-478.

Yu C-L, Tsai M-H, Stacey DW. Serum stimulation of NIH 3T3 cells induces the production of lipids able to inhibit GTPase-activating protein activity. <u>Mol. Cell. Biol.</u> 1990; 10:6683-6689.

pp75: A NOVEL TYROSINE PHOSPHORYLATED PROTEIN THAT HERALDS DIFFERENTIATION OF HL-60 CELLS

Ilana Bushkin[+], Jesse Roth[†], Daphna Heffetz[+] and Yehiel Zick[+] [+]Department of Chemical Immunology, The Weizmann Institute of Science, Rehovot 76100, Israel and [†]The Diabetes Branch, NIDDK, National Institutes of Health, Bethesda, MD 20892, USA.

INTRODUCTION

The promyelocytic (HL-60) leukemia cells are an excellent model system to study cellular differentiation since they undergo morphological changes in response to various differentiation agents which result in a phenotype with the characteristics of a monocyte/macrophage or a granulocyte (Collins, 1987). Alterations in protein tyrosine phosphorylation were implicated as playing an important role in the induction and maintenance of the differentiated phenotype of HL-60 cells (Frank and Sartorelli, 1988) as several specific oncogene products which possess protein tyrosine kinase (PTK) activity are expressed during early stages of myeloid differentiation (Sariban et al. 1985; Gee et al. 1986; Yu and Glazer, 1987). Hence, Potential substrates for these PTKs could play an important role in regulating the differentiation process. To observe changes in phosphotyrosine metabolism at the substrates level, HL-60 cells were treated with a combination of H_2O_2 and vanadate to inhibit the intracellular protein tyrosine phosphatase (PTPase) activity (Heffetz et al. 1990). Such treatment enabled us to enhance markedly protein tyrosine phosphorylation and detect a novel 75 kDa protein that undergoes enhanced tyrosine phosphorylation during early stages of differentiation of these cells.

RESULTS and DISCUSSION

Protein Tyrosine Phosphorylation During Differentiation of HL-60 Cells -

Immunoblotting of total HL-60 cell extracts with antibodies directed against P-Tyr residues has been employed to detect proteins that undergo preferential tyrosine phosphorylation in the course of differentiation of HL-60 cells into monocytes, granulocytes and macrophages. These differentiation pathways were induced upon incubation of HL-60 cells with 0.5 μM 1:25(OH)$_2$VitD$_3$, 0.5 mM Bt$_2$cAMP, or their combination, respectively (Mangelsdorf et al. 1984; Collins, 1987; Bushkin and Zick, 1990). We found that a 20 min. incubation of HL-60 cells with a combination of 2 mM H_2O_2 and 0.1 mM vanadate to inhibit intracellular PTPase activity enabled us to detect several proteins that underwent enhanced tyrosine phosphorylation.

In undifferentiated HL-60 cells, H_2O_2/vanadate augmented the phosphorylation of a single protein, pp53, that was considered to be a differentiation-independent tyrosine phosphorylated protein (Fig. 1, lane A). In HL-60 cells, treated with 1:25(OH)$_2$VitD$_3$ for 3 days, several additional phosphotyrosine-containing proteins (e.g. pp75, pp88, pp95, pp115) were found to undergo H_2O_2/vanadate-stimulated tyrosine phosphorylation (Fig. 1, lane F). The major one was pp75 whose phosphorylation was already increased 2 hr after induction, reached maximal levels on day three and remained unaltered during days 3-6 of the differentiation process. By contrast, the extent of differentiation, assessed by monitoring the reduction of intracellular nitroblue tetrazolium (NBT) into dark blue formazan deposit, caused by superoxide (O_2-) production in the differentiated pheno-type (Chaplinski and Niedel, 1982) was not detected earlier than 24 hours, and lagged behind through the course of differentiation (Fig. 1).   In Bt_2cAMP-treated HL-60 cells, induced to differentiate into granulocytes, we detected an initial increase in the P-Tyr content of pp75, which was fol-lowed by a dramatic reduction in its P-Tyr content through day two and its complete disappearance on day three (not shown).   Phosphorylation of pp75 occurred exclusively on tyrosine residues, as immunoblotting with anti P-Tyr antibodies was inhibited by 10 μM P-Tyr, but not by 1 mM P-Ser or P-Thr (not shown).

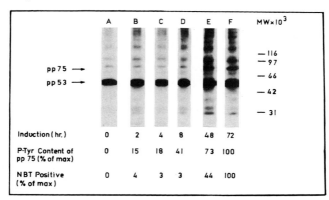

Fig. 1:   Time course of the effect of $1:25(OH)_2VitD_3$ on the induction of NBT positive cells and on increased P-Tyr content of pp75.   HL-60 cells ($0.5x10^6$/ml) were maintained uninduced or were induced with 0.25 μM $1:25(OH)_2VitD_3$ for the indicated hours.   NBT positive cells   were scored for deposits of reduced black formazan.   Phosphotyrosine content of pp75   (in H_2O_2/vanadate-treated cells) was determined by immunoblotting of cell extracts with anti P-Tyr antibodies.   The bands corresponding to pp75 were scanned by a densitometer, and the intensity of phosphorylated pp75 was plotted as the percent of maximum effect observed with $1:25(OH)_2VitD_3$.

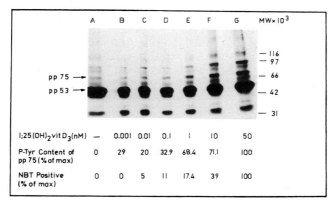

Fig. 2:   Effect of increasing concentrations of $1:25(OH)_2VitD_3$ on the induction of NBT positive cells and increased P-Tyr content of pp75. HL-60 cells were maintained uninduced or were induced to differentiate with the indicated concentration of $1:25(OH)_2VitD_3$ for three days.   NBT positive cells   and phosphotyrosine content of pp75   were determined as described in the Legend to Figure 1.

When HL-60 cells were treated with $1:25(OH)_2VitD_3$ for three days, the fraction of cells that scored a positive reaction with NBT increased in a dose-dependent manner, with half maximal effect detected in cells treated with $5x10^{-8}$ M $1:25(OH)_2VitD_3$ (Fig. 2). In contrast, the increase of the differentiation marker pp75 was evident at much lower concentrations with maximal effects already detected at $2x10^{-9}$ M $1:25(OH)_2VitD_3$. Hence, the induction of rapid tyrosine phosphorylation of pp75 emerges as a prominent early event of the response of the cells to differentiation agents. pp75 phosphorylation is not simply due to the growth arrest that accompanies terminal differentiation since inhibitors of cellular growth, such as adriamycin, fail to stimulate tyrosine phosphorylation of this protein (not shown). pp75 could also be labeled in cells that are more advanced in their differentiation state like the human leukemia cells U-937 and terminally differentiated monocyte-derived macrophages from peripheral blood. These findings suggest that the phosphorylation of pp75 is not a specific response of malignant cells induced to differentiate, but appears to be an integral part of the normally differentiated phenotype.

Subcellular Localization of pp75 - To examine the subcellular localization of pp75, cytosolic and particulate fractions from undifferentiated and $1:25(OH)_2VitD_3$-induced HL-60 cells were prepared. In the absence of $1:25(OH)_2VitD_3$ pp75 was barely detected in the cytosol or particulate fractions (Table I). Induction of differentiation by $1:25(OH)_2VitD_3$ increased the P-Tyr content of pp75 in the cytosol fraction of the differentiated cells, but no significant amounts of tyrosine phosphorylated pp75 could be detected in the particulate fraction. These data suggest that pp75 in HL-60 cells is localized mainly in the cytosol. It should be noted that unlike pp75, pp53 was localized both to the cytosolic and the particulate fractions.

Table I: Subcellular localization of pp75 in HL-60 cells. Undifferentiated HL-60 cells or cells induced to differentiate, for three days with 0.25 μM $1:25(OH)_2VitD_3$ were incubated at 37°C for 20 min with 2 mM H_2O_2 and 0.1 mM vanadate. At the end of incubation cells were homogenized, and fractionated into cytosol and particulate fractions. Immunoblotting with anti P-Tyr antibodies was carried out and the P-Tyr content of pp53 and pp75 was determined by scanning densitometry.

	Densitometry (arbitrary Units)			
Treatment	Cytoplasmic		Particulate	
$1;25(OH)_2vitD_3$	pp53	pp75	pp53	pp75
-	46.6	2.3	7.6	1.3
+	33.9	7.9	11.1	1.0

In vitro tyrosine phosphorylation of pp75 - To exclude the possibility that treatment with H_2O_2 and vanadate may activate tyrosine kinases with relaxed substrate specificity, we made an attempt to determine whether pp75 could be tyrosine phosphorylated _in vitro_ by its corresponding kinase in an H_2O_2/vanadate-independent manner. As seen in Fig. 3, incubation of cytosol derived from uninduced (control) HL-60 cells, with divalent cations and ATP had no effect on protein tyrosine phosphorylation of pp75. In contrast, cytosol derived from $1:25(OH)_2VitD_3$-induced HL-60 cells possessed a protein tyrosine kinase which phosphorylated pp75 in a time dependent manner. **Tyrosine phosphorylation of pp53 could already be detected in cytosols** derived from control cells, however its phosphorylation was markedly increased in cytosols derived from $1:25(OH)_2VitD_3$ treated cells. These data indicate that pp53 and pp75, as well as the enzymatic activities that catalyze their phosphorylation, are localized, at least in part, in the cytosol.

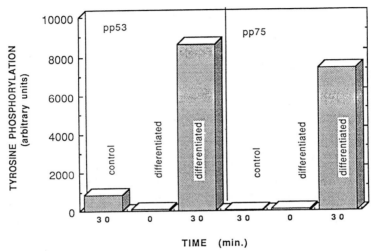

Fig. 3: **In vitro phosphorylation of pp75 in HL-60 cytosolic fraction.** HL-60 cells were maintained uninduced (control) or were induced for 3 days with 0.25 μM $1:25(OH)_2VitD_3$ (differentiated), washed twice in RPMI and frozen in liquid nitrogen. Cytosols were obtained and phosphorylated in vitro in the presence of 0.25 mM ATP, 2 mM $MnCl_2$ and 10 mM Mg acetate. The phosphorylated proteins were resolved by means of SDS-PAGE. Immunoblotting with anti P-Tyr antibodies and determination of the P-Tyr content of pp53 and pp75 were carried out as described in the Legend to Fig. 1.

These findings raise the intriguing possibility that pp75 may be a tyrosine kinase capable of undergoing autophosphorylation. This hypothesis is supported by the fact that HL-60 cells contain a cytosolic tyrosine kinase of 70 kDa (Kraft and Berkow, 1987). Although only slight changes in this enzyme activity were reported during differentiation into retinoic acid-derived granulocytes and TPA-derived macrophages, further studies will be needed to establish the correlation between pp75 described here, and the enzyme previously described.

If indeed pp75 functions as a protein tyrosine kinase, then our results indicate that either its amount or its activity are markedly increased during early stages of both monocytic and granulocytic differentiation. As such, pp75 phosphorylation could function as a 'common differentiation signal', that together with additional lineage-specific signals herald the differentiation of HL-60 cells. Such a conclusion is supported by recent observations that four growth factors that trigger the differentiation of HL-60 cells, G-CSF, GM-CSF, TNF-α and IFN-γ rapidly stimulate a transiet serine and tyrosine phosphorylation of a 75 kDa protein when incubated with permeabilized HL-60 cells in the presence of [γ-^{32}P] ATP (Evans et al. 1990). Since TNF and IFN-γ, like 1:25(OH)$_2$VitD$_3$, promote monocytic differentiation of HL-60 cells, whereas G-CSF and GM-CSF, like Bt$_2$cAMP, induce granulocytic differentiation, these studies support the notion that phosphorylation of pp75 may be associated with the induction of a common differentiation signal both towards the granulocytic and monocytic lineages. The data provided in the present work suggests that the content of pp75 and/or the activity of the kinase that catalyzes its phosphorylation are persistently elevated during the course of monocytic differentiation of HL-60 cells, suggesting that differentiation agents may induce both transient (Evans et al. 1990) and long-term effects (present study) on pp75. Further studies are therefore necessary to unravel the role of pp75 during the acute and chronic phases of the differentiation process.

References

Bushkin I, Zick Y (1990) Alterations in insulin receptor kinase activity during differentiation of HL-60 cells. Biochem Biophys Res Commun 172:676-682

Chaplinski TJ, Niedel JE (1982) Cyclic nucleotide-induced maturation of human promyelocytic leukemia cells. J Clin Invest 70:953-964

Collins SJ (1987) The HL-60 promyelocytic leukemia cell line; proliferation, differentiation, and cellular oncogene expression. Blood 70:1233-1244

Evans J, Mire-Sluis A, Hoffbrand V, Wickremasinghe G (1990) Binding of G-CSF, GM-CSF, tumor necrosis factor-α, and γ-interferon to cell surface receptors on human myeloid leukemia cells triggers rapid tyrosine and serine phosphorylation of a 75-kD protein. Blood 75:88-95

Frank DA and Sartorelli AC (1988) Alterations in tyrosine phosphorylation during the granulocytic maturation of HL-60 leukemia cells. Cancer Res 48:52-58

Gee CE, Griffin J, Sastre L, Miller LJ, Springer TA, Piwnica-Worms H, Roberts TM (1986) Differentiation of myeloid cells is accompanied by increased levels of pp60^{c-src} protein and kinase activity. Proc Natl Acad Sci USA 83:5131-5135

Heffetz D, Bushkin I, Dror R, Zick Y (1990) The insulinomimetic H$_2$O$_2$ and vanadate stimulate protein tyrosine phosphorylation in intact cells. J Biol Chem 265:2896-2902

Kraft AS, Berkow RL (1987) Tyrosine kinase and phosphotyrosine phosphatase activity in human promyelocytic leukemia cells and human polymorphonuclear leukocytes. Blood 70:356-362

Mangelsdorf DJ, Koeffler HP, Donaldson CA, Pike SW, Haussler MR (1984) 1:25-Dihydroxy vitamin D$_3$-induced differentiation in human promyelocytic leukemia cell line (HL-60): receptor-mediated maturation to macrophage-like cell. J Cell Biol 98:391-398

Sariban E, Mitchell T, Kufe D. (1985) Expression of the c-fms proto-oncogene during human monocytic differentiation. Nature 316:64-66

Yu G, Glazer RI (1987) Purification and characterization of p^{93fes} and p^{60src}-related tyrosine protein kinase activities in differentiated HL-60 leukemia cells. J Biol Chem 252:17543-17548

PRELIMINARY BIOCHEMICAL STUDIES OF A DROSOPHILA HOMOLOG OF p60^{c-src}

S.J. Kussick and J.A. Cooper
Fred Hutchinson Cancer Research Center
1124 Columbia Street
Seattle, WA 98104
USA

INTRODUCTION

A number of homologs of mammalian proto-oncogenes have been identified in Drosophila (for review see Shilo, 1987). These include homologs of genes encoding cytoplasmic and transmembrane tyrosine kinases, p21^{ras}, *raf*, and nuclear oncoproteins. Genetic experiments in flies (Henkemeyer, *et al.*, 1987; Bishop and Corces, 1988; van den Heuvel *et al.*, 1989; Elkins *et al.*, 1990) suggest that, rather than regulating cell growth, Drosophila proto-oncogenes may principally be involved in specifying cell morphology or regulating cell-cell interactions during development.

The Drosophila homolog of p60^{c-src} (D*src*) cloned by Hoffman, *et al.* (1983) and by Simon *et al.* (1985) is located at chromosomal region 64B, and encodes a protein (p62^D) whose predicted amino acid sequence is 47% identical to that of chicken p60^{c-src} and equally similar to the other vertebrate *src* family members. The catalytic and SH3/SH2 domains are highly conserved between D*src* and vertebrate *src* family members, while the amino and carboxyl termini of the molecules diverge. A number of residues important for p60^{c-src} function are conserved in p62^D, including the Gly2 myristylation site, the Lys295 ATP binding site, the Tyr416 autophosphorylation site, and the site of negative regulation of enzymatic activity at Tyr527. D*src* is expressed primarily in neural tissue and gut-associated smooth muscle of Drosophila embryos, larvae and pupae, and is rare in adults. We have expressed p62^D in mammalian fibroblasts and begun characterizing p62^D in these cells and in the Drosophila Schneider 2 cell line. Here we report a preliminary comparison of the enzymatic properties of this molecule with those of chicken p60^{c-src}.

RESULTS

Because we hoped to study p62^D in a physiologically relevant setting we tested the Drosophila Schneider 2 cell line (SL2) for endogenous expression of the kinase. Since this cell line was derived from dissociated Drosophila embryos (Schneider, 1972), it is not clear what tissue type(s) it represents. NP40 (1%) lysates of Schneider cells were immunoprecipitated with an antipeptide antiserum to the 13 C-terminal amino acids of p62^D (kindly supplied by M. Simon) or with a non-immune serum, and incubated with [γ^{32}P-ATP] in immune complex kinase assays. SDS-PAGE

NATO ASI Series, Vol. H 56
Cellular Regulation by Protein Phosphorylation
Edited by L. M. G. Heilmeyer, Jr.
© Springer-Verlag Berlin Heidelberg 1991

on a 12.5% gel revealed a doublet running at the same molecular weight as p62^D immunoprecipitated from Rat2 fibroblasts expressing a cDNA encoding p62^D (also supplied by M. Simon) driven by a retroviral long terminal repeat promoter (see Figure 1). Both bands of the Schneider cell doublet contain PTyr, as do the bands from fibroblast immunoprecipitates (the upper bands of each doublet are shown in Figure 2). The immunoprecipitate from Schneider cells phosphorylates a slightly-lysine rich histone preparation (histone 7S, Sigma) on tyrosine *in vitro* (see Figure 1 and 2C), as does the immunoprecipitate from Rat2 cells (phosphoamino acid analysis not shown).

Further evidence that Schneider 2 cells express p62^D comes from two-dimensional tryptic phosphopeptide mapping of Schneider cell or fibroblast immunoprecipitates labelled *in vitro* with γ^{32}P-ATP (see Figure 3). When separated by thin layer electrophoresis at pH 8.9, and by chromatography in a buffer of 37.5% n-butanol, 7.5% acetic acid, 25% pyridine, and 30% water, most of the major tryptic phosphopeptides of the 62 kd proteins from these two cell types comigrate. The few spots not shared by the two samples may be due to different proteins in the two cell types which cross react with

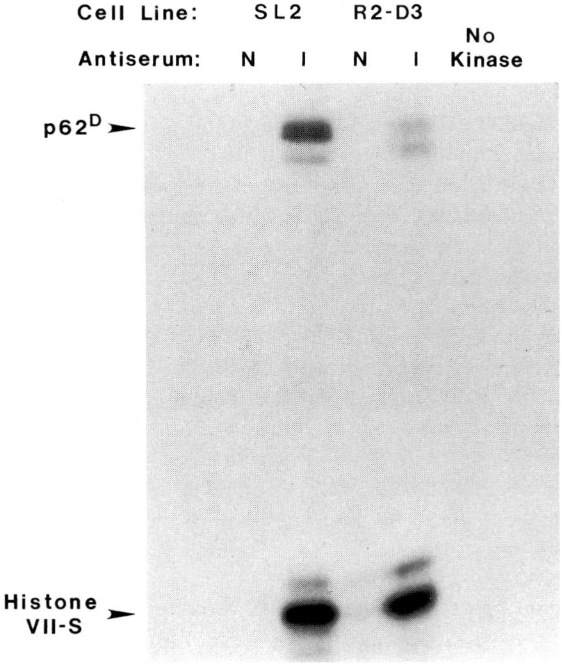

Figure 1: A 62 kd protein can be immunoprecipitated from Schneider cells with an antibody to the C-terminus of p62^D. SL2, Schneider 2 cell line. R2-D3, Rat 2 cells expressing Dsrc cDNA. N, non-immune serum. I, immune serum. "No kinase" denotes substrate histone 7S alone in kinase cocktail. Gel was alkali treated prior to auto-radiography.

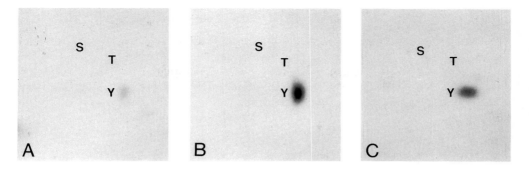

Figure 2: The 62 kd proteins from both Schneider and R2-D3 cells are labelled on tyrosine *in vitro*, as is the substrate histone 7S. Labelled proteins were eluted from an SDS gel, TCA precipitated, hydrolyzed at 110°C for one hour in 5.7 M HCl. Amino acids were separated by electrophoresis in two dimensions (pH 1.9 and 3.5), and identified by ninhydrin stainable phosphoamino acid standards. A, histone 7S labelled by Schneider cell immunoprecipitate; B, p62^D from Schneider cells; C, p62^D from Rat 2 cells. S, phosphoserine position; T, phosphothreonine position; Y, phosphotyrosine position.

the polyclonal antibody and are labelled *in vitro*. Phosphoamino acid analysis of the principal spots of the maps from fibroblast p62^D revealed that all contain phosphotyrosine. Since multiple spots are also seen when *in vitro* labelled p62^D is digested with chymotrypsin or thermolysin (data not shown), the data suggest that p62^D, unlike p60^{c-src}, autophosphorylates on multiple tyrosine *in vitro*. The vast majority of p60^{c-src} autophosphorylation *in vitro* occurs at a single residue: tyrosine 416 (Smart *et al.*, 1981).

An additional difference between p62^D and p60^{c-src} is the failure of the Drosophila enzyme to be stimulated by antibody binding to its C-terminus (data not shown), while C-terminal antibody binding to p60^{c-src} can stimulate its kinase activity approximately ten-fold, presumably by preventing the C-terminus from inhibiting kinase activity (Cooper and King, 1986). In this experiment p62^D immunoprecipitated from Schneider cells with an antiserum against its N-terminus was then incubated with buffer alone, with the antipeptide antiserum against the C-terminus of p62^D, or with the N-terminal antiserum. There was no difference in autophosphorylation of p62^D or in histone 7S phosphorylation under these three different conditions, suggesting that C-terminal antibody binding does not stimulate p62^D kinase activity *in vitro*.

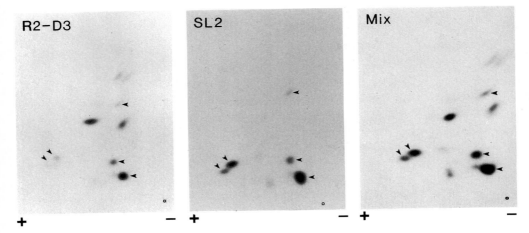

Figure 3: Two dimensional tryptic phosphopeptide maps of p62 immunoprecipitated from Schneider cells and fibroblasts are very similar, suggesting the protein in Schneider cells is the product of the 64B locus. Proteins were labelled *in vitro*, eluted from an SDS gel, TCA precipitated, and digested exhaustively with trypsin. Peptides were separated in two dimensions as described in the text. Small circle at the lower right denotes the origin.

CONCLUSION

As part of our initial characterization of p62^D, we have shown that the molecule is expressed in Schneider cells. Because the phosphopeptide maps of p62^D precipitated from Schneider cells and from rat fibroblasts expressing a cDNA clone of D*src* are very similar, we conclude that the protein in Schneider cells is very probably the product of the cloned mRNA from D*src* 64B. p62^D has tyrosine kinase activity whether immunoprecipitated from Schneider cells or from mammalian fibroblasts expressing a cDNA encoding p62^D. p62^D thus behaves like vertebrate p60^{c-src} in its ability to autophosphorylate on tyrosine *in vitro*. However, p62^D appears to behave differently from p60^{c-src} in several ways: 1) it phosphorylates an exogenous, histone VIIS, which p60^{c-src} recognizes poorly *in vitro* (p60^{c-src} data not shown); 2) it autophosphorylates on multiple tyrosines *in vitro*; 3) its kinase activity *in vitro* is not stimulated by C-terminal antibody binding. We are continuing our investigation of the functional regulation of p62^D by making substitution mutations at the p62^D homologs of amino acids known to be crucial to p60^{c-src} function: Lys295 and Tyr527. Future studies of these constructs expressed in Schneider cells and in transgenic flies should provide additional insight into the function of p62^D.

REFERENCES

Bishop JG, Corces VG. (1988) Expression of an activated *ras* gene causes developmental abnormalities in transgenic Drosophila melanogaster. Genes Dev. 2:567-577.

Cooper JA, King CS. (1986) Dephosphorylation or antibody binding to the carboxy terminus stimulates pp60^{c-src}. Mol. Cell. Biol. 6:4467-4477.

Elkins T, Zinn K, McAllister L, *et al*. (1990) Genetic analysis of a Drosophila neural cell adhesion molecule: interaction of fasciclin I and Abelson tyrosine kinase mutations. Cell 60:565-570.

Henkemeyer MJ, Gertler FB, Goodman W, Hoffmann FM. (1987) The Drosophila Abelson proto-oncogene homolog: Identification of mutant alleles that have pleiotropic effects in late development. Cell 51:821-826.

Hoffmann FH, Fresco LD, Hoffmann-Falk H, Shilo B-Z. (1983) Nucleotide sequences of the Drosophila *src* and *abl* homologs: Conservation and variability in the *src* family oncogenes. Cell 35:393-401.

Schneider I. (1972) Cell lines derived from late embryonic stages of *Drosophila melanogaster*. J Embryol Exp Morph 27:353-65.

Shilo BZ. (1987) Proto-oncogenes in Drosophila melanogaster. Trends in Genetics 3:69-72.

Simon MA, Drees B, Kornberg T, Bishop JM. (1985) The nucleotide sequence and the tissue-specific expression of Drosophila c-*src*. Cell 42:831-840.

Smart JE, Opperman AP, Czernilofsky AF, *et al*. (1981) Characterization of sites for tyrosine phosphorylation in the transforming protein of Rous sarcoma virus (pp60^{v-src}) and its normal cellular homolog (pp60^{c-src}). Proc Natl Acad Sci USA 78:6013-17.

van den Heuvel, Nusse R, Johnston P, Lawrence PA. (1989) Distribution of the *wingless* gene product in Drosophila embryos: a protein involved in cell-cell communication. Cell 59:739-749.

PHOSPHOINOSITIDE KINASES AND EGF RECEPTOR ACTIVATION IN PLASMA MEMBRANES FROM A431 CELLS

B. Payrastre, M. Plantavid, M. Breton, E. Chambaz[*] and H. Chap
INSERM - Unité 326
Hôpital Purpan
31059 Toulouse Cedex
France

ABSTRACT: When plasma membranes isolated from A431 cells were incubated in the presence of EGF, ^{32}P incorporation from $[\gamma^{32}P]$ ATP in inositol lipids was significantly enhanced. These lipids were identified as PtdIns(4)P and PtdIns(4,5)P$_2$ by H.P.L.C. analysis. Further experiments indicated that both PtdIns 4-kinase and PtdIns(4)P 5-kinase are probably regulated by the EGF-receptor tyrosine kinase. Results concerning the PtdIns 3-kinase are discussed.

INTRODUCTION: There is now growing evidence that polyphosphoinositides display various functions in signal transduction of many growth factors, hormones and neurotransmitters. First, it is well known that phosphatidylinositol 4,5-bisphosphate (PtdIns(4,5)P$_2$) plays a central role as the major substrate of phospholipases C and a source of the two second messengers, inositol 1,4,5-trisphosphate and diacylglycerols (Berridge, 1987 ; Nichizuka, 1988). Second, PtdIns(4,5)P$_2$ by itself probably contributes to the regulation of cytoskeleton reorganisation during cellular activation since specific interactions have been described between this inositol lipid and some cytoskeletal proteins such as profilin and gelsolin (Lassing et Lindberg, 1988 ; Goldschmidt et al., 1990). Two lipid kinases, the phosphatidylinositol (PtdIns) 4-kinase and the phosphatidylinositol 4-phosphate (PtdIns(4)P) 5-kinase, are involved in its synthesis whereas phosphatases and phospholipases C participate to its catabolism. However while phospholipase C activation is well documented, little is known about regulation of the other enzymes, both kinases and phosphatases.

On the other hand, new polyphosphoinositides phosphorylated on the 3-position of the inositol ring, have been recently subject to intense study (Whitman et al., 1985 ; Kaplan et al., 1986). Another lipid kinase, a PtdIns 3-kinase has been identified and now partially purified (Whitman et al., 1987 ; 1988 ; Auger et al., 1989 ; Morgan et al., 1990). This enzyme has been described as associated to the PDGF receptor as well as to the oncogene product, pp60[src]. Its activation might be involved in the mi-

---

Footnote: [*]INSERM - Unité 244, BACE, CENG, BP85X, 38041 Grenoble Cedex, France.

NATO ASI Series, Vol. H 56
Cellular Regulation by Protein Phosphorylation
Edited by L. M. G. Heilmeyer, Jr.
© Springer-Verlag Berlin Heidelberg 1991

togenic signal of PDGF. However, until now, there is no evidence for a precise role of these new inositol phospholipids ; indeed, they are poor substrates of the purified phospholipases C (Serunian et al., 1989).

The EGF receptor, like PDGF receptor, possesses an intrinsic and essential tyrosine protein kinase activity (Carpenter, 1987 ; Yarden et Ullrich, 1988). On the other hand, an increase in PtdIns turn-over has been reported upon EGF treatment (Sawyer et Cohen, 1981). More recently, it has been demonstrated that, at least in cells overexpressing EGF receptor, EGF stimulates PtdIns(4,5)P_2 hydrolysis, by a mechanism probably involving both tyrosine-phosphorylation and translocation of phospholipase Cγ1 (Hepler et al., 1987 ; Wahl et al., 1987 ; Meisenhelder et al., 1989 ; Payrastre et al., 1990 ; Todderud et al., 1990). We report here that EGF is also able to enhance PtdIns(4)P and PtdIns(4,5)P_2 synthesis in plasma membranes of A431 cells, overexpressing EGF receptor. Additional experiments suggest that this stimulation is due to a regulation of PtdIns 4-kinase and PtdIns(4)P 5-kinase activities by the EGF-receptor tyrosine kinase. Furthermore, we could not demonstrate in this model any increase in the 3-phosphorylated inositol phospholipid synthesis although immunoprecipitation studies indicate some connections between EGF receptor and PtdIns 3-kinase.

MATERIALS and METHODS: Plasma membranes from A431 cells were isolated as previously described (Payrastre et al., 1988). The phosphotyrosyl protein phosphatase (PTPase) from A431 membranes was purified according the procedure of Bütler et al. (1989). Lipid kinase assays, H.P.L.C. analysis of inositol lipids and immuno-isolation of phosphotyrosyl proteins were carried out as described by Payrastre et al (Biochem. J., 1990).

RESULTS and DISCUSSION

EGF stimulation of inositol-lipid labelling from [γ-^{32}P]ATP

When plasma membranes from A431 cells were preincubated under precise experimental conditions, in the presence of vanadate and ATP, EGF increased ^{32}P labelling of PtdIns P and PtdIns P_2 from [γ-^{32}P]ATP (Table 1).

Table 1: Stimulation by EGF of phosphoinositide phosphorylation in A431 plasma membranes, from [γ-^{32}P]ATP.

	PtdIns P	PtdIns P$_2$
	(% of control values)	
+ EGF (200 ng/ml)	134 ± 4 (p<0.001, n=6)	164 ± 19 (p<0.05, n=5)

Since, in isolated plasma membranes, phospholipase C activation following EGF addition was very low (Payrastre et al., 1990) and inositol-lipid phosphatase activities were negligible (results not shown), the data presented in Table 1 suggest that PtdIns and PtdIns(P) kinases are possible targets of EGF receptor. Vanadate alone, a potent inhibitor of phosphotyrosyl-protein phosphatases, was also able to enhance, at a less extent, inositol-lipid labelling as described (Payrastre et al., Biochem. J., 1990).

H.P.L.C. Analysis of the [^{32}P] labelled inositol phospholipids

Comparison of the two H.P.L.C. profiles (Fig. 1) indicates that, upon EGF treatment, there was only an increase in PtdIns(4)P and PtdIns(4,5)P$_2$ labelling with only appearance of trace amounts of PtdIns(3,4)P$_2$. Therefore, EGF-receptor activation leads to the stimulation of PtdIns 4-kinase. From these results it is difficult to state if PtdIns(4)P 5-kinase is also stimulated since only a simple increase in PtdIns(4)P might explain the increase in PtdIns(4,5)P$_2$ labelling.

Decrease in inositol-lipid labelling from [γ-^{32}P]ATP after plasma-membrane treatment with a specific PTPase

A431 plasma membranes were incubated first with a purified PTPase either in the presence of vanadate as inhibitor (inhibited PTPase) or in its absence (active PTPase) and lipid kinase activities were then assayed, after adjustment of vanadate concentrations. Table 2 indicates that PtIns(4)P and PtdIns(4,5)P$_2$ labelling was highly decreased when plasma-membrane proteins were previously dephosphorylated by the PTPase. These results emphasize the role of tyrosyl phosphorylation in lipid kinase activation.

Figure 1: <u>H.P.L.C. profiles of the deacylated inositol phospholipids from total lipid extracts.</u>

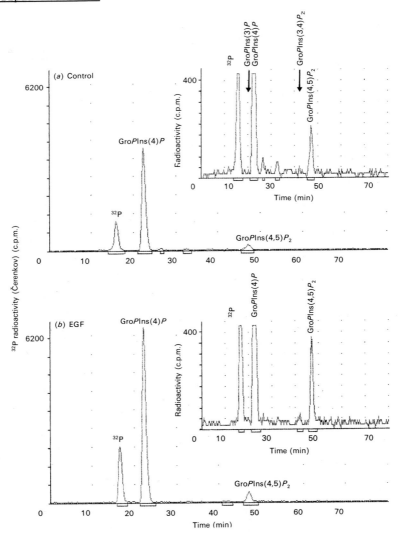

(a) Control plasma membranes and (b) plasma membranes stimulated by 200 ng of EGF/ml (from Payrastre et al., Biochem. J., 1990).

Table 2: Activities of PtdIns 4-kinase and PtdIns(4)P 5-kinase after dephosphorylation of plasma-membrane proteins by a specific PTPase.

Preincubation	PtdIns(4)P	PtdIns(4,5)P$_2$
	p.mol.^{32}P incorporated/mg protein/10 min.	
+ inhibited PTPase	174	5
+ active PTPase	78	1

Immuno-isolation of lipid kinases from A431-cell plasma membranes

Lipid kinase activities were assayed in proteins immunopurified from plasma membranes using a sepharose-linked antiphosphotyrosine antibody. Higher activities were present when plasma membranes were stimulated by EGF and H.P.L.C. analysis indicated that both PtdIns 4- and PtdIns(4P) 5-kinases were immuno-isolated (results not shown). However, a high PtdIns 3-kinase activity was also present.

The immuno-isolation of the three lipid kinases might indicate that they are all regulated by EGF receptor, either by a direct tyrosyl/phosphorylation or by some association with phosphotyrosyl proteins as well as with the autophosphorylated EGF receptor. However, we could not measure any significant increase in PtdIns(3)P or PtdIns(3,4)P$_2$ labelling when plasma membranes were incubated with EGF whereas PtdIns(4)P and PtdIns(4,5)P$_2$ radioactivities were higher. Such a dissociation between lipid kinase regulation and their behaviour towards antiphosphotyrosine antibody remains to be elucidated. Further experiments in whole cells are now in progress.

REFERENCES

Auger, K.R., Serunian, L.A., Soltoff, S.P., Libby, P., Cantley, L.C. (1989) Cell 57: 167-175.

Berridge, M.J. (1987) Annu. Rev. Biochem. 56: 159-193.

Bütler, M.T., Ziemiecki, A., Groner, B., Freis, R.R. (1989) Eur. J. Biochem. 185: 475-483.

Carpenter, G. (1987) Annu. Rev. Biochem. 56: 881-914.

Goldschmidt-Clermont, P.J., Machesky, L.M., Baldassare, J.J., Pollard, T.D. (1990) Science 247: 1575-1578.

Hepler, J.R., Nakahata, N., Lovenberg, T.W., DiGuiseppi, J., Herman, B., Earp, H.S., Harden, T.K. (1987) J. Biol. Chem. 262: 2951-2956.

Kaplan, D.R., Whitman, M., Schaffhausen, B., Raptis, L., Garcea, R.L., Pallas, D., Roberts, T.M., Cantley, L. (1986) Proc. Nat. Acad. Sci. 83: 3624-3628.

Lassing, I., Lindberg, U. (1988) Exp. Cell. Res. 174: 1-15.

Meisenhelder, J., Suh, P.G., Rhee, S.G., Hunter, T. (1989) Cell 57: 1109-1122.

Morgan, S.J., Smith, A.D., Parker, P.J. (1990) European J. Biochemistry 191: 761-767.

Nichizuka, Y. (1988) Nature (London) 334: 661-665.

Payrastre, B., Plantavid, M., Etievan, C., Ribbes, G., Carratero, C., Chap, H., Douste-Blazy, L. (1988) Biochem. Biophys. Acta 939: 355-365.

Payrastre, B., Plantavid, M., Chap, H. (1990) Biochem. Biophys. Acta, in press.

Payrastre, B., Plantavid, M., Breton, M., Chambaz, E.M., Chap, H. (1990) Biochem. J. 272 : 665-670.

Sawyer, S.T., Cohen, S. (1981) Biochemistry 20: 6280-6286.

Serunian, L.A., Haber, M.T., Fukui, T., Kim, J.W., Rhee, S.G., Lowenstein, J.M., Cantley, L.C. (1989) J. Biol. Chem. 264: 17809-17815.

Todderud, G., Wahl, M.I., Rhee, S.G., Carpenter, G. (1990) Science 249: 296-298.

Whitman, M.R., Kaplan, D.R., Schaffhausen, B., Cantley, L., Roberts, T.M. (1985) Nature (London) 315, 239-242.

Whitman, M., Kaplan, D., Roberts, T., Cantley, L. (1987) Biochem. J. 247: 165-174.

Whitman, M., Downes, C.P., Keeler, M., Keller, T., Cantley, L. (1988) Nature 332: 644-646.

Yarden, Y., Ullrich, A. (1988) Ann. Rev. Biochem. 57: 443-478.

PHOSPHORYLATION OF SYNAPTOPHYSIN BY THE c-src ENCODED PROTEIN TYROSIN KINASE pp60c-src

Angelika Barnekow
Dept. Exp. Tumorbiology
University of Muenster
Badestr. 9
D-4400 Muenster
Germany

A. INTRODUCTION

The phosphorylation and dephosphorylation of proteins plays an essential role in the regulation of cellular proliferation and differentiation processes. Within the central nervous system (CNS) protein phosphorylation is considered to be involved in various aspects of neuronal function including neurotransmitter uptake, storage, metabolism and release. A possible regulatory role of tyrosine phosphorylation in neurotransmission is supported by the recent observation that synaptophysin, a major synaptic vesicle protein is phosphorylated by the c-src encoded protooncogene product pp60c-src (Barnekow et al., 1990). Expression of pp60c-src, the first well defined protooncogene product, is developmentally regulated and tissue-specific, with neuronal tissues displaying high amounts of the c-src encoded pp60c-src kinase activity (Barnekow and Bauer, 1984; Schartl and Barnekow, 1984). In the CNS pp60c-src is preferentially expressed in regions characterized by a high content of grey matter and elevated density of nerve terminals (Sudol, 1988). The physiological function of pp60c-src is still unclear and specific target proteins need to be identified in order to obtain a better understanding of the role this protein plays in cellular differentiation processes.

B. METHODS AND RESULTS

Using rat synaptic vesicles purified through the step of chromatography on controlled-pore glass beads, highest amounts of pp60c-src were found in the vesicle fraction, also containing the highest amounts of synaptophysin (p38), a well characterized integral membrane glycoprotein (Hell et al., 1988; Barnekow et al., 1990). Based on the predicted amino acid sequence in conjunction with immunological and biochemical experiments a

NATO ASI Series, Vol. H 56
Cellular Regulation by Protein Phosphorylation
Edited by L. M. G. Heilmeyer, Jr.
© Springer-Verlag Berlin Heidelberg 1991

model for the membrane topology of synaptophysin has been proposed (Johnston et al., 1989). The data predict that the protein traverses the vesicle lipid bilayer four times with both the amino and carboxyterminal ends facing the cytoplasm. Within the carboxyterminal tail, p38 contains 9 potential tyrosine phosphorylation sites (Table 1).

Table 1: SEQUENCE OF THE CYTOPLASMIC DOMAIN OF RAT SYNAPTOPHYSIN

synaptic vesicle membrane -- KE		TGWAAPFMRA	PPGAPEKQPA
PGDAYGDAGY	GQGPGGYGPQ	DSYGPQGGYQ	PDYGQPASGG
GGYGPQGDYG	QQGYGQQGAP	TSFSNQM-COOH -- cytoplasm	

The co-localization and co-purification of pp60^{c-src} and synaptophysin and the tyrosine-rich cytoplasmic tail of p38 suggest a specific interaction of synaptophysin as a potential physiological substrate for the endogenous tyrosine kinase pp60^{c-src} present in synaptic vesicles.

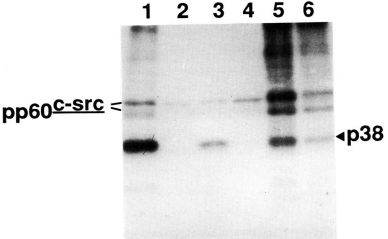

Fig.1: PHOSPHORYLATION OF SYNAPTOPHYSIN _in vitro_ AND IN _intact_ SYNAPTIC VESICLES BY pp60^{c-src}

pp60^{c-src} was immunopurified from c-src transfected 3T3 cell extracts using Mab 327 (Lipsich et al., 1983). Lane 1: 5μg vesicle extract was added to pp60^{c-src} before the addition of 5μCi gamma- ^{32}P-ATP for 10 min. at 25^{0}C and stopped by sample buffer as described earlier(Barnekow et al., 1990). The proteins were separated on a 12.5% polyacrylamide gel and detected by autoradiography; lane 2: same preparation in the absence of vesicle extract; lane 3: 0.1 μg synaptophysin, purified as described in Navone et al., 1986, added to immunopurified pp60^{c-src} before the phosphorylation reaction was started; lane 4: immunopurified pp60^{c-src} added to synaptophysin, immunoprecipitated from 5 μg synaptic vesicle extract by Mab C7.2 (Navone et al., 1986); lane 5: 5 μg intact synaptic vesicles added to immunopurified pp60^{c-src}; lane 6: 5 μg intact synaptic vesicles phosphorylated in the absence of exogenously added pp60^{c-src}.

In vitro, immunoprecipitated pp60^{c-src} phosphorylated p38 in synaptic vesicle extracts (Fig. 1, lane 1). With a preference for ZnCl$_2$ before MnCl$_2$ or MgCl$_2$ the phosphorylation reaction was found to be time - and pH - dependent (Fig. 2, Fig. 3).

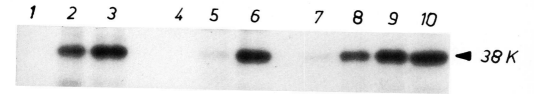

Fig. 2: **ION-AND TIME-DEPENDENT PHOSPHORYLATION OF SYNAPTOPHYSIN BY IMMUNOPURIFIED pp60^{c-src} in vitro**
Phosphorylation of synaptophysin (p38) was performed as described in Fig. 1 in the presence of 1= 50 mM Mg^{2+}; 2 = 50 mM Mn^{2+}; 3 = 50 mM Zn^{2+}; 4 = 0.5 mM Zn^{2+}; 5 = 5 mM Zn^{2+}; 6 = 50 mM Zn^{2+} ; 50 mM Zn^{2+} for 1 min = 7; for 3 min. = 8; for 5 min. = 9; for 10 min. = 10.

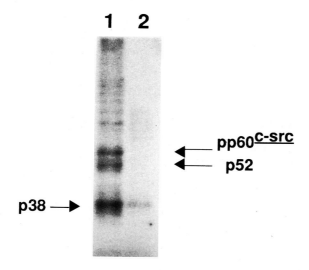

Fig. 3: **pH - DEPENDENT PHOSPHORYLATION OF SYNAPTOPHYSIN BY IMMUNOPURIFIED pp60^{c-src} in vitro**
Phosphorylation of synaptophysin (p38) was performed as described in Fig. 1. Lane 1 = pH 7.5; lane 2 = pH 4.5; p52 = indicates a pp60^{c-src} specific degradation product.

When purified synaptophysin was added to immunopurified pp60^{c-src} a distinct phosphorylation of p38 occured (Fig. 1, lane 3). In the presence of Mab C7.2, a monoclonal antibody against synaptophysin directed against the tyrosine-rich cytoplasmic tail of synaptophysin (Navone et al., 1986), the degree of phosphorylation of p38 decreases drastically (Fig. 1, lane 4). These results suggest a blocking of the potential phosphorylation site(s) in the synaptophysin molecule by Mab C7.2. Incubation of <u>intact</u> synaptic vesicles with gamma-^{32}P-ATP resulted in phosphorylation of synaptophysin (Fig. 1, lane 6). The degree of phosphorylation was further increased in the presence of exogenously added pp60$^{\underline{c-src}}$ (Fig. 1, lane 5).

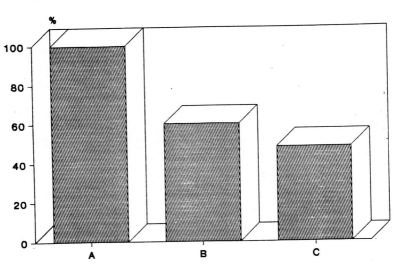

Fig. 4: **PHOSPHORYLATION OF SYNAPTOPHYSIN IN SYNAPTIC VESICLE EXTRACTS AFTER PREVIOUS ELIMINATION OF pp60$^{\underline{c-src}}$**
A: synaptic vesicle extract was phosphorylated in the presence of gamma-^{32}P-ATP.
B + C: pp60$^{\underline{c-src}}$ was eliminated from synaptic vesicle extracts using Mab 327 (B) or GD11(C)[Parsons et al., 1984)] before the extract was phosphorylated. Quantitation was performed by densitometric scanning of the ^{32}P-labeled band of synaptophysin on the autoradiographs. The data were normalized to the radioactivity found in the presence of pp60$^{\underline{c-src}}$ (100 %).

To prove the direct interaction of pp60$^{\underline{c-src}}$ and synaptophysin, synaptic vesicle extracts were incubated either with Mab 327 or Mab GD11 to remove the vesicular-bound pp60$^{\underline{c-src}}$. After separation of the pp60$^{\underline{c-src}}$ immuncomplexes from the incubation mixture, the phosphorylation reaction was started. As can be seen from Fig. 4, block B and C a significant reduction in synaptophysin phosphorylation can be observed compared to the sample shown in Fig. 4, block A, from which pp60$^{\underline{c-src}}$ has not been removed.

2-dimensional phosphoamino acid analyses of synaptophysin phosphorylated by pp60$^{\underline{c-src}}$ <u>in vitro</u>, in intact synaptic vesicles or after removal of pp60$^{\underline{c-src}}$ from vesicle extracts gave the following results. Synaptophysin is phosphorylated by pp60$^{\underline{c-src}}$ <u>in vitro</u> exclusively in tyrosine residues. In intact vesicles, synaptophysin reveals a major phosphorylation of tyrosine residues, most probably within the tyrosine-rich carboxyterminal tail, and a minor phosphorylation of serine. In the absence of pp60$^{c\text{-}src}$ revealed only the serine - but no significant tyrosine-specific phosphorylation.

C. CONCLUSION

pp60$^{c\text{-}src}$ expressed in neurons phosphorylates synaptophysin in synaptic vesicle extracts and in intact synaptic vesicles. Using highly purified synaptic vesicles from rat brain, we present evidence that synaptophysin, a major constituent of the synaptic vesicle membrane protein which is thought to play a key role in the exocytosis of small synaptic vesicles and possibly small clear vesicles in neuroendocrine cells, is a substrate for the endogenous protein tyrosine kinase pp60$^{c\text{-}src}$. The specific interaction of the kinase and the substrate synaptophysin suggests a possible regulatory role for this tyrosine kinase in signal transduction and intercellular communication.

ACKNOWLEDGEMENTS

This work was performed in collaboration with R. Jahn, MPI f. Psychiatrie and M. Schartl, Genzentrum MPI f. Biochemie, Martinsried, FRG. I thank E. Ossendorf for excellent technical assistance, J. Brugge and S. Parsons for the monoclonal antibodies Mab 327 and GD11 and D. Shalloway for the c-src transfected NIH3T3 cells. This work was supported by a grant from Deutsche Forschungsgemeinschaft (Ba 876/1-1 to A.B.).

REFERENCES

Barnekow A, Bauer H (1984) The differential expression of the cellular src gene product pp60$^{c\text{-}src}$ and its phosphokinase activity in normal chicken cells and tissues. Biochim Biophys Acta 782: 94-102

Barnekow A, Jahn R, Schartl M (1990) Synaptophysin: a substrate for the protein tyrosine kinase pp60$^{c\text{-}src}$ in intact synaptic vesicles. Oncogene 5: 1019-1024

Hell JW, Maycox PR, Stadtler H, Jahn R (1988) Uptake of GABA by rat brain synaptic vesicles isolated by a new procedure. EMBO J 7: 3023-3029

Lipsich LA, Lewis AJ, Brugge JS (1983) Isolation of monoclonal antibodies that recognize the transforming proteins of avian sarcoma viruses. J Virol 48: 352-360

Navone F, Jahn R, Di Gioia G, Stukenbrok H, Greengard P, De Camilli P (1986) Protein p38: an integral membrane protein specific of small clear vesicles of neurons and neuroendocrine cells. J Cell Biol 103: 2511-2527

Parsons SJ, McCarley DJ, Ely CM, Benjamin DC, Parsons T (1984) Monoclonal antibodies to the Rous sarcoma virus pp60$^{c\text{-}src}$ react with enzymatically active cellular pp60$^{c\text{-}src}$ of avian and mammalian origin. J Virol 51: 272-282

Schartl M, Barnekow A (1984) Differential expression of the celluar src gene during vertebrate development. Develop Biol 105: 415-422

Sudol M (1988) Expression of proto-oncogenes in neural tissues. Brain Res Rev 13: 391-403

ALTERED THYMOCYTE DEVELOPMENT INDUCED BY AUGMENTED EXPRESSION OF p56^{lck}

K.M. Abraham, S.D. Levin, J.D. Marth[*], K.A. Forbush, and R.M. Perlmutter
Howard Hughes Medical Institute
Departments of Immunology, Medicine and Biochemistry
University of Washington School of Medicine
Seattle, WA 98195 USA

ABSTRACT

The *lck* gene encodes a lymphocyte-specific membrane associated protein tyrosine kinase (p56^{lck}) that is implicated in T cell signal transduction by virtue of its physical association with CD4 and CD8 coreceptor molecules. To examine the role of this tyrosine kinase in thymocyte development, transgenic animals were produced which overexpress either wild-type p56^{lck}, or an activated form of the *lck* kinase (p56^{lckF505}) under the control of the *lck* proximal promoter. The primary defect observed in *lck* transgenic animals is an alteration in normal thymopoiesis. In addition, animals expressing high levels of *lck* transgenes exhibit rapid thymoma development. These results suggest that regulation of p56^{lck} activity is a critical feature of normal thymocyte development, and genetic mechanisms that are capable of altering this endogenous level of *lck* activity can lead to oncogenic transformation.

INTRODUCTION

The *lck* gene was originally identified as a result of its overexpression in the transformed lymphoid cell line LSTRA (Marth et al., 1985). The observation that *lck* mRNA and its protein product (p56^{lck}), are modulated in response to lymphocyte stimulation (Marth et al., 1987), and the subsequent identification of its physical association with CD4 and CD8 lymphocyte coreceptors (Rudd et al., 1988; Veillette et al., 1988) have led to the hypothesis that p56^{lck} regulates lymphocyte signaling events leading to cellular activation (Perlmutter 1989). In order to investigate the functional role of p56^{lck} during T cell development, we have overexpressed p56^{lck} specifically in thymocytes using transgenic technology. The phenotypes of the transgenic animals obtained are detailed below.

***

[*]Biomedical Research Centre, University of British Columbia, Vancouver, B.C., Canada

NATO ASI Series, Vol. H 56
Cellular Regulation by Protein Phosphorylation
Edited by L. M. G. Heilmeyer, Jr.
© Springer-Verlag Berlin Heidelberg 1991

RESULTS

The pLGF and pLGY transgene constructs used to overexpress p56^{lck} contain 11.2kb of murine genomic sequence including 1.0kb 5' of the proximal transcription start site. A portion of exon 12 sequence was obtained from the murine *lck* cDNA encoding either the wild type protein with tyrosine at position 505 (the pLGY construct), or a mutant form with phenylalanine replacing tyrosine at this position (pLGF). The polyadenylation signal for these constructs is provided by 3' sequences obtained from the human growth hormone gene (hGH). Thus, the majority of the transgene sequence is derived from the normal genomic sequence of *lck*, in effect introducing additional *lck* alleles into the mouse germline.

pLGF/pLGY Transgenes

Figure 1. Expression constructs used to overexpress p56^{lck} in thymocytes of transgenic mice.

Northern blot analysis of RNA obtained from tissues of animals bearing the pLGF transgene illustrates that transgene-encoded transcripts accumulate to highest levels in the thymus, but are not detectable in either non-lymphoid tissues such as the heart, kidney or liver, or in peripheral lymphoid tissues (A). In addition, thymocytes obtained from tumor-bearing pLGY and pLGF transgenic mice also contain increased levels of p56^{lck} protein when analyzed in *lck*-specific immunoblots (B).

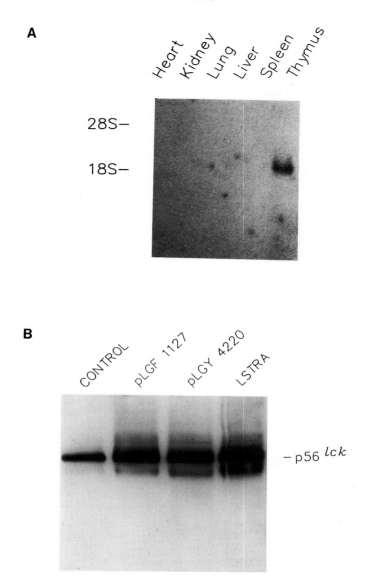

Figure 2. (A) Ten μg of total RNA was recovered from the indicated tissues, separated on formaldehyde-agarose gels, blotted and subsequently analyzed for transgene expression by hybridization with an hGH probe. Arrows indicate the migration of eukaryotic ribosomal RNAs. (B) 50μg of thymocyte whole cell lysates from pLGF or pLGY tumors, a littermate control, or the LSTRA cell line were analyzed by immunoblotting using p56^{lck}-specific antiserum. Reproduced from Abraham et al., 1991.

During normal development, thymocytes acquire surface expression of the T cell antigen receptor and the associated chains of the CD3 complex (reviewed in Clevers et al., 1988). Animals bearing the pLGF transgene invariably exhibit abnormalities in lymphocyte maturation as evidenced by an inability to produce thymocytes bearing this TCR/CD3 complex. In addition, thymocyte development can be monitored by the precisely timed sequential acquisition of CD8, CD4, and CD3 surface markers during fetal life (reviewed in von Boehmer, 1988). Analysis of pLGF and pLGY transgenics indicates that the normal progression of thymocytes through these developmental stages is dramatically delayed. Thus, it is likely that *lck* may play an important role during thymocyte development.

Accordingly, animals expressing extremely high levels of the pLGF transgene exhibit a profound arrest in thymocyte differentiation, and subsequently develop thymic tumors of an immature phenotype, being $CD3^-$, and $CD4^-CD8^{lo}$ or $CD4^-CD8^-$.

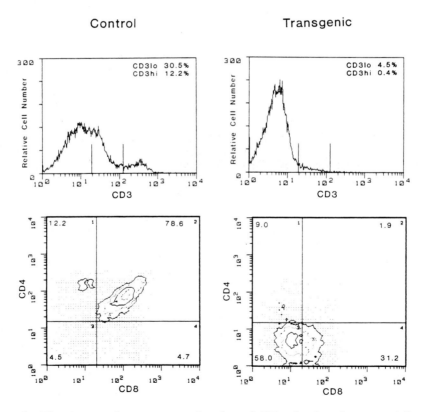

Figure 3. Thymocytes from a tumor-bearing pLGF animal and a normal littermate control were isolated, stained for surface CD3, CD4 and CD8 molecules and subsequently analyzed by flow cytometry.

CONCLUSIONS

1. Overexpression of *lck* in the thymocytes of transgenic animals can result in thymoma development.

2. Transgenic animals overexpressing either wild-type or F505 mutant forms of p56^{lck} in the thymus exhibit a defect in the production of mature CD3+ thymocytes.

3. This defective development, characteristic of *lck* overexpression, can in part be attributed to a delay in the kinetics of normal thymopoiesis.

REFERENCES

Abraham, K.M., Levin, S.D., Marth J.D., Forbush, K.A., and Perlmutter, R.M. (1991). Thymic tumorigenesis induced by overexpression of p56^{lck}. Manuscript submitted.

Clevers, H., Alarcon, B., Wileman, R., and Terhorst, C., (1988). The T cell receptor/CD3 complex: A dynamic protein ensemble. Ann Rev. Immunol. 6: 629-662.

Marth, J.D., Peet, R., Krebs, E.G., and Perlmutter, R.M. (1985). A lymphocyte-specific protein tyrosine kinase is rearranged and overexpressed in the murine T cell lymphoma LSTRA. Cell 43: 393-404.

Marth, J.D., Lewis, D.B., Wilson, C.B, Gearn, M.E., Krebs, E.G., and Perlmutter, R.M. (1987). Regulation of pp56*lck* during T-cell activation: functional implications for the *src*-like protein kinases. EMBO J 6: 2727-2734.

Perlmutter, R.M. (1989). T cell signalling. Science 245: 345.

Rudd, C.E., Trevillyan, J.N., Wong, L.L., Dasgupta, J.D., and Schlossman, S.F. (1988). The CD4 receptor is complexed to a T-cell specific tyrosine kinase (pp58) in detergent lysates from human T lymphocytes. Proc. Natl. Acad. Sci. USA 85: 5190-5194.

Veillette, A., Bookman, M.A., Horak, E.M., and Bolen, J.B. (1988). The CD4 and CD8 T cell surface antigens are associated with the internal membrane tyrosine-protein kinase p56^{lck}. Cell 55: 301-308.

von Boehmer, H. (1988). The developmental biology of T lymphoccytes. Ann. Rev. Immunol. 6: 309-326.

VII.   PROTEIN PHOSPHATASES

REGULATION AND REGULATORY ROLE OF THE INACTIVE ATP,Mg-DEPENDENT PROTEIN PHOSPHATASE (PP-1I)

J.R. Vandenheede, P. Agostinis and J. Van Lint
Afdeling Biochemie, Fakulteit der Geneeskunde
Katholieke Universiteit te Leuven
Campus Gasthuisberg
B-3000 Leuven
Belgium

Introduction

The ATP,Mg-dependent protein phosphatase family comprizes a number of enzymes with a wide species and tissue distribution (Ballou LM, Fischer EH, 1989; Cohen P, 1989; Vandenheede et al., 1989). All forms isolated contain the same 38 kDa catalytic subunit whose activity can be inhibited by two heat-stable proteins named inhibitor-1 and inhibitor-2. This inhibition has been used to identify this catalytic subunit in higher molecular weight complexes that contain phosphorylase phosphatase activity and to label these as type-1 enzymes. The 38 kDa catalytic subunit is therefore also referred to as PP-1C.

The _in vivo_ function of anyone member of this phosphatase family seems to be well defined, since the potential activity of the catalytic subunit is intricately controlled by regulatory proteins which determine its substrate specificity as well as its cellular localisation (Cohen P, 1989; Dent et al., 1990).

One particular characteristic of PP-1C is that it can be made to interconvert to an inactive conformation upon incubation with inhibitor-2. This inactivation reaction is specific for inhibitor-2 and is a time dependent process, not related to the phenomenon of (instantaneous) inhibition mentioned above. The inactive [PP-1.Inhibitor-2] produced in this way is remarkably similar to the inactive ATP,Mg-dependent phosphatase purified from rabbit skeletal muscle and other sources, which was isolated as a complex of an inactive catalytic subunit (F_C) and a

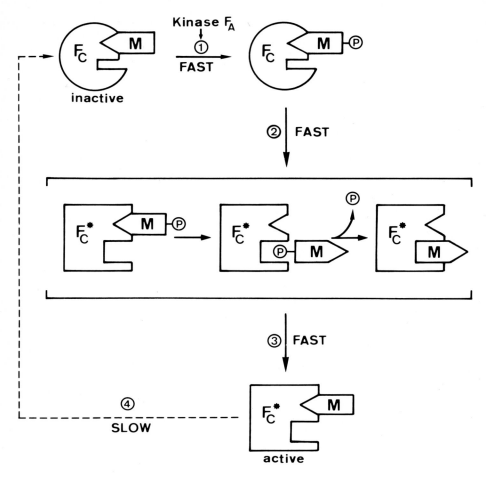

Figure 1. Reversible activation of the F_CM-enzyme
F_C : catalytic subunit in its inactive conformation
F^*_C: catalytic subunit in its active conformation
(identical to PP-1C)
M : modulator subunit (identical to inhibitor-2)

modulator subunit (M). Both the reconstituted [PP-1.Inhibitor-2] complex and the purified F_CM-enzyme are activated in a kinase dependent process which does not dissociate the two subunits (Vandenheede et al., 1989). The modulator protein has been identified as inhibitor-2 and the inactive F_C-catalytic subunit, after activation, has a substrate specificity identical to PP-1C.

The reversible activation of the rabbit skeletal muscle enzyme and its potential role in metabolism has been the subject of intensive studies by several research groups, and will be the main topic of this report.

The reversible activation of the inactive ATP,Mg-dependent phosphatase (F_CM)

The inactive F_CM-enzymes purified from rabbit skeletal muscle contains stoichiometric amounts of 38 kDa inactive catalytic subunit and 32 kDa modulator subunit.

The activation of the enzyme is a multistep process initiated by the kinase F_A-mediated phosphorylation of the M-subunit (Fig. 1, step 1). The activating kinase is identical to glycogen synthase kinase 3 (GSK-3).

The phospho-modulator induces a conformational change in the F_C-protein which activates the enzyme (step 2). The phospho-enzyme then undergoes an autocatalytic dephosphorylation which allows for the binding of exogeneous protein substrates such as phosphorylase a (step 3). In the absence of the kinase F_A and ATP,Mg, the active phosphatase complex reverts back the original, inactive conformation (t ½ = 20 min)(step 4).

The kinase F_A-mediated activation of the enzyme does not dissociate the modulator subunit, and yet, the active enzyme can be inhibited by the addition of exogeneous free modulator (inhibitor-2) protein. This indicates that one catalytic subunit may be able to bind two molecules of M. The time dependent inactivation of the free catalytic subunit by the modulator protein occurs at concentrations of M which do not cause an instantaneous inhibition of the enzyme activity. This suggests that the formation of inactive F_CM-enzyme reflects the binding of one M at a "high affinity" binding site on the catalytic subunit, and that inhibition of the active enzyme is the result of M attaching itself at a different, low-affinity site.

The fact that the activated enzyme auto-dephosphorylates suggests that the phospho-modulator is bound at the active site of the catalytic subunit. The obligatory dephosphorylation step could therefore simply mean that the modulator has to be removed out of the active site to allow for the binding of exogeneous protein substrates. Since the inactivation process specifically involves the "high affinity" binding site, it would be likely that the slow but spontaneous reversal of the kinase F_A-mediated activation is the direct result of an intramolecular translocation of the modulator subunit after its dephosphorylation at the active center to this "high affinity" site. This would then imply that the reversible activation of the F_CM-enzyme is made possible by the transient removal of the modulator from its "high affinity" site by making it a substrate for the phosphatase. It is conceivable that the active center of the enzyme constitutes the "low affinity" inhibitor-site for the modulator protein.

The fast auto-dephosphorylation step can explain why there is no correlation between the level of phosphate incorporated in the modulator and the phosphatase activation level reached.

Multisite phosphorylation of the modulator protein

The crucial role of the kinase F_A as the activating enzyme for the inactive ATP,Mg-dependent phosphatase is illustrated in Figure 1. Phosphorylation of the F_CM-enzymes occurs exclusively on threonine-72 of the modulator subunit, and this phosphorylation is the first step in the _in vitro_ activation of the phosphatase.

One disturbing observation has come from studies on the _in vivo_ phosphorylation state of the modulator protein, which showed that although the protein is phosphorylated at multiple serine residues (ser-86, -120 and -121), no phosphate could be detected at threonine-72 (Holmes et al., 1987). This would either suggest that the inactive ATP,Mg-dependent phosphatase is not activated _in vivo_, or argue that the enzyme is present in the cell exclusively under the activated dephospho-form (Fig. 1). As will be substantiated later in this report, we favor the second explanation.

Since at least three different serine residues were found to be phosphorylated _in vivo_, one has been searching for other kinases that will phosphorylate the modulator protein. Earlier studies on the kinetics of phosphorylation of proteins by kinase F_A (GSK-3) had clearly shown that prior phosphorylation of its potential substrates by casein kinase-2 (CK-2) makes these proteins better targets for kinase F_A (GSK-3) mediated phosphorylation (Picton et al., 1982; DePaoli-Roach et al., 1983). Soon thereafter, A.DePaoli-Roach reported on the synergistic _in vitro_ phosphorylation of the modulator protein by casein kinase-2 and kinase F_A (DePaoli-Roach AA, 1984). It was observed that CK-2 by itself did not activate the F_CM-enzyme, but it phosphorylated the M-subunit, making the phosphatase a better substrate for subsequent activation by kinase F_A. Since CK-2 has been implicated in quite a number of developmental processes such as protein synthesis, cell differentiation and even tumor promotion, its stimulatory role in the kinase F_A-mediated activation of the phosphatase may suggest the involvement of the F_CM-enzyme in these cellular events.

Casein kinase-2 phosphorylates _in vitro_ serines-86, -120 and -121 of the modulator (Holmes et al., 1986) which are also the sites phosphorylated _in vivo_. However, whereas serine-86 is the most extensively phosphorylated site _in vivo_, its _in vitro_ rate of phosphorylation by CK-2 is rather slow. This leaves open the possibility that the _in vivo_ phosphorylation of serine-86 is not exclusively catalyzed by CK-2.

Another likely candidate for phosphorylating this serine would be casein kinase-1, which has a substrate specificity similar to CK-2, and indeed, a casein kinase-1 mediated phosphorylation of the modulator subunit was reported next (Agostinis et al., 1987). Like CK-2, casein kinase-1 made the modulator protein a better substrate for subsequent phosphorylation by kinase F_A, but in contrast to CK-2, previous phosphorylation of the F_CM-enzyme by CK-1 blocked the activation of the phosphatase by kinase F_A. Although a complete sequence analysis of the CK-1 phosphorylation sites is not available yet, it is evident from peptide mapping that CK-1 and CK-2 have different preferences for the modulator sites. It would be logical to assume that the synergistic phosphorylation with kinase F_A is due to a common site (or sites) recognized by both CK-1 and CK-2, whereas the inhibition of the kinase F_A-mediated activation of the phosphatase may point to the existence of a (still unidentified) site specifically phosphorylated by CK-1.

Surprisingly enough, when the F_CM-enzyme was first fully activated by kinase F_A, subsequent phosphorylation by CK-1 did not inhibit the phosphatase activity towards phosphorylase _a_ as substrate and the phosphate introduced into the modulator subunit was autocatalytically removed.

Origin and regulatory role of the inactive F_CM-phosphatase

Although the F_CM-enzyme has been purified to homogeneity from several sources, the inactive phosphatase is not easily detected in fresh tissue extracts or cytosols (Vandenheede et al., 1989). It has moreover been observed that during the isolation of the enzyme from rabbit skeletal muscle an active phosphorylase phosphatase form is gradually being inactivated with each purification step (Vandenheede et al., 1982). Measurements of the phosphorylase phosphatase activity in crude tissue preparations has pointed to the fact that virtually 50% of the potential phosphatase activity is cytosolic in nature, and that this soluble enzyme activity is partially inhibited. This last observation has lead to the discovery of the heat and acid stable phosphatase inhibitors in the mid-seventies (Huang and Glinsmann, 1975; Lee et al., 1976). Since the inhibitor-2 has been identified as the regulatory or modulator subunit of the inactive phosphatase, it was reasonable to investigate whether this inhibitor could be present in fresh tissue cytosols in a complex with active phosphatase catalytic subunits. This lead to the discovery of a latent type of phosphorylase phosphatase which seemed to contain more than stoichiometric amounts of inhibitor-2 over catalytic subunit (Vandenheede et al., 1989). Although the characterization of this latent phosphatase is still quite incomplete, it could be represented as an "activated" $[F_C*M]$-enzyme (Fig. 1) with a second modulator bound at its active site. Since the catalytic subunit of the latent enzyme is in the active conformation, there is no reason to expect the associated modulators to be phosphorylated at threonine-72 (the kinase F_A specific site) seen the rapid auto-dephosphorylation of the enzyme upon activation. Purification of the latent phosphatase, or incubation at 30°C produces the inactive $F_C M$-enzyme which seems to have lost the second modulator unit.

As to the potential role of the cytosolic latent phosphatase, we can envisage three likely functions. First of all, the latent enzyme can be looked upon as a pro-enzyme from where phosphatase activity can be generated in response to a specific stimulus. This would require the removal of the "inhibitory" modulator unit if substrates like phosphorylase a are to be dephosphorylated.

Secondly, we can also look at the latent phosphatase as a potential source of modulator, to inactivate catalytic subunits which are translocated from specific cellular localities (where they were complexed to targetting subunits) into the cytosol. Evidence for such a translocation mechanism has been provided (Cohen P, 1989; Dent et al., 1990).

A third possibility is however that the physiological substrate for the latent phosphatase and for the activated F_CM-enzyme is a yet unknown protein, whose dephosphorylation is not impaired by the presence of the second modulator. This would be the case if the physiological substrate has a higher affinity for the active site of the enzyme than the second modulator unit. Myelin basic protein (MBP) is one example of a substrate for the activated F_CM-enzyme, whose dephosphorylation is not inhibited by the modulator protein (inhibitor-2) at physiological concentrations of substrate and inhibitor (Yang S-D, Fong F-L, 1987).

Acknowledgements

J.R. Vandenheede is a Research Director and P. Agostinis a Senior Research Assistant of the "Nationaal Fonds voor Wetenschappelijk Onderzoek". J. Van Lint is the recipient of a fellowship of "Levenslijn". This work was supported by grants from the "Nationaal Fonds voor Geneeskundig Wetenschappelijk Onderzoek".

References

Ballou LM, Fischer EH (1986) Phosphoprotein phosphatases. In:Boyer P, Krebs EG (eds) The Enzymes, Academic Press, New York, 17(A):311-361

Cohen P (1989) The structure and regulation of protein phosphatases. Ann Rev Biochem 58:453-508

Agostinis P, Vandenheede JR, Goris J, Meggio F, Pinna LA, Merlevede W (1987) The ATP,Mg-dependent protein phosphatase: regulation by casein kinase-1. FEBS Lett 224:385-390

Dent P, Lavoinne A, Nakielny S, Caudwell FB, Watt P, Cohen P (1990) The molecular mechanism by which insulin stimulates glycogen synthesis in mammalian skeletal muscle. Nature 348:302-308

DePaoli-Roach AA, Ahmad Z, Camici M, Lawrence JC Jr, Roach PJ (1983) Multiple phosphorylation of rabbit skeletal muscle glycogen synthase. J Biol Chem 258:10702-10709

DePaoli-Roach AA (1984) Synergistic phosphorylation and activation of ATP-Mg-dependent phosphoprotein phosphatase by F_A/GSK-3 and casein kinase II ($PC_{0.7}$). J Biol Chem 259: 12144-12152

Holmes CFB, Kuret J, Chisholm AAK, Cohen P (1986) Identification of the sites on rabbit skeletal muscle protein phosphatase inhibitor-2 phosphorylated by casein kinase-II. Biochim Biophys Acta 870:408-416

Holmes CFB, Tonks NK, Major H, Cohen P (1987) Analysis of the in vivo phosphorylation state of protein phosphatase inhibitor-2 from rabbit skeletal muscle by fast-atom bombardment mass spectrometry. Biochim Biophys Acta 929:208-219

Huang FL, Glinsmann WH (1975) Inactivation or rabbit muscle phosphorylase phosphatase by cyclic AMP-dependent kinase. Proc Natl Acad Sci USA 72:3004-3008

Lee EYC, Brandt H, Capulong ZI, Killilea SD (1976) Properties and regulation of liver phosphorylase phosphatase. Adv Enz Regul 14:467-490

Picton C, Woodgett J, Hemmings B, Cohen P (1982) Multisite phosphorylation of glycogen synthase from rabbit skeletal muscle. FEBS Lett 150:191-196

Vandenheede JR, Yang S-D, Merlevede W (1981) Rabbit skeletal muscle protein phosphatase(s). J Biol Chem 256:5894-5900

Vandenheede JR, Agostinis P, Staquet S, Van Lint J (1989) The inactive ATP,Mg-dependent protein phosphatase. Origin, role and regulation. Adv Prot Phosphatases 5:19-36

Vandenheede JR, Staquet S, Merlevede W (1989) Identification and partial characterization of a latent ATP,Mg-dependent protein phosphatase in rabbit skeletal muscle cytosol. Mol Cell Biochem 87:31-39

Yang S-D, Fong Y-L (1985) Identification and characterization of an ATP.Mg-dependent protein phosphatase from pig brain. J Biol Chem 260:13464-13470

OKADAIC ACID FROM LABORATORY CULTURES OF A DINOFLAGELLATE ALGA: EFFECTS ON PROTEIN PHOSPHORYLATION IN C3H10T1/2 FIBROBLASTS

H. Angel Manjarrez Hernandez, Lynda A. Sellers and Alastair Aitken
Laboratory of Protein Structure
National Institute for Medical Research
Mill Hill London NW7 1AA
United Kingdom

INTRODUCTION

Okadaic acid is a polyether carboxylic acid-containing compound of Mr 802, first isolated from species of the black sponge, Halichondria okadai and H. melanodocia (Tachibana et al., 1981). The compound is probably produced by the benthic marine dinoflagellate, Prorocentrum lima and may be absorbed by the sponge after feeding on this alga. Through progression of the food chain, okadaic acid has been shown to accumulate in the digestive glands of bivalves (Murakami et al., 1982) and is the major cause of diarrhetic shellfish poisoning (Hartshorne et al., 1989). Okadaic acid has also been shown to have potent tumour promoting activity (Suganuma et al., 1988), but unlike other promoters such as the phorbol ester TPA, okadaic acid does not activate protein kinase C (PKC) but is a specific inhibitor of protein phosphatases types 1 (PP1) and 2A (PP2A) (Cohen et al., 1989).

In the present study we have extracted from P. lima, inhibitory activity to protein phosphatases types 1 and 2A, and tested its effect on the phosphorylation levels of the major '80' kDa substrate of PKC in quiescent C3H10T1/2 mouse fibroblasts (Mahadevan et al., 1987). Partial purification by thin layer chromatography was made and monoclonal anti-bodies were used to determine if the inhibitory activities are due to the presence of okadaic acid and/or related compounds.

METHODS

Algal culture. P. lima (from North East Pacific culture collection of algae, isolate number 514) was grown unialgally in a synthetic sea water medium, "ultramarine" synthetic sea salts (Waterlife Research Industries Ltd., West Drayton, U.K.) enriched with ES-1 trace element supplement. The algae were cultured in aerated 5-litre conical glass flasks at $18^{\circ}C$

under natural sunlight. The cells were harvested when the growth reached the stationary phase.

Extraction of phosphatase inhibitors. The algae (6g wet weight) were collected by filtration through muslin gauze and extracted at room temperature with 5ml acetone followed by methanol. This was repeated three times and the combined extracts were dried in a rotary evaporator. The residue was redissolved in ether (5ml), back-extracted with water three times and the residue obtained after evaporation stored at $-20^{o}C$ under nitrogen.

Chromatography (tlc). Thin layer chromatography (tlc) of the algal residue dissolved in chloroform was carried out on precoated plates of Silica gel (Whatman, PE Sil 0.25mm) with the solvent system, chloroform; methanol; 6M acetic acid (90:9.5:0.5). Slices of the tlc plate (of 1 or 0.5cm widths) were extracted with the same solvent and evaporated to dryness under nitrogen.

Cell culture. C3H10T1/2 fibroblast cells were grown in Dulbecco's modified Eagle medium (DMEM) supplemented with 10% v/v foetal calf serum (FCS). Cells were maintained at $37^{o}C$ in a humidified atmosphere of air enriched with 5% CO_2. The media was changed every 2/3 days and the cells were reseeded on reaching confluence.

Monoclonal antibodies to okadaic acid. These were obtained by culturing mouse hybridoma cells supplied by UBE Industries Ltd., Yamaguchi-ken 755, Japan, covered by European patent, application number 88309441.9.

RESULTS AND DISCUSSION

The marine dinoflagellate alga, P. lima has yielded potent inhibitors of protein phosphatase types 1 and 2A (Fig. 1). A 10^{-5} dilution of P. lima extract resulted in 50% inhibition of phosphatase 2A under standard assay conditions (using 0.3 mU/ml of enzyme), but was approximately 100-fold less potent at inhibiting phosphatase type 1, consistent with the effects of authentic okadiac acid. Phosphatase 2A inhibitory activity, determined on extracts from strips of the tlc plate, was contained in two distinct fractions, one of which (fraction 5) had a very similar Rf to okadaic acid (0.53, Murakami et al., 1982). Fraction 10 contained the main proportion of the PP2A inhibitory activity (95% of the total) and had an Rf of 0.8. This more hydrophobic fraction could contain a derivative of okadaic acid such as dynophysistoxin or a methyl ester, both of which have been previously identified in a variety of marine organisms (Yasumoto et al., 1985).

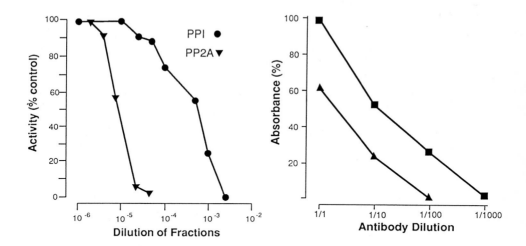

Fig. 1. Effect of P. lima extract on the activities of protein phosphatases. Assays were carried out with the purified catalytic subunits of phosphatases PP1 and PP2A (Cohen et al., 1989) (from rabbit skeletal muscle) using glycogen phosphorylase as substrate. The phosphatase concentration in the assays was 0.3 mU/ml.

Fig. 2. Okadaic acid antibody response of tlc fractions. This shows the titration of 1/1000 dilutions of fraction 5 (■) and fraction 10 (▲) against dilut- ions of monoclonal antibody. The ELISA tests were done in 96 well microtitre plates with antibody partially purified from the hybridoma supernatant using 50% ammonium sulphate precipitation.

Of the two distinct inhibitory activities obtained from tlc, fraction 5 (similar to okadaic acid) showed much stronger cross-reactivity with the okadaic acid antibodies (Fig. 2), although most of the PP2A inhibitory activity was contained in fraction 10. This may be explained by the anti- bodies being more selective for okadaic acid than its derivatives.

Treatment of C3H10T1/2 fibroblasts with fraction 5 gave an almost identical pattern of phosphorylation to that produced by a sample of authentic okadaic acid. However, after treatment with tlc fraction 10, there was increased phosphorylation of a 60 kDa protein (Fig. 3). This could be identical to the 60 kDa fragment of nucleolin that has previously been shown to be phosphorylated after treatment with okadiac acid in primary human fibroblasts (Issinger et al., 1988). In that study the state of phosphorylation of the 60 kDa protein was not greatly increased after treatment with TPA, consistent with our findings, and it has been suggested that casein kinase II and not PKC is responsible for the in vivo phosphorylation of this protein. In the present study, however, there was no significant increase in phosphorylation of this protein in mouse embryo C3H10T1/2 fibroblasts with fraction 5 (similar to okadaic acid in tlc) and with authentic okadaic acid.

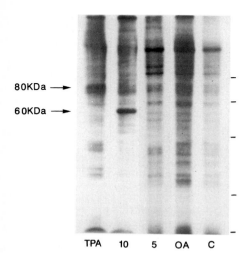

80KDa ➞

60KDa ➞

TPA   10   5   OA   C

Fig. 3. Autoradiography of SDS PAGE of C3H10T1/2 phosphoproteins. The lanes contained samples from cells treated for 2 h with tlc fractions 5 and 10 from P. lima; the phorbol ester (TPA) and authentic okadaic acid (OA). The fifth lane (C) represents unstimulated cells incubated with ^{32}P only. The position of the molecular weight markers (94, 67, 43, 30 and 20.1 kDa, from top to bottom) are indicated by the bars. The '80' kDa substrate for PKC and the 60 kDa protein are indicated by the arrows. The amount of each tlc fraction had a similar potency in the in vitro PP2A assay. By comparison with known okadaic acid concentrations this was equivalent to 10-20nM.

In contrast to TPA stimulation, okadaic acid or either of the tlc fractions resulted in a significant increase in phosphorylation of only one phosphoform of the 80 kDa substrate of kinase C (Fig. 4).

In this study we have shown that compounds closely resembling okadaic acid, may be produced in the laboratory with very little expenditure and can be used successfully for investigating the effects of altered levels of protein phosphatases and kinases in cell culture. The phosphatase inhibitor fractions from P. lima are currently being purified in our laboratory and their structures will be compared to those of okadaic acid and its known derivatives.

ACKNOWLEDGEMENTS - A.M.H. was supported by a studentship from the National University of Mexico (UNAM). The hospitality and assistance of Professor Philip Cohen during a visit by AMH to the University of Dundee, is much appreciated.

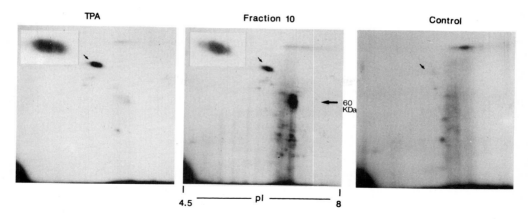

Fig. 4. Two-dimensional SDS PAGE of C3H10T1/2 phosphoproteins. Cells were incubated for 2 h in the presence of ^{32}P, with either TPA or fraction 10 and proteins subjected to two-dimensional SDS PAGE. The autoradiograph of the gels is shown together with a control. The position of the '80' kDa substrate of PKC is indicated by the small arrow, and the inset shows an enlargement of this area. The position of the 60 kDa protein is also shown.

REFERENCES

Cohen P, Klumpp S, Schelling DL (1989) FEBS Lett 250:296-300
Issinger OG, Martin T, Richter WW, Olson M, Fujiki H (1988) EMBO J 7:1621-1626
Mahadevan LC, Aitken A, Heath J, Foulkes JG (1987) EMBO J 6:921-926
Murakami Y, Oshima Y, Yasumoto T (1982) Bull Jpn Soc Sci Fish 48:69-72
Suganuma M, Fujiki H, Suguri H, Yoshizawa S, Hirota M, Nakayasu M, Ojika M, Wakamatsu K, Yamada K, Sugimura T (1988) Proc Natl Acad Sci USA 85:1768-1771
Tachibana K, Scheuer PJ, Tsukitani Y, Kikuchi H, Van Engen D, Clardy J, Gopichand Y, Schmitz FJ (1981) J Am Chem Soc 103:2469-2471
Yasumoto T, Murata M, Oshima Y, Sano M, Matsumoto GK, Clardy J (1985) Tetrahedron 41:10191025
Hartshorne DJ, Ishihara H, Karaki H, Ozaki H, Sato K, Hori M, Watabe S (1989) Adv Protein Phosphatases 5:219-231

REGULATION OF PROTO-ONCOGENE EXPRESSION AND RATE OF PROTEIN SYNTHESIS BY THE TUMOR PROMOTER OKADAIC ACID

Axel Schönthal
Cancer Center 0636
University of California at San Diego
La Jolla, CA. 92093, USA

KEYWORDS/ABSTRACT: okadaic acid; c-fos proto-oncogene; protein phosphatases. Okadaic acid is a non-phorbol-ester-type tumor promoter that specifically inhibits phosphoserine/phosphothreonine specific protein phosphatases. We show that treatment of cells with okadaic acid reduces the rate of protein synthesis. In parallel expression of the c-fos proto-oncogene is elevated. This latter effect can be further enhanced by simultaneous addition of anisomycin, a potent protein synthesis inhibitor.

INTRODUCTION

Phosphorylation events are major regulatory mechanisms of signal transduction pathways that control cell growth and differentiation (for a review, see Watson, 1988). Less is known about dephosphorylation of phosphoproteins in these processes. In this study we examined the effects of inhibition of two types of serine/threonine specific phosphoprotein phosphatases, PP-1 and PP-2A (for a review, see Cohen, 1989), on the expression of the c-fos proto-oncogene. For this purpose we treated cells with okadaic acid, a non phorbol ester type tumor promoter (Tachibana et al., 1981; Hakaii et al., 1986; Suganuma et al., 1988) that has recently been identified as a specific inhibitor of PP-1 and PP-2A activity (Bialojan and Takai, 1988; Hescheler et al., 1988). The c-fos proto-oncogene is a member of the group of immediate early genes. Its expression is induced very rapidly and transiently after serum-stimulation of starved cells or treatment of cells with different agents such as the phorbol ester tumor promoter TPA (12-O-tetradecanoyl-phorbol-13-acetate) (for reviews, see Herrlich and Ponta, 1989; Schönthal, 1990; Cooper, 1990). Recently we and others have demonstrated that c-fos is also induced by okadaic acid (Schönthal et al., 1990; Kim et al., 1990).

NATO ASI Series, Vol. H 56
Cellular Regulation by Protein Phosphorylation
Edited by L. M. G. Heilmeyer, Jr.
© Springer-Verlag Berlin Heidelberg 1991

Here we show that this induction is greatly enhanced in the presence of the protein synthesis inhibitor anisomycin. While treatment of intact cells with okadaic acid alone results in a reduced rate of protein synthesis, anisomycin and okadaic acid act synergistically to induce c-fos. These data suggest that protein synthesis inhibition by okadaic acid is not the sole mechanism by which it leads to elevated c-fos expression.

RESULTS

The accumulation of c-fos mRNA in response to okadaic acid is slow with a maximum around three hours. Eight hours after okadaic acid administration c-fos mRNA levels are still strongly elevated (Schönthal et al., 1990; see also Fig. 2 and 3). A similar elevation of c-fos mRNA expression has also been observed after treatment of cells with protein synthesis inhibitors (Greenberg et al., 1986; this paper). We therefore analyzed whether okadaic acid has an effect on protein synthesis. This was determined by pulsing the cells with ^{35}S methionine and measuring the incorporation of radioactivity into newly translated protein in the presence or absence of okadaic acid. As shown in Figure 1 okadaic acid concentrations of 500 nM or higher reduce protein synthesis to 69 and 60 % of the control level, respectively. This effect is weak in comparison to anisomycin which reduces protein synthesis to 1 %.

Since it has been shown earlier that inhibition of protein synthesis induces expression of the c-fos proto-oncogene (Greenberg et al., 1986), we analyzed whether the effects of okadaic acid on c-fos expression were due solely to its effects on protein synthesis. The response of c-fos gene expression after simultaneous treatment of cells with okadaic acid and anisomycin was analyzed. Since anisomycin alone inhibited protein synthesis by 99%, an induction of c-fos mRNA by okadaic acid in the presence of anisomycin would not likely be due to a further inhibition of translation by okadaic acid. Figure 2 shows that okadaic acid and anisomycin together have a strong synergistic effect and induce c-fos expression much stronger than either agent alone.

In another set of experiments we tested whether treatment of cells with sub-optimal concentrations of anisomycin (0.05-20 μM, which inhibits protein synthesis to 4-65 % of control levels) elevated c-fos mRNA expression with similar kinetics as seen with okadaic acid. As shown in Figure 3 the c-fos mRNA induction kinetics in response to either okadaic acid or different concentrations of anisomycin are quite different. After treatment of cells with low concentrations of anisomycin elevated c-fos mRNA levels are maximal at one hour. In contrast, after treatment with okadaic acid c-fos mRNA is low at one hour but is strongly elevated at three and six hours. These results suggest

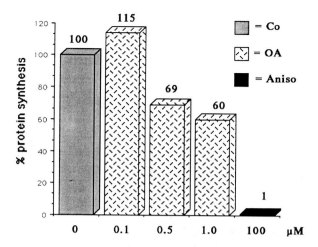

Figure 1: Logarithmically growing NIH3T3 cells were treated with the indicated concentrations of okadaic acid (OA) or anisomycin (Aniso) for 4 hours. Control cells (Co) received the solvent DMSO. Then the medium was changed to methionine-free medium supplemented with ^{35}S-methionine. 30 minutes later cells were harvested and the amount of incorporated ^{35}S-methionine determined. The amount of incorporation in control cells was set to 100 %. The values shown are the average of four experiments.

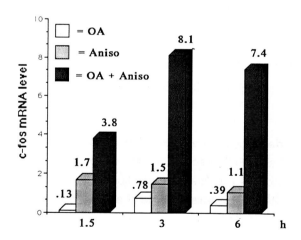

Figure 2: Logarithmically growing NIH3T3 cells were treated with either 500 nM okadaic acid (OA), or 100 µM anisomycin (Aniso), or both, and harvested at the indicated times. Poly A+ RNA was isolated and analyzed for c-fos mRNA expression by Northern hybridization with a fos specific probe. The filters were exposed to film and the amount of c-fos mRNA expression determined by scanning the autoradiographs. The values shown are the average of two independent experiments.

THE USE OF ß-CYCLODEXTRIN IN THE PURIFICATION OF PROTEIN PHOSPHATASE G FROM RAT LIVER

S. Wera, M. Bollen and W. Stalmans
Afdeling Biochemie, Faculteit Geneeskunde
Katholieke Universiteit Leuven
B-3000 Leuven
Belgium

The rate of glycogen synthesis in the liver is determined by the activity of glycogen synthase, which is activated by dephosphorylation of multiple Ser residues. The major enzyme involved in the activation of hepatic glycogen synthase is the glycogen-bound 'protein phosphatase G' (Bollen et al., 1988). This type-1 or AMD protein phosphatase also possesses phosphorylase phosphatase activity which, however, is considerably inhibited by subunit(s) that can be selectively destroyed by trypsin. Protein phosphatase G has recently been purified and was found to contain two non-catalytic polypeptides (161 and 54 kDa), the smaller one being similar or identical to α-amylase (Wera et al., 1991). The key to the purification was an affinity chromatography on immobilized ß-cyclodextrin.

Cyclodextrins are cyclic carbohydrates consisting of 6, 7 or 8 α-1,4-linked D-glucosyl units (α-, ß- and ɣ-cyclodextrin, respectively). Their appearance is a truncated cone with a hydrophobic core and a hydrophilic outside (Saenger, 1983). In plant chemistry cyclodextrins are known as inhibitors of starch-degrading enzymes. All three cyclodextrins inhibit potato phosphorylase competitively with starch (Staerk and Schlenk, 1967). Exo-type starch hydrolases (ß-amylase and glucoamylase) are also strongly inhibited, whereas α-amylases can slowly hydrolyze cyclodextrins (Suetsugu et al., 1974). Immobilized cyclodextrins display marked selectivity; Vretblad (1974) showed that α-cyclodextrin-Sepharose retained specifically ß-amylase but not α-amylase, which in turn showed high affinity for ß-cyclodextrin-Sepharose (Silvanovich & Hill, 1976).

We report here on the selective interaction of ß-cyclodextrin with protein phosphatase G from rat liver.

EXPERIMENTAL

Materials. The source of materials was as described by Wera et al. (1991). Cyclodextrins were coupled to epoxy-activated Sepharose 6B. The catalytic subunit of the type-1 (AMD) protein phosphatase was prepared from rabbit skeletal muscle, and glycogen synthase *b* from dog liver. Protein phosphatase G was partially purified from rat liver by chromatography of the isolated glycogen-protein fraction on heparin-Sepharose and subsequent precipitation of the holoenzyme with polyethylene glycol. Subcellular fractions were prepared from rat livers. The concentration of these fractions is expressed with respect to the liver tissue from which they were derived; e.g. a 1% cytosol corresponds to the cytosol of 1 mg liver in a volume of 100 µl.

   Assays. Enzyme activities were determined as described by Wera et al. (1991). The amylo-1,6-glucosidase activity of the debranching enzyme was measured by its ability to incorporate labelled glucose into glycogen. Synthase phosphatase activity was determined from the rate of activation of purified liver synthase *b*. Phosphorylase phosphatase was assayed with ^{32}P-labelled phosphorylase *a* as substrate; the activity was either measured as such (spontaneous activity) or after preincubation with trypsin, which destroys the inhibitory subunits (total activity). One unit of phosphorylase phosphatase dephosphorylates one unit of phosphorylase *a* per min at 25°C. One unit of phosphorylase *a* converts 1 µmol of substrate into product per min at 25°C.

RESULTS AND DISCUSSION

   When a freshly prepared liver extract is incubated, phosphorylase is progressively inactivated and glycogen synthase becomes activated by the action of protein phosphatases. We observed that the addition of ß-cyclo-dextrin (0.1 mg/ml) inhibited strongly the activation of glycogen synthase,

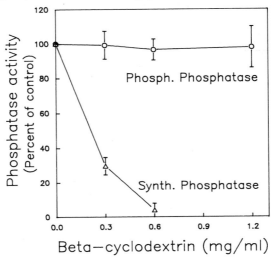

Fig. 1. *The effect of ß-cyclodextrin on partially purified protein phosphatase G.* The indicated protein phosphatase activities were determined in the presence of the indicated concentrations of ß-cyclodextrin. The results are means ± S.E.M. for three experiments.

although it did not affect the inactivation of phosphorylase (not illustrated). Further experiments showed that the cyclic maltoheptaose, even at 1 mg/ml, did not inhibit purified glycogen synthase from dog liver. Fig. 1 illustrates the effects of ß-cyclodextrin on partially purified protein phosphatase G from rat liver. The glycogen-synthase phosphatase activity of this enzyme was very sensitive to ß-cyclodextrin, half-maximal inhibition being reached at about 0.2 mg/ml, and complete inhibition at 0.6 mg/ml. In contrast, the phosphorylase phosphatase activity of the same enzyme was not affected by ß-cyclodextrin, even at 1.2 mg/ml.

These findings prompted us to investigate the value of ß-cyclodextrin coupled to Sepharose beads in the purification of various glycogen-bound enzymes. For this purpose, an enzyme-glycogen fraction was isolated from rat liver and passed slowly through a column of ß-cyclodextrin-Sepharose (Fig. 2). Essentially all the phosphorylase, glycogen synthase and amylo-1,6-glucosidase activities passed through the column, and subsequent elution with either 2 M NaCl or ß-cyclodextrin (1 mg/ml) did not release significant amounts of enzyme. In contrast, the column retained about one third of the glycogen-bound protein-phosphatase activity, which could be quantitatively eluted with 2 M NaCl.

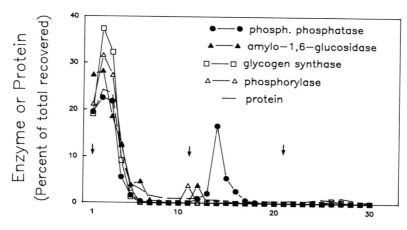

Fig. 2. Chromatography of an isolated enzyme-glycogen complex on ß-cyclodextrin-Sepharose. The liver glycogen fraction (600%) from 10 rats was recirculated for 30 min at room temperature through a column containing 10 ml of ß-cyclodextrin-Sepharose. The column was then washed at 4°C with standard buffer containing successively (arrows) 50 mM NaCl (fractions 1-10), 2 M NaCl (fractions 11-20) and 1 mg/ml of ß-cyclodextrin (fractions 21-30). The fractions (3 ml) were assayed for glycogen synthase, phosphorylase, debranching enzyme, total phosphorylase phosphatase, and protein as indicated.

The fate of protein phosphatase G is shown in more detail in Fig. 3. The phosphorylase-phosphatase activity of the native protein phosphatase G was severalfold increased by a treatment with trypsin (as shown by the enzyme not retained by ß-cyclodextrin-Sepharose). In contrast, the enzyme eluted by 2 M NaCl appeared to be the catalytic subunit of protein phosphatase G, since its phosphorylase phosphatase activity was not increased by trypsin treatment (Fig. 3), and since the eluted enzyme could not be re-bound to glycogen (not shown). Furthermore, some non-catalytic part of the enzyme appeared to be retained on the column; indeed, after subsequent re-equilibration of the column in buffer without salt, separately purified type-1 (AMD) catalytic subunit was completely retained on the column, and the 'reconstituted' holoenzyme could be eluted with free ß-cyclodextrin (Fig. 3). The binding of the type-1 catalytic subunit was quite specific, since it did not occur with the type-2A (PCS) enzyme, and since freshly prepared ß-cyclodextrin-Sepharose did not retain any free catalytic subunit (not shown).

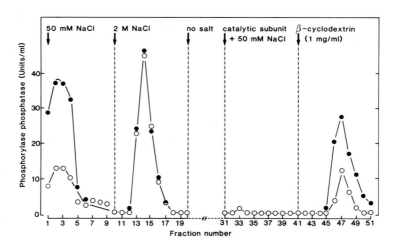

Fig. 3. Binding and reconstitution of protein phosphatase G on ß-cyclodextrin-Sepharose. The liver glycogen fraction (600%) from 10 rats was recirculated for 30 min at room temperature through a column containing 10 ml of ß-cyclodextrin-Sepharose. The column was transferred to the cold and washed with standard buffer containing successively 50 mM NaCl and 2 M NaCl. After re-equilibration of the column in standard buffer, 300 units of purified type-1 (AMD) phosphatase catalytic subunit were applied, and the column was then washed with standard buffer containing successively 50 mM NaCl and 1 mg/ml of ß-cyclodextrin. Fractions of 3 ml were collected and assayed for spontaneous (o) and total (●) phosphorylase phosphatase activity. The amount of catalytic subunit applied corresponds to the total phosphorylase phosphatase activity released by 2 M NaCl.

α- or γ -cyclodextrin-Sepharose were unable to retain protein phosphatase G or any other assayed enzyme (debranching enzyme, glycogen synthase, phosphorylase) present in a liver glycogen fraction (not shown).

The reconstituted holoenzyme obtained as illustrated in Fig. 3 was concentrated by precipitation with polyethylene glycol and displayed all the typical characteristics of the native protein phosphatase G (Bollen et al., 1988). Therefore, one or either non-catalytic subunit (see the introduction) confers to the catalytic subunit a series of properties, including high affinity for glycogen, marked resistance to inhibition by modulator protein (inhibitor-2), activation by trypsin (tenfold with the concentrated enzyme), and recognition of glycogen synthase as substrate. Furthermore, the synthase-phosphatase activity displayed regulatory properties which are specific for the liver enzyme; i.e. allosteric inhibition by physiological concentrations of phosphorylase a and of Ca^{2+}, and synergistic action with a cytosolic synthase phosphatase (Stalmans et al., 1987). The distinct properties of the holoenzymes from liver and muscle cannot reside in the catalytic subunit, since the reconstitution of the holoenzyme was performed with the catalytic subunit isolated from skeletal muscle. Further work will be required to establish the exact function of the non-catalytic subunits.

Acknowledgements - M. Evens and N. Sente provided expert technical assistance. This work was supported by Grant 3.0021.89 from the Belgian Fund for Medical Scientific Research and by a Concerted Research Action of the 'Vlaamse Executieve'. M. Bollen is a 'Bevoegdverklaard Navorser' of the National Fund for Scientific Research (Belgium).

REFERENCES

Bollen M, Vandenheede JR, Goris J, Stalmans W (1988) Characterization of glycogen-synthase phosphatase and phosphorylase phosphatase in subcellular liver fractions. Biochim Biophys Acta 969:66-77

Saenger W (1983) Stereochemistry of circularly closed oligosaccharides: cyclodextrin structure and function. Biochem Soc Trans 11:136-139

Silvanovich MP, Hill RD (1976) Affinity chromatography of cereal α-amylase. Anal Biochem 73:430-433

Staerk J, Schlenk H (1967) Interaction of potato phosphorylase with cycloamyloses. Biochim Biophys Acta 146:120-128

Stalmans W, Bollen M, Mvumbi L (1987) Control of glycogen synthesis in health and disease. Diabetes Metab. Rev. 3:127-161

Suetsugu N, Koyama S, Takeo K, Kuge T (1974) Kinetic studies on the hydrolyses of α-, ß-, and γ -cyclodextrins by Taka-amylase A. J Biochem 76:57-63

Vretblad P (1974) Immobilization of ligands for biospecific chromatography via their hydroxyl groups. The cyclohexaamylose-ß-amylase system. FEBS Lett 47:86-89

Wera S, Bollen M, Stalmans W (1991) Purification and characterization of the glycogen-bound protein phosphatase from rat liver. J Biol Chem 266: in press

Characterization of a Human T-cell Protein Tyrosine Phosphatase Expressed in the Baculovirus System

N.F. Zander, J.A. Lorenzen, G. Daum, D.E. Cool* and E.H. Fischer

Department of Biochemistry and *Howard Hughes Medical Institute; University of Washington; Seattle, WA

Phosphorylation of proteins on tyrosyl residues has been implicated in signal transduction and the control of cell growth, proliferation, differentiation and transformation. This process is regulated by the interplay of protein tyrosine kinases and protein tyrosine phosphatases (PTPases, review : Tonks and Charbonneau, 1989). A PTPase of 35 kDa (PTPase 1B) was previously purified to homogeneity from human placenta and characterized (Tonks et al., 1987a,b). Subsequently, a related PTPase of 48 kDa was cloned from human T-cells (Cool et al, 1989). The T-cell PTPase mainly differed from purified placenta PTPase 1B in having an additional 98 residues at the C-terminus. To address the function of this region, a truncated enzyme (TCΔC11.PTPase) was generated by inserting a stop codon behind Arg 317 (Cool et al., 1990). Both forms of the enzyme were expressed in

NATO ASI Series, Vol. H 56
Cellular Regulation by Protein Phosphorylation
Edited by L. M. G. Heilmeyer, Jr.
© Springer-Verlag Berlin Heidelberg 1991

the baculovirus system using recombinant *Autographa californica* nuclear polyhedrosis virus (Ac-NPV) as the vector (Summers and Smith, 1988). In this manuscript, we describe the high level expression of both forms of the PTPase, their purification and their enzymatic properties.

Spodoptera frugiperda (Sf9) cells were cotransfected with recombinant plasmid DNA (open reading frames of TC.PTPase and TCΔC11.PTPase cloned into pVL 941) and Ac-NPV wild type DNA. After virus purification, cells were infected with recombinant virus, harvested 72 h post infection and sequentially extracted in low salt buffer (25 mM imidazole pH 7.2, 2 mM EDTA, 0.1% (v/v) β-mercaptoethanol, protease inhibitors), low salt buffer containing 0.5% Triton X-100 and high salt buffer containing 0.5 % Triton X-100 and 0.6M KCl. The full-length PTPase could be extracted from the particulate fraction only with detergent and salt, whereas the truncated form was soluble in low salt buffer alone. Figure 1A shows a Western Blot of Sf9 cell extracts; both forms of the protein could be detected with peptide antibody 8172 (Cool et al., 1990). Limited trypsinolysis of either form gave rise to a fully active 33 kDa tryptic fragment.

For purification of the full-length enzyme, the Triton/KCl extract was precipitated by adding ammonium sulfate to 20% saturation. After centrifugation, the pellet was resuspended and subjected to FPLC Superose 12 column chromatography. The enzyme eluted as a

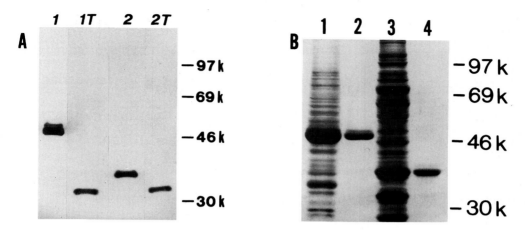

Figure 1 : Expression and Purification of TC.PTPase and TCΔC11.PTPase.
A.Western Blot of Sf9 cell extracts. Protein (0.5 μg) was subjected to SDS-PAGE, blotted onto nitrocellulose and probed with anti-peptide antibody 8172. Lanes 1: extracts of cells expressing the full-length enzyme; lanes 2: extracts of cells expressing the truncated form; lanes T: extracts after limited trypsinolysis.
B. SDS-PAGE of cell extracts and purified proteins. Lane 1: 10 μg extract of cells expressing TC.PTPase; lane 2: 1 μg purified TC.PTPase; lane 3: 20 μg extract of cells expressing TCΔC11.PTPase; lane 4 : 1 μg purified TCΔC11.PTPase.

single peak of activity with an apparent Mr of 220 kDa, indicating aggregation, insertion in detergent micelles or asymmetry of the molecule. The peak fractions contained three bands when analyzed by SDS-PAGE: a doublet at 48 kDa (not resolved in figure 1B) and a much less intense band at 50 kDa. However, all three bands corresponded to different species of the PTPase since they cross-reacted with three different peptide antibodies.

For purification of the truncated form, the low salt buffer extract was applied directly to a Sephadex G75 superfine column. The enzyme eluted in a symmetric peak at its expected molecular weight, although a minor trailing band can be seen. This protein was recognized by peptide antibody 8172 and, therefore, corresponded to a different species of the truncated enzyme.

The reason for the molecular heterogeneity of the expressed proteins is still not clear. It was not due to phosphorylation as incubation of the cells in the presence of $^{32}P_i$ did not lead to phosphate incorporation into any of the bands.

Table I : Specific activities of PTPase forms [U/mg][1]

Substrate:	RCML	MBP
TC.PTPase	850	10,300
TCΔC11.PTPase	26,000	4,700

[1] activities measured at substrate saturation in the presence of 5 mM EDTA; 1 U = 1 nmol/min phosphate release

The truncated form of the enzyme displayed a specific activity of 26,000 U/mg toward RCML. In contrast, the full-length enzyme was far less active (850 U/mg) toward this substrate suggesting that enzyme activity was repressed by its C-terminal portion. Either limited trypsinolysis of the purified protein (which removes the C-terminal segment) or generation of the truncated form by site-directed mutagenesis led to a 30 fold increase in PTPase activity. The low activity of the TC.PTPase was not a general phenomenon but depended on the nature of the substrate. With MBP as the substrate, the full length enzyme displayed double the activity (10,300 U/mg) of the truncated form (4,700 U/mg). These data suggest that MBP interacts with the C-terminal segment facilitating access to the catalytic site.

Both forms of the T-cell enzyme were inhibited by micromolar concentrations of vanadate, molybdate and zinc. Nanomolar concentrations of heparin inhibited the enzymes; however, this effect could only be observed with RCML as the substrate. Polycationic compounds activated the full-length enzyme (spermine up to sevenfold, unphosphorylated MBP up to threefold), whereas activation of the truncated form by these compounds was at most 30%, indicating interaction of polycations with the C-terminal segment of the molecule.

References

Cool DE, Tonks NK, Charbonneau H, Walsh KA, Fischer EH, Krebs, EG (1989) cDNA isolated from a human T-cell library encodes a member of the protein-tyrosine-phosphatase family. Proc.Natl.Acad.Sci. USA 86: 5257-5261.

Cool DE, Tonks NK, Fischer EH, Krebs, EG (1990) Expression of a human T-cell protein-tyrosine-phosphatase in baby hamster kidney cells. Proc.Natl.Acad.Sci.USA 87:7280-7284.

Summers MD, Smith, GE (1988) A Manual of Methods for Baculovirus Vectors and Insect Cell Culture Procedures. Texas Agricultural Experiment Station Bull. No. 1555.

Tonks NK, Charbonneau H (1989) Protein tyrosine dephosphorylation and signal transduction. Trends in Biochem.Sci. 14:497-500.

Tonks NK, Diltz CD, Fischer, EH (1987a) Purification of the Major Protein-tyrosine-phosphatses of Human Placenta. J.Biol.Chem. 263: 6722-6730.

Tonks NK, Diltz CD, Fischer EH (1987b) Characterization of the Major Protein-tyrosine-phosphatases of Human Placenta. J.Biol.Chem. 263:6731-6737.

INHIBITION OF TYROSINE PROTEIN PHOSPHATASES FROM MUSCLE AND SPLEEN BY NUCLEIC ACIDS AND POLYANIONS

C. Stader, S. Dierig, N. Tidow, S. Kirsch, and H. W. Hofer
Faculty of Biology, University of Konstanz,
P.O. Box 5560, W-7750 Konstanz, Germany

The reversion of tyrosine phosphorylation is potentially involved in the regulation of the cell cycle and signal transduction (cf. to the contributions by E. Fischer and G. Draetta, this volume). Structural relationship was detected between a 38 kDa protein tyrosine phosphatase (PTP) isolated from human placenta (PTP-1B, Tonks et al., 1988) and the catalytic domains of membrane proteins like CD45 (Tonks et al., 1990), LAR (Streuli et al., 1988), and LCP (Matthews et al., 1990) which possess tyrosine phosphatase activity. Clones encoding for a major soluble PTP of $M_r \approx 50,000$ have been obtained from human and rat cDNA libraries (Chernoff et al., 1990; Brown-Shimer et al., 1990; Guan et al., 1990). These clones included the complete code for PTP-1B indicating that PTP-1B was a proteolytic fragment of a larger precursor. We isolated PTPs from the cytosolic fraction of porcine spleen (designated PTP-S) and two soluble enzymes (designated PTP-M1 and PTP-M2) from rat muscle and compared their properties with respect to the inhibition by various polyanions.

Purification and preliminary molecular characterization of PTP from porcine spleen and rabbit muscle

Purified a soluble PTP-S exhibited a single protein band on SDS-PAGE corresponding to an estimated molecular weight between 52,000 and 57,000, depending on the type of marker proteins used for calibration. This M_r was in good agreement with that described by Chernoff et al. (1990) for the *in vitro* translation product of the full-length clone of a major human PTP. Chromatography of PTP-S on a Superose 12 column indicated a smaller M_r of 40,000. The apparent discrepancy between the M_r data obtained under denaturing and native conditions may be due to interaction of the protein with the Superose 12 matrix, an effect which had been observed with protein tyrosine kinases before (Batzer et al., 1990).

The migration of PTP-M1 on SDS-PAGE was the same as of PTP-S, whereas the migration of PTP-M2 was in agreement with an estimated M_r 48,000. The native M_r of both PTP from muscle was the same as determined for PTP-S.

PTP-M2 from rat muscle and PTP-S cross-reacted on Western blots with an antiserum raised against PTP in the laboratory of Prof. E. Fischer. This fact also indicates that the PTP-M2 and PTP-S belong to the same family of PTPs as the PTPs from lymphocytes and human placenta.

NATO ASI Series, Vol. H 56
Cellular Regulation by Protein Phosphorylation
Edited by L. M. G. Heilmeyer, Jr.
© Springer-Verlag Berlin Heidelberg 1991

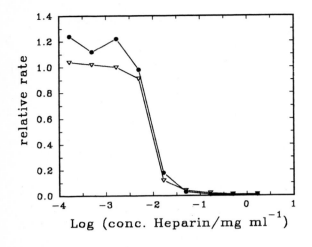

Log (conc. Heparin/mg ml^{-1})

Fig. 1

Inhibition of PTP-M1 (triangles) and PTP-M2 (circles) from skeletal muscle by heparin. Phosphatase activity was assayed with modified lysozyme (Tonks, 1988a) phosphorylated on tyrosine residues by a purified lienal tyrosin-specific protein kinase (Batzer et al., 1990).

The search for physiological inhibitors of PTP-S and PTP-M

The PTPs from muscle and spleen were strongly inhibited by heparin (cf. Fig. 1) as previously observed for PTP-1B by Tonks et al. (1988b).

In an attempt to find a physiological equivalent for heparin we used an approach similar to that which had previously led to the isolation of an inhibitor protein by Ingebritsen (1989). Fractions from spleen extracts inhibiting the PTP activity were eluted by high ionic strength from a DEAE-chromatography column. The inhibitory activity was stable to prolonged heating (60 min at 95 °C) and its elution profile from a Superose 12 column corresponded to a $M_r < 100,000$. The material obtained fromthe gel filtration column exhibited a ratio of UV absorbance at 260 nm and 280 nm of nearly 2 (cf. Fig. 2).

Fig. 2.

UV absorption spectrum of the heat-stable inhibitor preparation of PTPs obtained from porcine spleen as described in the text. The spectrum was measured in the fraction exerting the strongest inhibition (Fraction 21) after chromatography on Superose 12 (pH 7.0).

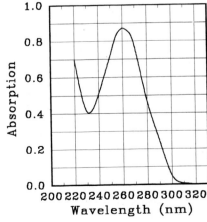

Wavelength (nm)

Fig. 3.

Effect of the treatment (2 h, ambient temperature) of the PTP inhibitor preparation from porcine spleen with proteases and nucleases (1 mg/ml). Chymotrypsin and trypsin were used before (hollow bars) and after (hatched bars) of heating to 95 °C).

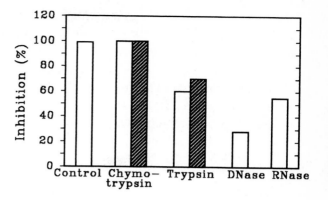

The inhibitor preparation from spleen retained its activity after treatment with highly purified chymotrypsin (cf. Fig. 3) whereas incubation with RNase and DNase distinctly reduced the inhibitory action on PTP. The partial loss of inhibitor activity after treatment with trypsin (and in a similar manner with proteases of bacterial origin)

Fig. 4.

Chromatography of the PTP inhibitor on a Superose 12 column (HR 15/30). The line indicates absorption at 280 nm, the hatched bars represent inhibitor activity.
a: Untreated sample.
b: Sample incubated with chymotrypsin (1 mg/ml) for 2 h before chromatography.
The samples were heated at 95 °C for 10 min prior to chromatography.

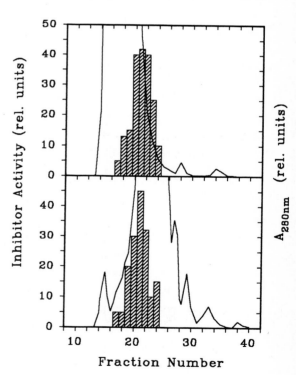

may be due to RNase contaminations in standard qualities of some commercial proteases, since heat-treatment did not reduce inhibition. As shown in Fig. 4, the molecular weight of the inhibitory components was not changed when the samples were treated with a highly purified chymotrypsin before gel filtration chromatography, whereas there was a distinct change in the pattern of UV-absorption in the eluate from the chromatography, indicating that there had been chymotrypsin-sensitive material in the sample. Analogous experiments performed with an inhibitor sample treated with RNase eliminated most of the inhibitory activity. In the light of these data it appeared more likely that the PTP-inhibiting material from spleen was nucleic acids rather than a protein, as previously reported for an inhibitor from brain (Ingebritsen, 1989).

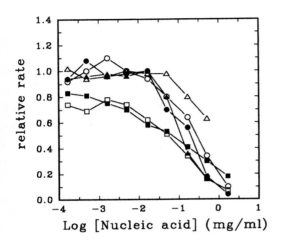

Fig. 5.

Inhibition of PTPs from muscle and spleen by commercial preparations of herring sperm DNA (triangles) and yeast RNA (circles) and by polyadenosine (squares). PTP assays were performed with modified lysozyme (cf. Legend of Fig. 1). The hollow symbols refer to experiments performed with PTP-M1, the full symbols to experiments done with PTP-M2.

Inhibition of PTPs by nucleic acids

The inhibition of PTPs from skeletal muscle by commercial preparations of herring sperm DNA and unfractionated yeast RNA is shown in Fig. 5. There were only minor differences in the efficiency of the nucleic acids with respect to their inhibitory potency on the PTPs. Since an almost identical effect was also obtained with polyadenosine, it was clear that the inhibition did not depend on a specific arrangement of bases within the nucleic acids but was most likely conferred by the presence of phosphate anions. The fact that chondroitin sulfate was an almost equipotent inhibitor of the PTPs as heparin and nucleic acids indicated that the presence of multiply charged polyanions was a sufficient requirement to inhibit PTPs.

A comparison of the inhibition of PTPs by low and high molecular weight anionic substances (cf. Table 1) suggested that, on the basis of charges per weight, the polyanionic substances were distinctly more efficient to inhibit the PTPs than low molecular weight phosphates or sulfates.

Table 1

	$IC_{50\%}$ (mg/ml)
ATP	8.47
Na_2SO_4	7.6
Na_2PO_4	9.8
Heparin	0.022
Chondroitin Sulfate	0.045
RNA	0.012
DNA	0.25

Conclusion

PTPs exhibit remarkable conservation of structure in the catalytic part of their molecules. This fact is also supported by the results of this investigation which revealed immunologic cross-reaction and, therefore, presumably a close structural relationship between PTPs from spleen and muscle and the catalytic moiety of a PTP which is supposedly a membrane bound protein (Cool et al., 1990). Nevertheless, the PTPs described here were isolated from cytosolic fractions and the reproducible contents of their activities in the tissues did not support the possible assumption that the solubilization of the enzymes was mediated by proteolytic attack on a membrane-bound precursor. It is more likely, that the enzymes isolated in our laboratory are the products of the recently cloned genes of soluble PTP which apparently is fairly ubiquitous in mammalian tissues.

The PTPs described here not only revealed striking structural similarities to enzymes derived from other sources but also similar kinetic behaviour. Though heparin itself is not likely to regulate PTP activity in situ, we tried to find more likely candidates. It turned out that other polyanions, especially nucleic acids, also possess affinity to the PTP proteins. In theory, the affinity should be sufficiently high to allow binding under physiological conditions and, therefore, offers a potential mechanism of regulation of PTP activity.

Acknowledgement

We are indebted to Dr. Günter Daum for valuable discussions and to Dr. Edmond Fischer and Dr. Norbert Zander for the gift of a polyclonal antiserum against protein tyrosine phosphatases.

References

Batzer A, Kirsch S, Hofer HW (1990) Characterization of two tyrosine-specific protein kinases from pig spleen. Substrate-specific effect of autophosphorylation. Eur. J. Biochem., 194, 251-258.

Brown-Shimer S, Johnson KA, Lawrence JB, Johnson C, Bruskin A, Green NR, Hill DE (1990) Molecular cloning and chromosome mapping of the human gene encoding protein phosphotyrosyl phosphatase 1B. Proc. Natl. Acad. Sci. USA 87:5148-5152.

Charbonneau H, Tonks NK, Walsh, KA, Fischer EH (1988) The leukocyte common antigen (CD45): a putative receptor-linked protein tyrosine phosphatase. Proc. Natl. Acad. Sci. USA 85:7182-7186.

Charbonneau H, Tonks NK, Kumar S, Diltz CD, Harrylock M, Cool DE, Krebs EG, Fischer EH, Walsh KA (1989) Human placenta protein tyrosine phosphatase: Amino acid sequence and relationship to a family of receptor like proteins. Proc. Natl. Acad. Sci. USA 86:5252-5256.

Chernoff J, Schievella AR, Jost CA, Erikson RL, Neel BG (1990) Cloning of a cDNA for a major human protein tyrosine phosphatase. Proc. Natl. Acad. Sci. USA 87:2735-2739.

Cool DE, Tonks NH, Charbonneau H, Fischer EH, Krebs EG (1990) Expression of a human T-cell protein-tyrosine-phosphatase in baby hamster kidney cells. Proc. Natl. Acad. Sci. USA 87:7280-7284.

Guan KL, Haun RS, Watson SJ, Gaehlen RL, Dixon JE (1990) Cloning and expression of a protein tyrosine phosphatase. Proc. Natl. Acad. Sci 87:1501-1505.

Ingebritsen TS (1989) Phosphotyrosyl-protein phosphatases. II. Identification and characterization of two heat-stable protein inhibitors. J. Biol. Chem. 264:7801-7808.

Matthews FJ, Cahir, ED, Thomas ML (1990), Identification of an additional member of the protein-tyrosine-phosphatase family: Evidence for alternative splicing the tyrosine phosphatase domain. Proc. Natl. Acad. Sci. USA. 87:4444-4448.

Streuli M, Krueger NX, Hall LR, Schlossman SF, Saito H (1988) A new member of the immunoglobulin superfamily that has a cytoplasmic region homologous to the leukocyte common antigen. J. Exp. Med. 168:1553-1562.

Tonks NK, Diltz CD, Fischer EH (1988a) Purification of the major protein-tyrosine-phosphatases from human placenta. J. Biol. Chem. 263:6722-6730.

Tonks NK, Diltz CD, Fischer EH (1988b) Characterization of the major protein-tyrosine-phosphatases from human placenta. J. Biol. Chem. 263:6731-6737.

Tonks NK, Diltz CD, Fischer EH (1990) CD45, an integral memembrane protein tyrosine phosphatase. Characerization of enzyme activity. J. Biol. Chem. 265:10674-10680.

VIII.   CONTROL
OF CELLULAR PROCESSES

BIOCHEMICAL REGULATION OF THE CDC2 PROTEIN KINASE

Giulio Draetta
Differentiation Programme
European Molecular Biology Laboratory
Postfach 10 2209
D - 6900 Heidelberg
Germany

cdc2 is the catalytic subunit of a universal cell cycle regulator

cdc2[1] and its homolog CDC28, are 34 kd protein kinases originally identified in the yeasts *S.pombe* and S. *cerevisiae,* respectively (for review see (Cross et al., 1989) (Nurse, 1990). In these organisms, mutations in *cdc2/28* arrest cells in either the G1 or the G2 phases of the cell cycle. In addition, some dominant mutations of *cdc2* cause cells to traverse the cell cycle at an advanced rate, confirming that this gene acts at a major control point of the cycle. Using complementation of the genetic defect both *cdc2* and *CDC28* were cloned. From their predicted protein sequences it appeared that *cdc2* and *CDC28* encode ser/thr protein kinases (Reed et al., 1985) (Simanis and Nurse, 1986). Indeed, in vitro assays demonstrated that cdc2 can phosphorylate histone H1 and other substrates on serine or threonine residues (with a primary sequence requirement: X - Ser/Thr - Pro - X). A major breakthrough came when it was found that cdc2 homologs are present also in higher eukaryotes (Draetta et al., 1987) (Lee and Nurse, 1987). Furthermore, it was found that cdc2 is a component of the Maturation Promoting Factor (MPF) (Dunphy et al., 1988) (Gautier et al., 1988), a soluble component of mitotic cells which is able, when injected in interphase cells, to drive their nuclei into mitosis. MPF activity can be easily assayed by microinjection of the large oocytes of *Xenopus*. These cells are arrested at the G2/M prophase of meiosis I. Upon microinjection of MPF, a burst of protein phosphorylation occurs and oocytes mature into eggs, which will remain arrested in metaphase until fertilization.

   cdc2 is a unique transducer able to respond to extracellular signals (availability of nutrients, growth factors), and to sense the occurrence of events such as cellular growth, DNA replication and mitosis. All these signals converge on (and activate) specific forms of cdc2, which differ in their state of phosphorylation and association with

---

[1] Throughout the text, italics will be used to refer to genes (eg. *cdc2*), while for their protein products regular characters willbe used (eg. cdc2).

NATO ASI Series, Vol. H 56
Cellular Regulation by Protein Phosphorylation
Edited by L. M. G. Heilmeyer, Jr.
© Springer-Verlag Berlin Heidelberg 1991

regulatory subunits. Although structural and biochemical studies on the cdc2 kinase are just beginning, numerous elements of its regulation have been elucidated so far. In this paper I will summarize the present knowledge on the biochemical regulation of such kinase.

Association with regulatory subunits

In 1988 it was found that a 62,000 Mr protein (p62) associates with cdc2 and becomes phosphorylated as cells enter mitosis (Draetta and Beach, 1988). Monomeric cdc2 is inactive as histone H1 kinase, unless it is bound in a high molecular weight complex with p62 (Brizuela et al., 1989). The pattern of accumulation of the cdc2/p62 complex prompted to investigate the possibility that cdc2 might associate with cyclins, proteins which accumulate and are selectively degraded at specific stages of the cell cycle. It had been previously demonstrated that injection of mRNA encoding for a clam cyclin induces maturation in *Xenopus* oocytes, suggesting that cyclins may be direct activators of MPF (Swenson et al., 1986). Indeed clam cyclins were found to associate with cdc2, and to form an active histone H1 kinase (Draetta et al., 1989). Furthermore it was found that some suppressors of cdc2 mutations in yeast encode proteins with structural homologies to the cyclins (Booher and Beach, 1987) (Booher et al., 1989). It was also demonstrated that human p62 is a cyclin of the B type (Pines and Hunter, 1989), and that a complex between cyclin B and cdc2 is the MPF purified from *Xenopus* eggs (Gautier et al., 1990).

Many cyclin genes have been cloned, and for some of the encoded proteins it has been shown that they physically interact with cdc2. On the basis of their sequence similarity and their pattern of accumulation cyclins have been divided into different subtypes. For example, the *S.pombe* cdc13 cyclin belongs to the B-type class, and a complex of cdc13 and cdc2 is maximally active at mitosis. In *S. cerevisiae* the cyclins encoded by the *CLN* genes accumulate at the G1/S transition of the cell cycle (Hadwiger et al., 1989). In human cells, cDNA encoding A- and B- type cyclins have been cloned (Pines and Hunter, 1989) (Wang et al., 1990). A 60,000 Mr polypeptide(p60) originally identified as a cellular protein which binds the E1A oncogene in Adenovirus- infected cells, was found to associate with cdc2 as well (Giordano et al., 1989). It is now known that p60 is the human homolog of cyclin A (Pines and Hunter, 1990). Its pattern of accumulation and binding to the cdc2 protein kinase differs from the one of cyclin B (p62, see above) in that it does bind cdc2 in S and G2 , but it is degraded before metaphase.

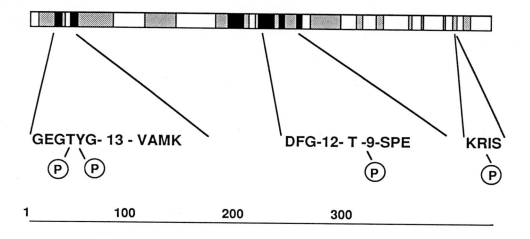

Fig.1 *Functional domains of the cdc2 protein*
Black areas indicate regions of homology with the protein kinase family. Grey areas indicate sequences   present in all cloned cdc2 homologs. Also shown are amino acid sequences surrounding mapped phosphorylation sites.

A further subunit of the cdc2 protein kinase is suc1, a 13 kd protein. The *suc1* gene was identified through a genetical analysis of cell cycle defects in *S. pombe* (Hayles et al., 1986). *suc1* is a suppressor of certain mutations in the *cdc2* gene. It was found that the suc1 protein can bind monomeric cdc2, as well as cdc2/cyclin B complexes (Brizuela et al., 1987). Therefore it is implied that some of the cdc2 mutants have a defect in the ability to bind suc1 and can be rescued by suc1 overexpression. Recently it has been found that the association between cdc2 and suc1 is cell-cycle regulated (M.Pagano and G. Draetta, in preparation; B.Ducommun and G. Draetta in preparation). The existence of an equilibrium between a suc1- free and a bound form of cdc2 is also suggested by the fact that large overproduction of suc1 leads to cell cycle arrest, probably by sequestering the cdc2 protein in a permanently bound form (G. Basi and G. Draetta, in preparation).

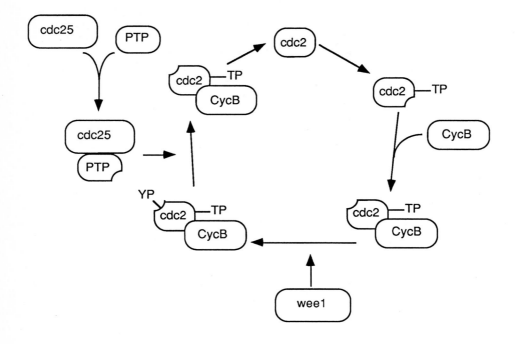

Fig.2 Model for the activation of the cdc2/cyclin B complex at the G2/M transition.
Monomeric cdc2 is inactive. Thr161 phosphorylation allows binding to cyclin B.
As the complex forms, the wee1 protein kinase induces phosphorylation of cdc2
on Tyr15, keeping the cyclin B/cdc2 complex inactive until DNA replication is
completed. cdc25,   following completion of DNA replication, activates a protein
tyrosine phosphatase. Dephosphorylation of cdc2   leads to full activation of the
histone H1 kinase, and starts mitosis.

cdc2 regulation by post-translational modifications

cdc2 is phosphorylated in vivo on serine, threonine and tyrosine residues. There is no
evidence for the occurrence of cdc2 autophosphorylation in vitro. Four major sites of
phosphorylation have been identified to date (Fig.1). In mammalian cells,
phosphorylation on serine   (Ser277) occurs specifically in the G1 phase of the cycle;
this phosphorylation site has not been identified in yeast (Krek and Nigg, 1991).   Tyr15
is located within the nucleotide binding domain of the cdc2 protein (Gould and Nurse,
1989); phosphorylation at this site occurs during the S and G2 phases of the cell cycle.
The tyrosine- phosphorylated cdc2, complexed with cyclin B, is inactive; at the end of

G2 dephosphorylation of Tyr15 leads to full activation of the kinase and entry into mitosis in yeast (Gould and Nurse, 1989) . Indeed, yeast strains carrying a mutated cdc2 with a phenylalanine in position 15 instead of tyrosine, are deregulated for cdc2 activation and proceed through the cycle at an advanced rate. Two sites of threonine phosphorylation have been identified. Thr14 is located within the ATP binding site (Krek and Nigg, 1991), while Thr167 (Thr 161 in the mammalian cdc2) is located in a position homologous to Thr197 and Tyr 416, putative autophosphorylation sites of cAMP-dependent protein kinase and p60^{c-src}, respectively (Booher and Beach, 1986). Recent evidence shows that phosphorylation of Thr161 is an absolute requirement for cyclin binding in vitro (B. Ducommun, P. Brambilla and G. Draetta, unpublished). It has been demonstrated that cyclin B binding to cdc2 increases the ability of cdc2 to undergo phosphorylation on tyrosine (Solomon et al., 1990).

A phosphorylation cascade must be involved in the mechanism of activation of the cdc2 kinase (Fig.2). Two *S. pombe* protein kinases, mik1 and wee1, are part of the upstream regulatory mechanism for cdc2 activation. Wee1 is a dose-dependent inhibitor of mitosis (Russell and Nurse, 1987); an increase in the amount of wee1 in the cell leads to cell cycle delay, which is proportional to the level of its expression. Less is known about mik1, but in the absence of both kinases tyrosine phosphorylation of cdc2 is abruptly lost and a complete alteration of cell cycle timing occurs with appearance of anucleated, multinucleated and fragmented cells (mitotic catastrophe)(K. Lundgren and D. Beach, person. commun.). This suggest that mik1 and wee1 are involved either directly or indirectly in the tyrosine phosphorylation of cdc2.

cdc25 is an activator of mitosis (Russell and Nurse, 1986). It has been found that yeast cells lacking the cdc25 gene product arrest in G2, with a preformed cdc2/cyclin B complex. The cdc2 protein in this complex contains high levels of phosphotyrosine. Lack of cdc25 can be suppressed by overexpression of a tyrosine phosphatase from human cells (Gould et al., 1990). Since there is no obvious structural similarity between cdc25 and tyrosine phosphatases, it is reasonable to hypothesize that cdc25 binds and activates the cdc2 tyrosine phosphatase . In fact a cdc2 protein lacking Tyr15, does not need cdc25 function. The presence of Thr 14 phosphorylation in mammalian cells renders the regulation even more complicate. In fact it is known that tyrosine dephosphorylation of the cdc2 complex isolated from G2- arrested cells does not lead to activation of its kinase activity (Morla et al., 1989). Probably Thr14 and Tyr15 dephosphorylation are needed for cdc2 activation in mammalian cells.

cdc2 targets

In Table I I have listed the cdc2 substrates identified so far, together with the functional effects brought by phosphorylation, when detected (for review see, (Lewin, 1990)). The first identified cdc2 substrate was Histone H1, the phosphorylation of which at mitosis has been known for a long time. The activation of the cdc2 kinase before mitosis leads to phosphorylation of a number of different proteins, all involved in the structural changes occurring in the cell at mitosis. It is interesting to find among the cdc2 substrates the cellular oncogene products p60^{c-src}, c-abl, the antioncogene products retinoblastoma and p53. They are all phosphorylated *in vitro* by cdc2 at sites which are phosphorylated *in vivo*.

It is clear that some of the cdc2 substrates are phosphorylated at times in the cell cycle other than mitosis. This, together with the genetical evidence from yeast confirming that the kinase activity of cdc2 is required during the G1/S as well as the G2/ M transition suggests that different cdc2/cyclin complexes, activated at discrete stages of the cell cycle, phosphorylate discrete sets of substrates. The cyclins might have a role in controlling the timing of activation of the kinase, but also in directing cdc2 to a specific subcellular location or in changing its substrate specificity.

Perspectives

The identification of cdc2 in higher eukaryotes, together with the finding that it is itself a target of tyrosine kinases (all of the known tyrosine kinases are encoded by either cellular oncogenes or growth factor receptors) has stimulated intensive investigation to try to link cdc2 activation to the activation of proliferation as a prerequisite for abnormal (cancerous) growth. It is obvious that cell cycle control proteins must be the target of regulatory signals which sense the outside environment and mediate the choice growth versus quiescence. Our hope is to be able to understand how permanent activation of this kinase can be responsible for cellular immortalization, and be the basis for further oncogenic events. In two occasions cdc2 has been found to be involved with oncogenes. Firstly, it was shown that the cyclin A subunit of cdc2 can bind the

Adenovirus E1A nuclear oncogene. Recent mapping studies (A.Giordano and B.R. Franza, Jr., pers. commun.) have shown that the binding sites for cyclin A on the E1A

Table I. cdc2 substrates

Substrate	Functional effects of phosphorylation	When in the cell cycle ?
RNApol II CTD	Activation of transcription?	G1,S
p53	Unknown	S
SV40 large T	Activation of DNA replication	S
RFA	Activation of DNA replication	S
p105Rb	Inactivation of Rb in late G1?	S,G2,M
p60c-src	Increased kinase activity	M
v-abl	Unknown	M
Histone H1	Chromatine condensation?	M
Nucleolin	Inhibition of rRNA transcription?	M
Vimentin	Reorganization of microfilaments	M
SWI5	Inhibition of nuclear localization	M
Myosin LC's	Inhibition of cytokinesis?	M
Lamins	Lamin depolymerization	M
Caldesmon	Morphological changes	M

polypeptide coincide with E1A domains necessary for immortalization. In addition the same authors have found that a fraction of the cdc2 protein is bound to cyclin A and E1A. Secondly, the cyclin A gene has been identified in the DNA sequences flanking the site of integration of the Hepatitis B Virus, in a hepatocellular carcinoma. It will be interesting to investigate the possible effects of this events on cyclin A stability at the mRNA and protein level, as well as on the activity of the cyclin A/cdc2 kinase during the cell cycle.

Further links with the growth control pathways have been established. Studies done in S.cerevisiae have shown that a gene involved in negative regulation of growth, FAR1, is dispensable in the absence of the CLN2 cyclin gene (Chang and Herskowitz,

1990). It has been suggested that FAR1 regulates the expression of this cyclin, the inactivation of which is required for growth arrest. All three CLN cyclins have to be inactivated for G1 phase arrest in S.*cerevisiae*. In the same yeast FUS3, a protein kinase which is also a negative regulator of growth, has been isolated. FUS3 is probably acting through an inhibition of the CLN3 (Elion et al., 1990). It has regions of homology to MAP2 kinase, a serine/threonine protein kinase which is involved in signal transduction in mammalian cells.

As we proceed with our investigations, more and more protein kinases and phosphatases are being discovered. Understanding the cdc2 activation cascade will represent a great challenge for the biochemist of the 90's, in many ways comparable to the work started in the 50's (and still going at full speed!) on the regulation of glycogen metabolism by phosphorylation. As of the immediate future a number of obvious questions remain open. It will be interesting, for example, to find out more about the role of the cyclin subunits in the regulation of the kinase; whether for example binding to the cyclins relieves any inhibition exerted by pseudo- substrate sequences in the cdc2 protein, or whether binding to the cyclins helps forming the substrate binding site. And more: the role of the suc1 protein is still unknown. The possibility that suc1 might target cdc2 to a tyrosine phosphatase is been actively investigated. cdc2, the cyclins and suc1 can be expressed as recombinant proteins, and attempts to study molecular structure are in progress. I hope to seeing more and more biochemists involved with cdc2 during the years to come!

References

Booher, R., Alfa, N. C., Hyams, J. S., and Beach, D. H. (1989). The fission yeast cdc2/cdc13/suc1 protein kinase: regulation of catalytic activity and cellular localization. Cell *58*, 485-497.

Booher, R., and Beach, D. (1986). Site-specific mutageneisis of cdc2+, a cell cycle control of the fission yeast *Schizosacharomyces pombe*. Mol. Cell. Biol. *6*, 3523-3530.

Booher, R., and Beach, D. (1987). Interaction between cdc13+ and cdc2+ in the control of mitosis in fission yeast; dissociation of the G1 and G2 roles of the cdc2+ protein kinase. EMBO J. *6*, 3441.

Brizuela, L., Draetta, G., and Beach, D. (1987). p13suc1 acts in the fission yeast cell division cycle as a component of the p34cdc2 protein kinase. EMBO J. *6*, 3507.

Brizuela, L., Draetta, G., and Beach, D. (1989). Activation of human CDC2 protein as a histone H1 kinase is associated with coplex formation with the p62 subunit. Proc. Natl. Acad. Sci. USA *86*, 4362-4366.

Chang, F., and Herskowitz, I. (1990). Identification of a gene necessary for cell cycle arrest by a negative growth factor of yeast: FAR1 is an inhibitor of a G1 cyclin, CLN2. Cell *63*, 999- 1011.

Cross, F., Roberts, J., and Weintraub, H. (1989). Simple and complex cell cycles. Annu. Rev. Cell Biol. *5*, 341-395.

Draetta, G., and Beach, D. (1988). p34 protein kinase, a human homolog of the yeast cell cycle control proteins encoded by cdc2+ and CDC28. Cancer Cells *6*, 259-263.

Draetta, G., Brizuela, L., Potashkin, J., and Beach, D. (1987). Identification of p34 and p13, human homologs of the cell cycle regulators of fission yeast encoded by cdc2+ and suc1+. Cell *50*, 319-325.

Draetta, G., Luca, F., Westendorf, J., Brizuela, L., Ruderman, J., and Beach, D. (1989). Cdc2 protein kinase is complexed with both cyclin A and B: evidence for proteolytic inactivation of MPF. Cell *56*, 829-838.

Dunphy, W. G., Brizuela, L., Beach, D., and Newport, J. (1988). The Xenopus cdc2 protein is a component of Xenopus MPF, a cytoplasmic regulator of mitosis. Cell *54*,

Elion, E. A., Grisafi, P. L., and Fink, G. R. (1990). *FUS3* encodes a *cdc2+/CDC28*-related kinase required for the transition from mitosis into conjugation. Cell *60*, 649-664.

Gautier, J., Minshull, J., Lohka, M., Glotzer, M., Hunt, T., and Maller, J. L. (1990). Cyclin is a component of maturation-promoting factor from *Xenopus*. Cell 487-494.

Gautier, J., Norbury, C., Lohka, M., Nurse, P., and Maller, J. (1988). Purified maturation-promoting factor contains the product of a Xenopus homolog of the fission yeast cell cycle control gene cdc2+. Cell *54*, 433-439.

Giordano, A., Whyte, P., Harlow, E., Franza, B. R. J., Beach, D., and Draetta, G. (1989). A 60 kd cdc2- associated polypeptide complexes with the E1A proteins in Adenovirus- infected cells. Cell *58*, 981-990.

Gould, K., and Nurse, P. (1989). Tyrosine phosphorylation of the fission yeast cdc2+ protein kinase regulates entry into mitosis. Nature *342*, 39-45.

Gould, K. L., Moreno, S., Tonks, N. K., and Nurse, P. (1990). Complementation of the mitotic activator, p80^{cdc25}, by a human protein-tyrosine phosphatase. Science *250*, 1573-1575.

Hadwiger, J. A., Wittenberg, C., Richardson, H. E., De Barros Lopes, M., and Reed, S. I. (1989). A family of cyclin homologs that control the G1 phase in yeast. Proc. Natl. Acad. Sci. USA *86*, 6255-6259.

Hayles, J., Beach, D., Durkacz, B., and Nurse, P. M. (1986). The fission yeast cell cycle control gene cdc2: isolation of a sequence suc1 that supreses cdc2 mutant function. Mol. Gen. Genet. *202*, 291-293.

Krek, W., and Nigg, E. (1991). Differential phosphorylation of vertebrate p34cdc2 kinase at the G1/S and G2/M transitions of the cell cycle: identification of major phosphorylation sites: EMBO J. , in press.

Lee, M., and Nurse, P. (1987). Complementation used to clone a human homologoue of the fission yeast cell cycle control gene cdc+2. Nature *327*, 31.

Lewin, B. (1990). Driving the cell cycle: M phase kinase, its partners, and substrates. Cell *61*, 743- 752.

Morla, A. O., Draetta, G., Beach, D., and Wang, J. Y. J. (1989). Reversible tyrosine phosphrylation of cdc2: dephosphorylation accompanies activation during entry into mitosis. Cell *58*, 193-203.

Nurse, P. (1990). Universal control mechanism regualting onset of M-phase. Nature *344*, 503-508.

Pines, J., and Hunter, T. (1989). Isolation of a human cyclin cDNA: evidene for cyclin mRNA and protein regulation in the cell cycle and for interaction with p34cdc2. Cell *58*, 833- 846.

Pines, J., and Hunter, T. (1990). Human cyclin A is adenovirus E1A- associated protein p60 and behaves differently from cyclin B. Nature *760*, 760- 763.

Reed, S. I., Hadwiger, J. A., and Lörincz, A. T. (1985). Protein kinase activity associated with the product of the yeast cell cycle gene CDC28. Proc. Natl. Acad. Sci. USA *82*, 4055-4059.

Russell, P., and Nurse, P. (1986). cdc25+ functions as an inducer in the mitotic control of fission yeast. Cell *45*, 145.

Russell, P., and Nurse, P. (1987). Negative regulation of mitosis by weel+, a gene encoding a protein kinase homolog. Cell *49*, 261.

Simanis, V., and Nurse, P. M. (1986). The cell cycle control gene cdc2+ of yeast encodes a protein kinase potentially regulated by phosphorylation. Cell *45*, 261- 268.

Solomon, M., Glotzer, M., Lee, T. H., Philippe, M., and Kirschner, M. W. (1990). Cyclin activation of p34cdc2. Cell *63*, 1013- 1024.

Swenson, K. I., Farrell, K. M., and Ruderman, J. V. (1986). The clam embryo protein cyclin A induces entry into M phase and resumption of meiosis in *Xenopus* oocytes. Cell *47*, 861-870.

Wang, J., Chenivesse, X., Henglein, B., and Brechot, C. (1990). Hepatitis B virus integration in a cyclin A gene in a hepataocellular carcinoma. Nature *343*, 555-557.

MOLECULAR AND BIOCHEMICAL CHARACTERIZATION OF THE MITOGEN-ACTIVATED S6 KINASE

George Thomas
Friedrich Miescher Institute
P.O. Box 2543
4002 Basel, Switzerland

INTRODUCTION

The approach generally taken to determine how growth factors induce cells to proliferate begins with purification of the growth factor and then using it to probe for its corresponding receptor. This has led to the finding that most growth factor receptors are tyrosine kinases and that binding of the growth factor to the receptor leads to an increase in this activity (for a review see Carpenter G, 1987; Yarden Y and Ulrich A, 1988). Results obtained from site-directed mutagenesis experiments further argue that the increase in this kinase activity is essential in conveying the mitogenic response (Carpenter G, 1987; Yarden Y and Ulrich A, 1988). Indeed, it is thought that the mitogenic response is propagated through a cascade of kinase reactions which are initiated by the tyrosine kinase receptor (Czech MP et al, 1988; Carpenter G and Cohen S, 1990). In our laboratory we have approached this question from inside the cell. We have begun with an obligatory step in the mitogenic response, the activation of protein synthesis, shown how this event may be triggered by the multiple phosphorylation of a ribosomal protein, S6, and attempted to trace the pathway back to the cell membrane describing the molecular components which are involved in the activation of this process (Kozma SC et al, 1989a; Hershey JWB, 1989).

Protein synthesis

The activation of protein synthesis is usually observed on sucrose gradients as a large shift of inactive 80S ribosomes into more heavily sedimenting actively translating polysomes. The inactive 80S ribosomes, unlike 80S monosomes, contain no attached mRNA, but instead are made up of naked 40S and 60S subunits (Hershey JWB,

NATO ASI Series, Vol. H 56
Cellular Regulation by Protein Phosphorylation
Edited by L. M. G. Heilmeyer, Jr.
© Springer-Verlag Berlin Heidelberg 1991

1989). This increase in polysome formation is accompanied by a shift of newly transcribed messenger RNA as well as a large pool of stored messenger RNA into the actively translating polysome population; in fact, about 80% of the increase in protein synthesis seen during the first eight hours of the mitogenic response is due to stored messenger RNA (Rudland PS, 1974). This differential use of messenger RNA prompted us to ask the question of whether alterations in messenger RNA expression also led to changes in the pattern of translation. Therefore, quiescent cells or serum-stimulated cells were pulse-labeled with ^{35}S-methionine and total cytoplasmic proteins were extracted and analyzed by NEPGE-2-dimensional gel electrophoresis. The observed patterns of translation from both quiescent and stimulated cells were quite similar, but with several notable changes (Thomas G et al, 1981). We were able to identify a number of those proteins whose levels were altered but, more importantly, we were able to show that about half of them were under translational control. Furthermore, one of these proteins, Q_{49}, which we have identified as elongation factor EF1-α (Thomas G and Thomas G, 1986), has been shown to be under selective translational control (H. Jefferies and G. Thomas, unpublished). This information will be very important in terms of establishing *in vitro* systems to test the importance of S6 phosphorylation in the activation of protein synthesis.

The increase in translational activity following the stimulation of quiescent cells to proliferate was shown very early on to be due to an increase in the rate of initiation of protein synthesis (Kozma SC et al, 1989; Hershey JWB, 1989). The components involved in this process are well described and we know many of their functions from both biochemical and genetic studies (Hershey JWB, 1989). Therefore, we argued that the activation of protein synthesis may serve as a useful model of how a growth factor acting on the cell surface turns on a mitogenic response within the cell. However, because of the large number of components involved we initially limited ourselves to looking at protein phosphorylation and dephosphorylation. This choice was made for two reasons: (1) we knew that phosphorylation reactions served as common mediators of intracellular regulatory signals, and (2) we knew that many of the translational components involved in initiation could be phosphorylated under a variety of growth conditions (Kozma SC et al, 1989a; Hershey JWB, 1989). This approach led to the finding that a single protein on the 40S ribosomal subunit, S6, became highly phosphorylated when quiescent cells were stimulated to proliferate (Thomas G et al, 1979).

Ribosomal Protein S6

To determine how much S6 was becoming phosphorylated following stimulation of quiescent cells to proliferate we took advantage of the fact that the more phosphory-lated this protein becomes the slower is its electrophoretic mobility on two-dimension-al polyacrylamide gels (Krieg J et al, 1988). For example, in quiescent cells S6 lies on a diagonal line with ribosomal proteins S2 and S4. However, thirty minutes after treatment with EGF most of the protein is largely relocated in 5 derivatives designated a - e, representing 1 to 5 moles of phosphate incorporated into S6. By 1 hour post-stimulation a large portion of the protein has shifted to derivatives e and d, representing the most highly phosphorylated forms of S6 (Olivier AR et al, 1988). This increase was shown to be due to phosphorylation since treatment of the fully phosphorylated form of S6 with alkaline phosphatase causes the protein to shift back to its native unphosphorylated state (Thomas G et al, 1979). At this point we focused on the question of whether S6 phosphorylation could be de-coupled from the activation of protein synthesis in vivo (Thomas G et al, 1980; Thomas G et al, 1982). Titration experiments were carried out with either individual growth factors or in the presence of inhibitors of either protein synthesis or S6 phosphorylation. In the case of cycloheximide, a potent inhibitor of protein synthesis, there was no effect on the rate or extent to which S6 became phosphorylated (Thomas G et al, 1980). In contrast, when S6 phosphorylation was blocked or slowed down with methylxan-thines, a corresponding effect on the initiation of protein synthesis was observed (Thomas G et al, 1980). This led to the speculation that the phosphorylation of S6 was somehow involved in recruiting inactive 80S ribosomes and stored mRNA into actively translating polysomes. To test this possibility extracts were prepared from quiescent cells which were stimulated for a short time, a time at which inactive 80S ribosomes were just beginning to be recruited into polysomes. Native 40S subunits, inactive 80S couples and polysomes were isolated and the extent of S6 phosphorylation was analyzed in each case (Thomas G et al, 1982; Duncan R and McConkey EH, 1982). The results show that the most highly phosphorylated ribosomes were present in polysomes and that 80S and 40S ribosomes were phosphorylated to a much lesser extent. The selection began with derivative c, was stronger for derivative d, and derivative e and was almost exclusively found in polysomes (Thomas G et al, 1982).

When the above experiments were carried out it was known that five moles of phosphate could be incorporated per mole of S6. However, it was not clear whether these five phosphates were incorporated into five distinct or many different sites or whether the phosphorylation was ordered or random. Over the last few years we have been able to show that there are (1) five distinct sites, (2) that they are clustered at the carboxy end of S6 and (3) that they apparently appear in a very specific order (see H.R. Bandi and G. Thomas, this volume). These results, together with the polysome shift experiments described above, suggested that the first two sites of phosphorylation, derivates a and b, may be involved in regulating the later sites of S6 phosphorylation and that the later sites of phosphorylation were those which had a direct effect on the activation of protein synthesis. This hypothesis is also consistent with both biochemical and genetic studies (Palin E and Traugh JA, 1987; Johnson SP and Warner JR, 1987). Furthermore, these findings have led us and others to speculate that the phosphorylation of S6 may somehow be involved in increasing the affinity of the 40S ribosome for messenger RNA (Kozma SC et al, 1989a; Hershey JWB, 1989). This hypothesis fits well with our knowledge about the location of S6 within the 40S ribosome. From crosslinking and protection experiments it has been argued that S6 resides in the mRNA-tRNA binding site (Nygård, O and Nilsson L, 1990). Recently we have prepared mono-specific antibodies against a synthetic S6 peptide containing all sites of phosphorylation. Together with U. Bommer and J. Stahl (Institute of Molecular Biology, Berlin) we are presently attempting by immunoelectromicroscopy (Lutsch G et al, 1990) to directly determine where the sites of S6 phosphorylation reside within the 40S ribosome (see H.R. Bandi and G. Thomas, this volume). We would now like to learn how important each of these sites is in protein synthesis, and to determine the mechanism which is responsible for controlling this event. Obviously, both of these studies would be greatly facilitated by identification of the kinase which controls S6 phosphorylation.

Mitogen-activated S6 kinase

The approach used to search for the activated kinase responsible for phosphory-lating S6 involved preparation of whole-cell extracts from either quiescent cells or cells treated with EGF. The extracts were then incubated in the absence or presence of 40S subunits. The pattern of protein phosphorylation in the absence of 40S subunits was quite similar for both quiescent and stimulated cells. However, addition of 40S subunits to these extracts led to a 30-fold increase in the amount of phosphate incorporated into a protein of M_r 30 kd. This protein was identified as ribosomal protein S6 by two-dimensional gel electrophoresis and shown to contain all the same tryptic-phosphopeptides observed *in vivo*, which indicated that the stimulated cell extract contained all the activity required for phosphorylating S6 in the intact cell (Novak-Hofer I and Thomas G, 1984). Detection of this activity required the presence of phosphatase inhibitors. Initially, β-glycerolphosphate was used as the phospha-tase inhibitor, but it was later found that many other phosphatase inhibitors could substitute for β-glycerolphosphate, including phosphoserine, p-nitrophenyl phosphate and pyrophosphate. The finding that phosphatase inhibitors were required to recover full kinase activity from cell extracts meant that S6 kinase phosphatases as well as S6 kinases had to be considered in the regulation of S6 phosphorylation. We therefore set out to identify these components several years ago. The strategy was to test the S6 kinase at each stage of the purification for its stability in the absence of phosphatase inhibitors. The kinase and phosphatase activities appeared to co-purify with each other through the first three successive steps of purification, high-speed centrifugation, followed by ammonium sulfate fractionation and then anion exchange chromatography. However, chromatography on a cation-exchange column separated these two activities, with the inactivator, the putative phosphatase, appearing in the flow-through and the kinase remaining bound to the column (Ballou et al, 1988a). The finding that the two activities appeared to co-fractionate with one another through the anion exchange column prompted us to reassay this column in the presence of an authentic phosphatase substrate, phosphorylase *a*. The results showed two poorly resolved peaks of phosphorylase *a* phosphatase activity, termed A and B, and that the kinase emerged between these two peaks. This column was then assayed with kinase as substrate and it was found that the major inactivating activity comigrated with the phosphatase in fraction B. Two phosphatases, 1 and 2A, are known to readily attack phosphorylase *a*. Furthermore, it is known that phosphatase 2A elutes later on an anion exchange column than does phosphatase 1. This result suggested that peak B represented a type 2A phosphatase, a conclusion substantiated by a number of additional findings, e.g., (1) peak B activity was not blocked by a specific inhibitor of phosphatase 1, (2) it selectively dephosphorylates the α subunit of phosphorylase kinase, and (3) the purified catalytic subunit of phosphatase 2A was 3- to 4-fold more efficient than that of phosphatase 1 in inactivating the kinase (Ballou et al, 1988a).

We used the knowledge that we could separate the phosphatase from the kinase on a cation exchange column as a first step in establishing a purification procedure for the isolation of the kinase. When the active fraction from the last step of purification was analyzed on silver stained SDS-PAGE a single band of M_r 70 kd was seen (Jenö P et al, 1988). During the latter stages of the purification we realized that an enzyme of about this molecular weight was becoming phosphorylated and was following the activity of the kinase (Jenö P et al, 1988), which suggested that the kinase itself may be autophosphorylated. To determine if this were the case, the ability of this enzyme to self-phosphorylate was tested against its ability to phosphorylate S6 in individual fractions through the last column of purification. The results showed that the two activities directly paralleled one another. Finally, we had assumed until then that this enzyme was responsible for phosphorylating all of the sites in S6 observed *in vivo*. To establish if this were indeed the case, we incubated unphosphorylated 40S ribosomes with γ^{32}P-ATP in the absence and the presence of the kinase. The results showed that in the presence of the kinase S6 could be driven to the most highly phosphorylated derivatives. Subsequent drying and autoradiography of this gel showed that only S6 became phosphorylated under these conditions. Finally, when the radioactively labeled material was excised and digested with trypsin, the same phosphopeptides as observed *in vivo* were obtained (Jenö P et al, 1988), indicating that this kinase is responsible for phosphorylating all the sites observed *in vivo*. Hence, from the data above, it was argued that the M_r 70 kd band is the kinase which is responsible for phosphorylating S6 in the intact cell.

Rat liver S6 kinase

Although we had identified the S6 kinase, there was insufficient material from cultured cells to generate either protein sequence data or polyclonal antibodies (Jenö P et al, 1989). Therefore, we searched for ways in which others had activated S6 phosphorylation in animals to provide much larger amounts of starting material for the purification of the enzyme. Most of these studies have been carried out in rat liver. In the first approach the animals have been starved for 24-36 hours and then either refed or given insulin. Both protocols are known to lead to an increase in protein synthesis and, in the case of insulin, an increase in S6 phosphorylation (Nielsen PJ et al, 1982). The second approach has been to inject cycloheximide into normally feeding animals, which is known to turn on a number of the early mitogenic responses, including S6 phosphorylation. In addition, we knew that vanadate, a potent phosphotyrosyl phosphatase inhibitor, could induce S6 phosphorylation in tissue culture cells (Novak-Hofer I and Thomas G, 1985) and thus reasoned that it may induce S6 phosphorylation in the intact animal. Therefore animals starved for 36 hrs were either refed or injected with insulin, and in a second set of experiments normally feeding animals were injected with either vanadate or cycloheximide. All treatments led to an increase in S6 phosphorylation, with the most potent being refeeding and cycloheximide injection (Kozma SC et al, 1989). Because it was simpler to handle a large number of animals employing cycloheximide we used this approach to determine whether we could induce an S6 kinase similar to the one observed in Swiss 3T3 cells. Comparison of kinase extracts from either serum-stimulated 3T3 cells or the livers of cycloheximide-injected rats on three distinct chromotographic columns revealed that both activities eluted in the identical position, indicating that the same enzyme was being activated in both systems. With that information in hand we worked out a large-scale purification procedure for the isolation of M_r 70 kd kinase (Kozma SC et al, 1989b), which we have recently modified (Lane HA and Thomas G, 1991). That this protein was equivalent to the 3T3 cell enzyme was shown by comparison of chyotrypsin or cyanogen bromide peptide maps of the two autophosphorylated enzymes (Kozma SC et al, 1989b). From approximately 1.5 kg of liver we isolated 70 to 90 µg of kinase.

Cloning the S6 kinase

The livers of cycloheximide-treated rat liver provided sufficient material to begin sequencing the protein. Tryptic peptides were separated by microbore reverse-phase HPLC, and 17 of them were sequenced. Three of the initial peptide sequences had evident homology with subdomains 6, 9 and 11 of the conserved catalytic domain of the serine/threonine kinases. On the basis of this observation degenerate oligonucle-otides representing sequences in domains 6 and 11 were synthesized and a 0.4 kb cDNA fragment was generated by PCR employing rat embryo cDNA as template (Kozma SC et al, 1990). The size of the DNA fragment was consistent with that predicted for a peptide extending from subdomains 6 to 11 and hybridized on Southern blots with a "Guessmer" derived from the peptide presumed to represent the S6 kinase subdomain 9. Subcloning and sequencing of the 0.4 kb cDNA fragment revealed that it encoded the S6 kinase peptide fragment presumed to represent subdomain 9. On the basis of its DNA sequence it was judged that this cDNA probe would hybridize specifically to the S6 kinase mRNA.

Northern blot analysis

The 0.4 kb cDNA fragment was used to probe a Northern blot of rat liver polyA + mRNA to obtain an idea of the size and number of mRNA transcripts. It was assumed from the low abundancy of the S6 kinase that the mRNA transcript would be rare, and thus require a cDNA probe of high specificity for Northern blot analysis and for screening cDNA libraries. We therefore turned to a recent method employing PCR to generate DNA probes of less than 0.5 kb having specific activities in the range of 0.5 to 1.0×10^{10} cpm/µg DNA (Schowalter PB and Sommer SS, 1989). The 0.4 kb cDNA probe labeled in this manner hybridized by Northern blot analysis to four transcripts of 2.5, 3.2, 4.0 and 6.0 kb (Kozma SC et al, 1990).

cDNA clones encoding S6 kinase

A rat liver cDNA library was screened with the radioactively labeled 0.4 kb DNA fragment described above. From 1.6×10^6 clones, 5 positive clones, with insert sizes of 1.4 to 28 kb were obtained. Each clone was partially sequenced employing an internal primer derived from the 0.4 kb cDNA fragment and found to contain the S6 kinase sequence. More extensive sequencing revealed that 3 were incomplete in the 5' coding region, and these were considered no further. By restriction map analysis and partial sequencing the two remaining clones appeared to be identical, reflecting the fact that the library had been amplified one time. An insert from one clone was sequenced and found to contain 2,805 nucleotides. The sequence encoding the S6 kinase contains 1506 nucleotides representing a protein of 502 amino acids and is terminated by two consecutive translational stop codons. The 5' untranslated region has a length of 133 nucleotides consistent with the average length found in most mRNA sequences. Sequencing from the 3' end of the gene did not reveal a polyA tail, suggesting that the cDNA clone had been initiated by random priming. Of the 17 peptide sequences, all but one could be aligned with the DNA sequence. Although the 3' untranslated region is not complete the size of the cDNA could roughly fit with the 3.2 kb transcript detected by Northern blot analysis. However, the greater abundance of the 2.5 kb mRNA led us to suspect that the cDNA might correspond to this transcript. Finally, when this clone was transcribed and translated *in vitro* it generated a polypeptide with a molecular weight identical to that of purified dephosphorylated S6 kinase, arguing that this clone represents the complete kinase (S. Kozma and G. Thomas, unpublished).

Comparison of catalytic domains

When the catalytic domains of the 65 reported protein kinase sequences are compared, absolute conservation of amino acids is seen in nine positions, and in five positions the same amino acid is present in 64 of 65 sequences. All of these conserved amino acids are present in the S6 kinase sequence. Finally, the motifs DLKPEN and G (T/F(XX)Y/F(XAPE) in subdomains 6 and 8 identify the S6 kinase as a member of the serine threonine kinase family. Of the five major subfamilies defined by amino acid sequence comparison, the S6 kinase appears to fall into the protein kinase C family. The most striking difference with this family is that the S6 kinase is the only member which contains tyrosine rather than phenylalanine in position 79. Within this subfamily the S6 kinase has 56% identity with the catalytic domain of S6 kinase 2_a, 49% with yeast protein kinase 1, and 44% with protein kinase C_ϵ (Kozma SC et al, 1990). However, these identities fall off sharply outside the catalytic domain, with the yeast protein kinase 1 having the highest overall homology of 41%. Together the results show that the S6 kinase is a member of the serine/threonine family of protein kinases most closely related to the group of kinases activated by second messengers.

Activation by direct phosphorylation

Having purified S6 kinase also allowed us to test whether the activity of the enzyme is directly controlled by phosphorylation. Thus, activated S6 kinase from either mouse or rat was incubated with phosphatase 2A in the absence and presence of a phosphatase inhibitor. In the absence of the inhibitor both enzymes were rapidly inactivated, whereas in the presence of the inhibitor both enzymes were stable over the incubation period (Kozma SC et al, 1989b). These results would suggest that phosphate is at least required to maintain the activity of the enzyme. To test whether phosphorylation leads to the activation of the kinase the enzyme was purified from quiescent and 60-min-serum-stimulated cells which were prelabeled with ^{32}P. No ^{32}P-labeled bands were detected in quiescent cells following the sizing step on gel filtration, whereas at this point a distinct band of M_r 70 kd could be resolved in stimulated cells (Ballou LM et al, 1988b). This band was purified to near homogeneity in the last step of purification, ATP-affinity chromatography. From its M_r 70,000 chromatographic behavior and ability to autophosphorylate, we argued that this protein band represented the kinase, indicating that phosphorylation leads to its activation (Ballou LM et al, 1988b). Furthermore, we showed that when this sample is treated with phosphatase 2A, the ^{32}P-phosphate is lost from the protein and in parallel the enzyme loses its ability to phosphorylate S6 *in vitro* (Ballou LM et al, 1988b). Phosphoamino acid analysis revealed that the protein was phosphorylated on serine and threonine, arguing that if this kinase lies on a phosphorylation cascade initiated by the tyrosine kinase of a receptor, then there must be at least one kinase between this enzyme and the receptor, namely, an S6 kinase kinase. In turn, this enzyme, although a serine-threonine kinase, would be activated by tyrosine phosphorylation (G. Thomas, this volume).

Future perspectives

In recent years the study of S6 phosphorylation has revealed a much more complex network of regulatory events than was initially perceived (Figure 1). Here I have outlined the present status of the field including the possible relevance of individual S6 phosphorylation sites to the existence of an S6 kinase cascade initiated by either growth factor receptor or an oncogene kinase. Future studies will shed additional light on the importance of these events in controlling the mitogen-induced increase in protein synthesis required for cell proliferation. To further elucidate the role of S6 phosphorylation in protein synthesis obvious genetic approaches will be employed, although it would be much more satisfying to apply biochemical approaches directly. To carry out such experiments would require two components, an active reconstituted system in protein synthesis and a messenger RNA which is known to be under selective translational control at this time. Both of these components have recently become available. We now have a reconstituted system which is very active in protein synthesis (Morley SJ and Hershey JWB, 1990) and have identified a protein which is under selective translational control following mitogenic stimulation of quiescent cells, EF-1α (H.B.J. Jefferies and G. Thomas, unpublished). Presently we are testing the effect of S6 phosphorylation on the expression of this mRNA in the reconstituted protein synthesizing system.

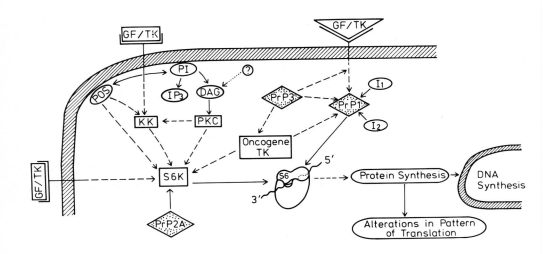

Figure 1. <u>Regulation of S6 phosphorylation.</u> □ = protein kinases: GF/TK (growth factor receptor-associated tyrosine kinase); oncogene-TK (cellular oncogene-associated tyrosine kinase); PKC (protein kinase C); KK (kinase kinase); S6K (S6 kinase). ◇ = protein phosphatases: PrP1 (serine/threonine protein phosphatase 1); PrP2A (serine/threonine protein phosphatase 2A); PrP3 (tyrosine protein phospha-tases). O = effectors: PI (phosphatidylinositol); IP$_3$ (inositol 1,4,5 trisphosphate); DAG (diacylglycerol); POS (phosphooligosaccharide); I_1 (inhibitor 1); I_2 (inhibitor $_2$). Solid lines indicate known events, broken lines indicate hypothetical events. (Reprinted with permission from Kozma et al, 1989a).

Acknowledgements: I would like to thank Drs. M. Dihanich and P.F. Jones for their critical reading of the manuscript, and C. Wiedmer for editing and typing expertise.

REFERENCES

Ballou LM, Jenö P and Thomas G (1988a) Protein phosphatase 2A inactivates the mitogen-stimulated S6 kinase from Swiss mouse 3T3 cells. J Biol Chem 263:1188-1194

Ballou LM, Siegmann M and Thomas G (1988b) S6 kinase in quiescent Swiss mouse 3T3 cells is activated by phosphorylation in response to serum treatment. Proc Natl Acad Sci USA 85:7154-7158

Carpenter G (1987) Receptors for Epidermal Growth Factor and Other Polypeptide Mitogens. In Ann Rev Biochem 56:818-914

Carpenter G and Cohen S (1990) Epidermal growth factor. J Biol Chem 265:7709-7712

Czech MP et al. (1988) Insulin receptor signalling. J Biol Chem 263:11017-11020

Duncan R and McConkey EH (1982) Preferential utilization of phosphorylated 40S ribosomal subunits during initiation complex formation. Eur J Biochem 123:535-538

Hershey JWB (1989) Protein phosphorylation controls translation rates. J Biol. Chem. 264:20823-20826

Jenö P et al. (1988) Identification and characterization of a mitogen-activated S6 kinase. Proc Natl Acad Sci USA 85:406-410

Jenö P et al. (1989) Purification and characterization of a 40S ribosomal protein S6 kinase from vanadate-stimulated Swiss 3T3 cells. J Biol Chem 264:1293-1297

Johnson SP and Warner JR (1987) Phosphorylation of the saccharomycecerevisiae equivalent of ribosomal protein S6 has no detectable effect on growth. Mol Cell Biol 7:1338-1345

Kozma SC, Ferarri S. and Thomas G (1989a) Unmasking a growth factor/oncogene-activated S6 phosphorylation cascade. Cell Signalling 3:219-225

Kozma SC et al. (1989b) A stimulated S6 kinase from rat liver: identity with the mitogen-activated S6 kinase of 3T3 cells. EMBO J 8:4125-4132

Kozma SC et al. (1990) Cloning of the mitogen-activated S6 kinase from rat liver reveals an enzyme of the second messenger subfamily. Proc Natl Acad Sci USA 87:7365-7369

Krieg J, Olivier AR and Thomas G (1988) Analysis of 40S ribosomal protein S6 phosphorylation during the mitogenic response. In Meth Enzymol 164:575-581

Lane HA and Thomas G (1991) Purification and properties of mitogen-activated S6 kinase from rat liver and 3T3 cells. In Methods of Enzymol 200, in press

Lutsch G et al (1990) Immunoelectron microscopic studies on the location of ribosomal proteins on the surface of the 40S ribosomal subunit from rat liver. Eur J Cell Biol 51:140-150

Morley SJ and Hershey JWB (1990) A fractionated reticulocyte lysate retains high efficiency for protein synthesis. Biochemie 72:259-264

Nielsen PJ et al. (1982) The phosphorylation of ribosomal protein S6 in rat tissues following cycloheximide injection, in diabetes, and after denervation of diaphram. J Biol Chem 257:12316-12321

Novak-Hofer I and Thomas G (1984) An activated S6 kinase in extracts from serum- and epidermal growth factor-stimulated Swiss 3T3 cells. J Biol Chem 259:5995-6000

Novak-Hofer I and Thomas G (1985) Epidermal growth factor-mediated activation of an S6 kinase in Swiss mouse 3T3 cells. J Biol Chem 260:10314-10319

Nygård O and Nilsson L (1990) Translational dynamics. Eur J Biochem 191:1-17

Olivier AR, Ballou LM and Thomas G (1988) Differential regulation of S6 phosphorylation by insulin and epidermal growth factor in Swiss mouse 3T3 cells: insulin activation of type 1 phosphatase. Proc Natl Acad Sci USA 85:4720-4724

Palin E and Traugh JA (1987) Phosphorylation of ribosomal protein S6 by cAMP-dependent protein kinase and mitogen-stimulated S6 kinase differentially alters translation of globin mRNA. J Biol Chem 262:3518-3523

Rudland PS (1974) Control of translation in cultured cells: continued synthesis and accumulation of messenger RNA in nondividing cultures. Proc Natl Acad Sci USA 71:750-754

Schowalter PB and Sommer SS (1989) The generation of radiolabeled DNA and RNA probes with polymerase chain reaction. Anal Biochem 177:90-94

Thomas G, Siegmann M and Gordon J (1979) Multiple phosphorylation of ribosomal protein S6 during the transition of quiescent 3T3 cells into early G_1 and cellular compartmentalization of the phosphate donor. Proc Natl Acad Sci USA 76:3952-3956

Thomas G et al. (1980) Regulation of 40S ribosomal protein S6 phosphorylation. In Swiss mouse 3T3 cells. Cell 19:1015-1023

Thomas G, Thomas G and Luther H (1981) Transcriptional and translational control of cytoplasmic proteins following serum stimulation of quiescent Swiss 3T3 cells. Proc Natl Acad Sci USA 78:5712-5716

Thomas G et al. (1982) The effect of serum, EGF, $PGF_{2\alpha}$ and insulin on S6 phosphorylation and the initiation of protein and DNA synthesis. Cell 30:235-247

Thomas G and Thomas G (1986) Translational control of mRNA expression during the early mitogenic response in Swiss mouse 3T3 cells: identification of specific proteins. J Cell Biol 103:2137-2144

Yarden Y and Ulrich A (1988) Growth Factor Tyrosine Kinases. In Ann Rev Biochem 57:443-478

PROTEIN PHOSPHORYLATION IN THE NERVOUS SYSTEM

Manfred W. Kilimann
Institut für Physiologische Chemie
(Abteilung für Biochemie Supramolekularer Systeme)
Ruhr-Universität Bochum
Universitätsstraße 150
W 4630 Bochum 1
Germany

Protein phosphorylation is involved in many mechanisms that contribute to neuronal plasticity, i.e. in the modulation of the information processing and storage properties of nerve cells resulting, ultimately, in the ability of the nervous system to learn. At the molecular and cellular level, this means that a signal passing through a neuron leaves behind a trace that affects the future behaviour of the cell. The time scale of such an effect may be very different: it may last for only a few seconds, or it may be lifelong.

Neuronal plasticity can be achieved by protein phosphorylation at various levels.

I. Proteins directly affecting neuronal signalling are subject to regulation by phosphorylation, such as proteins involved in

   - neurotransmitter metabolism
   - neurotransmitter release
   - neurotransmitter recognition
   - membrane excitability
   - formation of synaptic connections

II. The gene expression of such proteins may be regulated by the phosphorylation of transcription factors.

The purpose of this review is to illustrate, by one or a few examples for each case, how neuronal plasticity can be achieved by protein phosphorylation.

Modulation of neurotransmitter synthesis: tyrosine hydroxylase

Tyrosine hydroxylase (TH) catalyzes the first, rate-limiting step in the biosynthesis of catecholamines (dopamine, noradrenaline and adrenaline). Catecholaminergic cells are found in the adrenal medulla, the sympathetic part of the peripheral nervous system, and small but important cell populations of the central nervous system. It has been known for long that catecholamine release is followed by an increase of TH activity, which makes sense because it leads to the repletion of catecholamine stores. A long-term increase of the amount of TH protein, lasting hours to days, is achieved by a enhancement of TH gene transcription. An acute stimulation of TH activity, lasting for minutes, is achieved by phosphorylation of TH itself.

The N-terminal 40 amino acids of TH comprise a cluster of phosphorylatable serine residues upon which several signal pathways converge. Phosphorylation, with differential preferences for one or the other serine residue, can be stimulated in various experimental systems by depolarization, cAMP-elevating agents, phorbol esters or growth factors (nerve growth factor, NGF, or epidermal growth factor, EGF). Signal mechanisms likely to be involved in vivo are nicotinic cholinergic stimulation and calcium/calmodulin dependent protein kinases, certain neuropeptides and cAMP-dependent protein kinase (PKA), and muscarinic cholinergic stimulation and protein kinase C (PKC), respectively.

Selected literature: Zigmond, 1985; Campbell et al., 1986; Black et al., 1987; Haycock, 1990.

Modulation of neurotransmitter release: synapsin I

Synapsin I is present in almost all nerve terminals, where it binds with high affinity to the cytoplasmic side of small neurotransmitter vesicles. It also binds to, and bundles, actin filaments, and is thought to immobilize the vesicles by cross-linking them to the microfilament meshwork of the synaptic terminal. Both binding affinities are reduced by calcium/calmodulin-dependent phosphorylation and, less efficiently, by cAMP-dependent phosphorylation. Phosphorylation has been shown to occur in vivo in response to depolarization, serotonin, dopamine and noradrenalin.

Therefore, repeated stimulation of a synapse is imagined to lead to enhanced phosphorylation of synapsin I, resulting in the liberation of part of its immobilized neurotransmitter vesicle pool, and an increase of neurotransmitter exocytosis. In fact, it has been shown in vivo that injection of dephospho-synapsin I inhibits, whereas injection of calcium/calmodulin dependent protein kinase II stimulates synaptic conductivity.

Recent reviews: Hemmings et al., 1989; Bähler et al., 1990.

Modulation of neurotransmitter receptors

The nicotinic acetylcholine receptor, a ligand-gated ion channel, is phosphorylated by PKA, PKC, and a protein tyrosine kinase, on its subunits beta, gamma and delta. All phosphorylation sites are located in similar positions on the major intracellular loop of each subunit. All three kinases, or agents that stimulate them, have been shown to enhance receptor desensitization.

Desensitization of G protein-linked receptors like the beta-adrenergic receptor (ß-AR), the alpha-adrenergic receptor, and rhodopsin, is also modulated by phosphorylation. The ß-AR can be phosphorylated by PKA, PKC, a protein tyrosine kinase, and a second messenger-independent ß-AR kinase that phosphorylates only the ligand-occupied receptor. Multiple phosphorylation sites are located on the third cytoplasmic loop and the cyto-plasmic C-terminus of the protein. Thus, mechanisms are provided for different dimensions of desensitization and down-regu-lation, and for cross-talk between different receptors on a given cell.

Recent literature: Hausdorff et al., 1990; Huganir and Green-gard, 1990 (reviews); Valiquette et al., 1990).

Modulation of membrane excitability: ion channels

A large number of ion conductances in many systems have been shown to be influenced by protein kinases or agents that acti-vate them. The voltage-gated Na^+ channel is phosphorylated by PKA and PKC, and phosphorylation by PKA enhances channel inac-tivation. Phosphorylation of the dihydropyridine-sensitive Ca^{2+} channel by PKA enhances channel conductivity (review: Catte-rall, 1988), whereas PKA-catalyzed phosphorylation of the ino-sitol 1,4,5-trisphospate receptor, a ligand-gated Ca^{2+} channel, prevents ligand-induced opening of the channel (Suttapone et al., 1988). The slow afterhyperpolarization K^+ channel in mam-malian hippocampus, a brain structure important for memory, is modulated by a number of neurotransmitters/neuromodulators acting presumably through protein kinases (review: Nicoll, 1988). A particularly well-studied case is the short-term facilitation of the gill and siphon-withdrawal reflex of Aplysia, which occurs through the PKA-catalyzed phosphorylation and inhibition of K^+ channels in response to serotonergic stimula-tion (Shuster et al., 1985).

Axonal growth and synapse formation: GAP-43

Correct physical connections between neurons are a prerequisite for the proper functioning of a nervous system, and some learning events are associated with the formation of new synaptic contacts. GAP-43 (also known as B-50, pp46 and F1) is a phosphoprotein found in neurons and glia, whose expression is correlated with axonal growth and regeneration, and synaptogenesis. It is prominent in axonal growth cones. Expression of GAP-43 in nonneuronal tissue culture cells by gene transfer leads to the extension of cellular processes. Recent literature: Skene, 1989 (review); Zuber et al., 1989; da Cunha and Vitkovic, 1990.

Control of gene expression by phosphorylation of transcription factors

Expression of neuronal genes has been shown to be modulated e.g. by
- membrane electrical activity
- neurotransmitters
- neurotrophic factors,
and this modulation is likely to be important both in development and adaptive plasticity of the nervous system.

Genes responding to a stimulus by enhanced transcription can be grouped in two classes: immediate early response (IER) genes (response within minutes) and late response genes (response within hours, depending on protein synthesis). Many IER genes have been shown to be transcription factors.

A prominent example is the proto-oncogene, c-fos, whose mRNA is

induced rapidly and transiently in cell culture by depolarization, phorbol esters, forskolin, NGF or EGF, and in vivo by
seizure or massive sensory stimulation. Regulatory DNA elements
for transcription control of the c-fos gene reside in its 5'-
flanking sequences. An octanucleotide consensus sequence,
TGACGTCA, named CRE (cAMP-responsive element), is the binding
site for a transcription factor, CREB (CRE-binding protein).
CREB can be phosphorylated on a single serine residue by PKA
and a Ca^{2+}/calmodulin-dependent kinase; it inhibits transcription when it is dephosphorylated, and stimulates it in its
phosphorylated state. CREs confer cAMP-responsiveness to several other genes in whose cis-regulatory regions they are found
(somatostatin, proenkephalin, tyrosine hydroxylase genes).

PKC-stimulating agents and serum growth factors act on c-fos
transcription through a more upstream regulatory sequence, the
SRE (serum response element). The gene product, Fos, as a
heterodimer with a second proto-oncogene and IER gene product,
Jun, is itself a transcription factor. Both can also be regulated by phosphorylation. Recent literature: Sheng and Greenberg, 1990 (review), Wisden et al., 1990.

The gill and siphon-withdrawal reflex of aplysia provides an
example how a change in gene transcription can affect the
behaviour of a neuron. It was described above that short-term
facilitation (lasting minutes to hours) is achieved by cAMP-
dependent phosphorylation of K^+ channels. Repeated serotonergig
stimulation of the sensory neuron/motor neuron synapse leads to
long-term facilitation, lasting days to weeks. This can be prevented by inhibitors of transcription or translation. Apparently, the cAMP-dependent phosphorylation and activation of CREB,
which then stimulates the transcription of other gene(s) (whose
identity is as yet unknown), plays an important role in long-
term facilitation. Microinjection of CRE oligonucleotides into
the sensory neuron blocks long-term potentiation, presumably by
titrationg out the cellular CREBs (Dash et al., 1990).

REFERENCES

Bähler M, Benfenati F, Valtorta F, Greengard P (1990) The synapsins and the regulation of synaptic function. BioEssays 12:259-263

Black IB, Adler JE, Dreyfus CF, Friedman WF, LaGamma EF, Roach AH (1987) Biochemistry of information storage in the nervous system. Science 236:1263-1268

Campbell DG, Hardie DG, Vulliet PR (1986) Identification of four phosphorylation sites in the N-terminal region of tyrosine hydroxylase. J Biol Chem 261:10489-10492

Catterall WA (1988) Structure and function of voltage-sensitive ion channels. Science 242:50-61

da Cunha A, Vitkovic L (1990) Regulation of immunoreactive GAP-43 expression in rat cortical macroglia is cell type specific. J Cell Biol 111:209-215

Dash KD, Hochner B, Kandel ER (1990) Injection of the cAMP-responsive element into the nucleus of Aplysia sensory neurons blocks long-term facilitation. Nature 345:718-721

Hausdorff WP, Caron MG, Lefkowitz RJ (1990) Turning off the signal: desensitization of ß-adrenergic receptor function. FASEB J 4:2881-2889

Haycock JW (1990) Phosphorylation of tyrosine hydroxylase in situ at serine 8, 19, 31, and 40. J Biol Chem 265:11682-11691

Hemmings HC, Nairn AC, McGuinnes T, Huganir RL, Greengard P (1989) Role of protein phosphorylation in neuronal signal transduction. FASEB J 3:1583-1592

Huganir RL, Greengard P (1990) Regulation of neurotransmitter receptor desensitization by protein phosphorylation. Neuron 5:555-567

Nicoll RA (1988) The coupling of neurotransmitter receptors to ion channels in the brain. Science 241:545-551

Sheng M, Greenberg ME (1990) The regulation and function of c-fos and other immediate early genes in the nervous system. Neuron 4:477-485

Shuster MJ, Camardo JS, Siegelbaum SA, Kandel ER (1985) Cyclic AMP-dependent protein kinase closes the serotonin-sensitive K^+ channels of Aplysia sensory neurones in cell-free membrane patches. Nature 313:392-395

Supattapone S, Danoff SK, Theibert A, Joseph SK, Steiner J, Snyder SH (1988) Cyclic AMP-dependent phosphorylation of a brain inositol trisphosphate receptor decreases its release of calcium. Proc Natl Acad Sci USA 85:8747-8750

Wisden W, Errington ML, Williams S, Dunnett SB, Waters C, Hitchcock D, Evan G, Bliss TVP, Hunt SP (1990) Differential expression of immediate early genes in the hippocampus and spinal cord. Neuron 4:603-614

Valiquette M, Bonin H, Hnatowich M, Caron MG, Lefkowitz RJ, Bouvier M (1990) Involvement of tyrosine residues located in the carboxyl tail of the human ß-adrenergic receptor in agonist-induced down-regulation of the receptor. Proc Natl Acad Sci USA 87:5089-5093

Zigmond RE (1985) Biochemical consequences of synaptic stimulation: The regulation of tyrosine hydroxylase activity by multiple transmitters. Trends in Neurosci 1985:63-69

Zuber MX, Goodman DW, Karns LR, Fishman MC (1989) The neuronal growth-associated protein GAP-43 induces filopodia in non-neuronal cells. Science 244:1193-1195

MITOTIC CONTROL IN MAMMALIAN CELLS, POSITIVE AND NEGATIVE REGULATION BY PROTEIN PHOSPHORYLATION

Anne Fernandez and Ned Lamb.

Cell Biology Unit,
CRBM, CNRS-INSERM,
BP 5051,
34033, Montpellier Cedex,
France

INTRODUCTION

Of the numerous processes under the regulation of differential protein phosphorylation, mitotic transit has recently attracted intense examination with the isolation of a 34000 Mr kinase as an active component of the maturation promoting factor (MPF) in oocytes and its identification as an homolog of the yeast cell division cycle gene product cdc2, and the histone 1 kinase from mammalian cells (1, 2, 3). This kinase, now known as p34^{cdc2} is itself regulated by multiple dephosphorylation and phosphorylation and its association with two or more cyclin proteins (for reviews see 4, 5, 6). However, whilst more and more reports seemed to converge to make p34^{cdc2} an universal control switch driving the different events associated with mitotic induction, we found that microinjection of this kinase in a highly purified and active form was insufficient to induce interphase somatic mammalian cells to enter mitosis (7). The present report will detail aspects of the potential functions of this kinase and other regulatory molecules which bring further insights on the role of differential protein phosphorylation in the regulation of mammalian cell mitosis. Our approach has involved microneedle microinjection to examine the consequences on the progression through the cell cycle, cell morphology and cytoarchitecture of artificially elevating or inactivating different kinase and phosphatase pathways in synchronized mammalian cells. We have examined in particular a crucial point in mitosis, prophase entry, focussing on the role of p34^{cdc2} kinase, src kinase and the requirement of a distinct pathway involving inactivation of the cAMP-dependent protein kinase.

NATO ASI Series, Vol. H 56
Cellular Regulation by Protein Phosphorylation
Edited by L. M. G. Heilmeyer, Jr.
© Springer-Verlag Berlin Heidelberg 1991

PROPHASE ENTRY INVOLVES ACTIVATION OF p34cdc2 KINASE

The p34cdc2 kinase is an ubiquitously conserved kinase identified in all species so far examined (reviewed 8). Its inactivation in yeast is associated with a critical incapacity to divide and its overexpression following injection into oocytes is a sufficient inductive signal to cause unstimulated eggs to enter a pathway leading to maturation (Reviewed 5 and 6). With respect to mammalian cells, research on HeLa cells (reviewed 5) has shown that the protein kinase activity of p34cdc2 is associated with one or two cyclin proteins proposed to modulate its kinase activity (9,10) and activation of the kinase is associated with dephosphorylation of both tyrosine and threonine residues (11). Although p34cdc2 could be associated with mitotic entry, its functions in that process remained unclear, and apart from histone H1, p34cdc2 kinase seemed to phosphorylate few specific substrates (12), a finding confirmed by the identification of a consensus sequence for phosphorylation by p34cdc2 (S/TPXZ).

An attempt to examine the in vivo functions of p34cdc2 kinase in mammalian cells utilized microinjection of affinity purified antibodies (13). This study concluded that inhibiting p34cdc2 kinase in vivo (although the antibodies were non-inhibitory to the kinase activity in vitro) was sufficient to prevent cells from progressing through cytokinesis (ie. the separation of cells) but that nuclear division was not affected since the cells apparently became bi or tetra nuclear. We chose to examine the consequences of directly overexpressing active mitotic kinase in synchronized mammalian cells. The kinase, provided by Dr. Jean-Claude Labbe of Marcel Doree's group, was purified from starfish oocytes blocked in the second meiotic division (3). In vitro, this kinase phosphorylated histone H1 and was competent to induce maturation of arrested oocytes from a variety of species. The kinase consisted of a single protein band as analyzed by gel electrophoresis. In this respect, it differed from similar kinase activity purified by Maller's group (reviewed. 4) in lacking an associated protein of Mr 45000, now known to be cyclin B. The reason for this difference remains unclear, but similar effects to those described below can be obtained with cyclin-B containing p34cdc2 kinase, suggesting that once activated, p34cdc2 no longer requires the presence of cyclin-B to be effective.

We examined the consequences of elevating p34cdc2 kinase in vivo, through microinjection of rat embryo fibroblasts (REF-52). These cells are particularly amenable to synchronization and their long cell cycle time (25 hours) allowed us to select them far from mitotic stage, minimizing the possibilities of artifacts consequent from the natural cell cycle. Cells could be synchronized through serum starvation to generate G0 quiescent cells, G1 cells (0 to 15 hours after refeeding), S-phase cells undergoing DNA synthesis (15-20 hours, peaking at 18 hours) and G2 cells (20-28 hours). Microinjection of

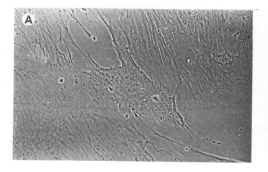

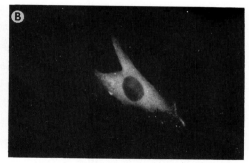

Figure 1. <u>Changes in cell morphology and microtubules organization after injection of p34^{cdc2}</u>. REF52 cells were injected with purified p34^{cdc2} (0.1 mg/ml) and fixed 30 minutes after in 3.7% formalin (panel A and B) or -20°C methanol (panel C), and subsequently stained for injected marker rabbit antibody (B), to stain for the injected cell, or for the microtubules using a monoclonal anti-tubulin antibody (panel C), using the procedures detailed before (7). Panel A and B shows the same injected cell (arrowed). Panel A is a phase contrast image showing the altered cell morphology and the feet-like extensions (arrowed) of the p34^{cdc2} injected cell. Panel C shows (arrowed) the complete disappearance of filamentous microtubule staining in the p34^{cdc2} injected cell.

purified active p34^{cdc2} kinase in those cells lead to a rapid loss of cell shape within 30 minutes of injection at all phases in the cell cycle except in S-phase. As shown in Fig. 1 these effects involved loss of cell substratum contact, with cells forming long feet like processes, strongly mimicking early mitotic prophase (7). Such changes could be induced with a variety of kinase concentrations and preparations, but were dependent on active p34^{cdc2} kinase. Since the level of endogenous intracellular p34^{cdc2} kinase is constant throughout the cell cycle (6), the observation that simple addition of purified active kinase is sufficient to induce some prophase-like effects imply that cycling factors must be responsible for activating the endogenous kinase at mitosis and/or restrain its activity during interphase. Accompanying these changes in cell morphology, the cellular cytoskeleton undergoes pronounced changes. As shown in Fig. 1 these include complete loss of microtubule organization and changes in actin filament networks. Immunofluorescence analysis localized p34^{cdc2} kinase in association with the microtubule organizing centers (13,14), leading to speculation that reorganization of microtubules at mitosis could stem from an effect at these sites. Alternatively, p34^{cdc2} kinase may modulate microtubule organization through activation of casein kinase 2, a kinase which has long been associated with microtubule regulation (15). In addition to these effects, microinjection of p34^{cdc2} appeared to induce limited chromatin condensation in G1 cells, which become pronounced in G2 cells, suggesting that this effect may imply the cooperation of another cell cycle dependent event (see below). Similar changes occurred in many cell lines tested, confirming that activation of p34^{cdc2} kinase plays an integral role in the induction of these events during mammalian cell mitosis. These data also clearly illustrated however, that p34^{cdc2} was itself insufficient to induce premature mitosis in the mammalian cell system since the injected cells did not show any signs of the formation of a mitotic microtubule spindle or the disassembly of their nuclear envelope. In light of these limitations, we began to examine other possible regulators which could cooperate with p34^{cdc2} kinase in the complete induction of mitosis.

IMPLICATION OF p60 c-src IN THE EFFECTS OF p34^{cdc2}

To investigate potential substrates for p34^{cdc2} kinase we performed in vivo metabolic labelling of injected cells. This technique which is extensively described elsewhere (16,17) involves injection of few cells followed by labelling in high specific activity radiolabel (in this case [^{32}P] orthophosphate). Changes in cellular phosphoproteins are followed after labelling, by two-dimensional electrophoresis. Such metabolic labelling revealed, amongst the numerous changes in phosphorylation which take place during the first 10 minutes, the increased phosphorylation of a protein of 60000 Mr which focussed to the same position as the endogenous p60 c-src protein (p60 c-src). The marked increase in phosphorylation of this protein suggested that indeed, in vivo, the p34^{cdc2} kinase either directly phosphorylated or induced the rapid phosphorylation of c-src proteins. Although little is known of the substrates for p34^{cdc2} kinase, independent lines of evidence implied c-src activation accompanied mitosis (18) and that site specific phosphorylation by p34^{cdc2} kinase was involved in this process (19,20). Since extensive data has described changes in cell shape and actin organization following overexpression of activated c-src or infection with viral v-src protein, p34^{cdc2} dependent activation of p60 c-src may provide a possible mechanism through which p34^{cdc2} kinase elicit the changes we observed in cell shape and morphology. To address this possibility, we examined first if overexpression of activated c-src protein induced similar changes in morphology and secondly if inhibiting c-src kinase in vivo could prevent these effects of p34^{cdc2}. Because of the difficulty in obtaining purified active src kinase, we resorted to overexpressing the kinase in REF-52 through microinjection of a plasmid construct encoding c-src under the regulation of a strong SV40 promoter (kindly provided to us by Kurt Ballmer-Hoffer, F.M.I., Basel). The activation of the c-src kinase requires dual phosphorylation/dephosphorylation events which can be brought about by Middle-T antigen, (from polyoma virus), leading to the constitutive activation of *c-src* kinase. This activation protocol provided us with an endogenous control since we could overexpress *c-src* protein in the presence or absence of similarly overexpressed middle-T to produce inactive or activated kinase respectively. Whilst gene injection is not new, its use has been curtailed

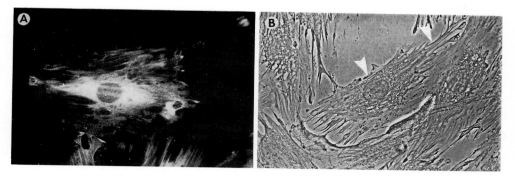

Figure 2. Overexpression of c-src; morphological changes in cells expressing c-src activated with middle-T. Panel A shows a cells injected with pSV40 c-src plasmid alone (1.0 mg.ml) fixed in Formalin 50 minutes after injection and stained for the expression of c-src with monoclonal anti-src antibodies. It shows that although src protein is overexpressed there is no alteration in cell morphology. Panel B shows a phase micrograph of a cell (arrowed) injected with both pSV40 c-src and pSV40 middle-T antigen plasmids. The injected cells overexpressing activated c-src shows pronounced changes in cells shape and loss of substratum contact (upper arrow).

by high mortality rates associated with nucleic acid injection. The main reason for this is that many experimenters are confounded by a prerequisite that genes must be injected into the nucleus, which is completely irrelevant since oligonucleotides injected into the cytoplasm translocate to the nucleus within minutes of injection. As shown in Figure 2, although within 45 minutes the cytoplasm of cells injected with *c-src* gene alone has become filled with *c-src* proteins, little or no change in morphology takes place implying that the synthesized protein is inactive in this intracellular environment. Similar results could be obtained in cells injected through the periods G1 and early G2. Plasmid DNA injection gives poor results in G0 or during S phase because cells are not transcriptionally very active and the plasmid DNA gets destroyed. When the same plasmid construct is injected together with a middle-T antigen plasmid construct, similar changes in morphology to those observed after injection of the p34^{cdc2} kinase occur. Nonetheless, these changes (Fig 2) differ from p34^{cdc2} in that they are not accompanied by changes in microtubule organization. Confirming that active *c-src* kinase is responsible for the effects on cell shape observed, we have seen no effect of the injection of middle T antigen plasmid alone, and coinjection of a monoclonal anti-*c-src* antibody prevented changes induced following c-src gene injection. The same monoclonal anti-src, when coinjected with p34^{cdc2} kinase, prevented many of the changes in cell shape and actin reorganization normally elicited by p34^{cdc2} kinase, whereas it did not for example affect the reorganization of the microtubules. Taking together, these data imply that a component in the mitotic activation by p34^{cdc2} kinase pathway includes direct modulation of p60 *c-src* activity.

THE POTENTIAL ROLES OF cAMP-DEPENDENT PROTEIN KINASE INHIBITION IN MITOTIC ENTRY

Whilst increases in protein phosphorylation accompanying mitotic entry have been extensively investigated, the processes involving protein dephosphorylation at this time are poorly understood. Two aspects interested us particularly, firstly the reported inhibitory role of A-kinase in oocyte maturation and secondly the eventual direct role of protein phosphatases in mitotic entry. As one of the first kinases to be purified active to homogeneity, the functions of the cAMP-dependent protein kinase (A-kinase) have been examined in a variety of systems and in various regulatory pathways including glycogen metabolism and muscle contraction (21). Three features of A-kinase are important to the present discussion, it is thought to normally exists as a double homodimer complex containing two catalytic 'C' subunits and two regulatory 'R' subunits which upon binding of cAMP to the latter, releases the free 'C' subunits that then proceed to catalyze phospho-transferase reactions. The 'R' have both an extremely high affinity constant for binding free 'C' in the absence of cAMP as well as carrying a sequence responsible for the inhibition of catalytic activity. A similar short kinase inhibitor sequence is present in another protein initially described by Walsh et al., (22), and subsequently purified to homogeneity by Demaille et al., (23). This small protein of 75 amino acids was described as protein kinase inhibitor, shortened to PKI, and is a highly specific inhibitor for A-kinase. The inhibitory capacity of the PKI protein is retained in a synthetic peptide derived from the N-terminal first 26 amino acids of the sequence (24). Since peptides of this size are highly unstable in vivo, we modified this peptide sequence through addition of a blocking group at the carboxy terminus and substitution of a D-amino acid in the arginine cluster at the peptide center. Whilst these changes did not appear to confer different inhibitory properties to PKI, they did increase its half life following injection into cells from seconds to 4-8 hours providing us with an effective in vivo tool (manuscript submitted).

Since their first description and purification, A-kinase 'C' and 'R' subunits and PKI have been used to examine in vivo functions of A-kinase. In oocytes it became clear that A-kinase was not involved in the increase in protein phosphorylation at mitosis but, instead, seemed to play an inhibitory role on maturation. Many examples exists which confirmed the early work of Maller and Krebs (25) and although the fine details differ between species, they converge to conclude that in most early embryonic systems, A-kinase plays an inhibitory role to maturation induction (26). In somatic mammalian cells, the functions of A-kinase in division are less clear. Early work described "reverse transformation" of certain transformed CHO cell lines following long treatments with drugs activating A-kinase (27). We observed contradictory immediate effects in cells injected with the catalytic subunit of A-kinase, correlated with changes in actin and vimentin organization and we have detailed the potential role of A-kinase dependent pathways in these processes (16,17). To examine the consequences on cellular dynamics of inhibiting endogenous A-kinase in living cells, we have used microinjection of the modified PKI peptide (PKi(m)), and more recently, of the PKI protein purified from producing bacteria, or a gene sequence encoding it (28)(both supplied to us by Dr. Richard Maurer). As shown in Figure 3, the most immediate and marked changes in cells observed after injection of the protein, peptide or gene is a rapid and extensive condensation of the chromatin.

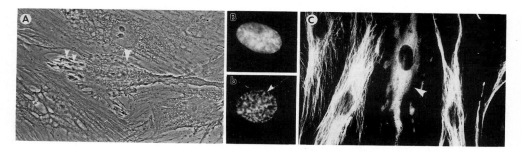

Figure 3. Changes in cell morphology, chromatin and microtubule organization following inhibition of cAMP-dependent protein kinase. Panel A shows a phase micrograph of a cell (arrowed) injected with PKI peptide during the G2 period and in comparison to the surrounding uninjected cells, shows a marked change in cell shape including the formation of phase dense feet-like structures at the periphery (double arrow). Panels B and b, show DNA staining (Hoescht) from cells uninjected and injected with PKI peptide. In comparison to the uninjected cell (panel B), the DNA in the injected cell (panel b) (it is the same cell as shown in panel A) shows heavily condensed chromatin. Panel C shows the complete disappearance of filamentous microtubule staining (as described for Figure 1) in the cell injected with PKI peptide (arrowed).

The extent of this effect, which occurred throughout the cell cycle, far exceeded that observed following injection of p34^{cdc2} kinase, was dependent on the presence of active PKI and could be specifically reversed by microinjection of A-kinase catalytic subunit. This results suggest that the decondensed state of the chromatin involves the continuous A-kinase dependent phosphorylation of some histones and/or other chromatin associated proteins, a concept which is consistent with previous reports that A-kinase is associated with transcriptionally active chromatin (29). In late G1 and throughout G2 the condensation of chromatin in PKI injected cells was accompanied by pronounced changes in cell shape (Fig. 3). Like with injection of p34cdc2, analysis of the cytoskeleton revealed the dramatic reorganization of the microtubules accompanied inhibition of A-kinase. The similarity of these effects and those observed with p34cdc2 kinase was intriguing, particularly since the most prominent effects occurred in late G2 when p34cdc2 kinase becomes activated. To investigate the potential relationship between these effects, we examined the effects of coinjecting p34cdc2 kinase with PKI. Cells injected with both components underwent highly pronounced morphological changes and heavy chromatin condensation. In addition to these effects, we observed the disassembly of the nuclear envelope, an event not induced with injection of either component alone. This latter observation that a combination of both p34cdc2 kinase and A-kinase inhibition lead to envelope disassembly strongly support the idea that this event requires both the activation of p34cdc2 kinase, leading to the phosphorylation of specific lamin sites, as suggested before (30,31,32), but also the inactivation of cAMP-dependent protein kinase resulting in the dephosphorylation of other lamin sites. We have indeed observed the marked dephosphorylation of lamin A and C in cells after injection of PKI, whilst microinjection of p34cdc2 results in the increased phosphorylation of all 3 lamins (manuscript submitted).

The marked effects of inhibiting A-kinase on chromatin and microtubule organization added to the effect on nuclear envelop when coupled with p34cdc2 injection strongly support the notion that A-kinase dependent phosphorylation is continuously involved in maintaining the interphase state of mammalian cells. As such, mitotic induction would involve both the inhibition of A-kinase and activation p34cdc2 kinase and other related kinase pathways. For example, with respect to the process of chromatin condensation inhibiting A-kinase was far more effective than injection of active p34cdc2, which would appear to contradictory to the dogma linking H1 phosphorylation (likely by p34cdc2), to the condensation of chromatin although such a link was never definitely demonstrated. In contrast, examples exist in tetrahymena where the macronuclei, which divide amitotically, contain H1 and p34cdc2 like activity, whilst the micronuclei, which undergo mitotic division and normal chromatin condensation, do not contain H1 like histones nor p34cdc2 activity, but instead the equivalent proteins (termed linkers) are rich in potential A-kinase consensus sites (David Allis, personal communication). Inhibition of A-kinase in the process of mitotic entry would also be consistent with the implied role of A-kinase in transcriptional regulation (33,34,35) since at mitosis gene transcriptional activity is at its lowest. In addition, PKI is an attractive candidate for such a role in down-regulating A-kinase at mitosis since it is highly susceptible to proteolytic degradation and protease activation is a key step involved after mitotic entry (at metaphase) in reversal processes such as cyclin degradation.

MAJOR EVENTS IN MITOTIC ENTRY INVOLVE SEVERAL DISTINCT AND COORDINATED CHANGES IN PHOSPHORYLATION

From the data overviewed in the present report we can begin to summarize some of the activation steps required for mitotic entry. Beforehand however, it is important to keep in mind that at least two levels of mitotic regulation must exist. One involving the series of events bringing about activation of the protein kinases, phosphatases or their specific inhibitors, and the second the direct action of coupled kinase and phosphatase pathways on the different events associated with mitotic induction. Activation of the p34^{cdc2} kinase is the subject of intense investigation. In addition to differential dephosphorylation, it involves the interaction of p34^{cdc2} kinase with other regulatory proteins. So far these include various cyclin proteins, a 13000 Mr protein equivalent to yeast p13 suc protein (c.f. 4,5,6,8) and possibly the phosphatase type-1 and 2A, c-src kinase and regulatory subunits of the cAMP-dependent protein kinase. These latter elements also acting to effect many of the p34^{cdc2} modulated changes.

The sequence leading to mammalian cell mitosis would involve two different classes of events, essentially activation of mitotic related pathways and the concomitant coordinate inactivation of counteracting interphase pathways. Various key morphological land marks have been previously extensively described and it is worth looking at two of these, chromatin condensation and nuclear envelope disassembly, to illustrate how cells combine the strategies outlined above. Although chromatin condensation is probably the most visible changes in cells as they enter mitosis, it is in reality an enigmatic process which begins during G1 and culminates only at mitosis. What is well established is that a histone 1 kinase activity homologous to p34^{cdc2} kinase becomes active at early prophase when major changes in chromatin condensation take place. However, this is clearly not the only step in this process since the chromatin condenses steadily throughout G2 in association with a gradual increase in H1 phosphorylation (36,37,38), at a time when p34^{cdc2} is still inactive. Furthermore, as outlined above, premature activation of p34^{cdc2} kinase induces only poor chromosome condensation in G1 cells whilst inhibition of A-kinase resulted in heavy chromatin condensation. In vitro, A-kinase has been shown to phosphorylate preferentially histone 2A, but its in vivo substrates are still unclear (21). Metabolic labelling has allowed us to identify a marked decrease in H2A phosphorylation following PKI injection (data not shown). H2A is inferred to play a role in the core nucleosome structure where its differential phosphorylation functions in gene transcriptional regulation. In this respect coupling A-kinase dependent phosphorylation of H2A to nucleosome core structure and gene transcriptional activation would be provide a consistent means to ensure that decreases in transcriptional activity were concomitant with entry of cells into mitosis. Alternatively changes in chromatin may result from changes in the phosphorylation of other DNA associated non histone proteins such as a dephosphorylation of the nuclear lamins, since there are links between the chromatin and particularly lamins A and C (which are on the inner surface of the nuclear lattice). In support of such a possibility, our data show that these two lamins become dephosphorylated upon inhibition of A-kinase through PKI injection. Other DNA associated proteins may also be involved through changes in their phosphorylation status, for example those associated with DNA coiling such as topoisomerases. In reality, cells probably use a combination of effects to bring about chromatin condensation with each separate process adding further regulatory capacity to the system, a hypothesis supported by the observation that chromatin condensation is maximal (forming chromosome like structures) when cells are injected with both PKI and p34^{cdc2}.

The data presented have also provided further insights into the mechanism driving another major event in the induction of mitosis, nuclear envelope disassembly. As mentioned above there is clear evidence that the nuclear envelope cannot be induced to disassemble in the presence of active p34^{cdc2} alone in vivo where this seems in some cases to be sufficient in vitro. Complete envelope loss in living cells appears to require the concomitant inactivation of A-kinase, raising the question of how does inactivating A-kinase allow the envelope to disassemble. An attractive hypothesis can be advanced. There is evidence that A-kinase induces the phosphorylation of sites on the nuclear lamins A and C (17) in vivo. Likewise, injection of A-kinase produces a marked increase in vimentin phosphorylation in vivo and, unlike the reported effects of other kinases which induced vimentin disassembly, A-kinase dependent phosphorylation of vimentin stabilized filament organization (17). Since the same consensus sequence site phosphorylated on vimentin is present on the nuclear lamins, A-kinase may maintain nuclear envelope organization through promoting nuclear lamin assembly via phosphorylation. Indeed, as mentioned before, we have observed the clear dephosphorylation of nuclear lamins A and C following injection of PKI (Lamb et al., manuscript submitted). Thus, like the condensation of chromatin the process of nuclear envelope disassembly appears to involve the coordinate action of at least two pathways: activation of p34^{cdc2} kinase and inactivation of cAMP-dependent phosphorylation. However, whilst both these pathways have shown separately or together a strong effect on the organization to the microtubules which in both cases undergo extensive disassembly, we have never observed the reorganization of the microtubules into a mitotic spindle, even when the two are coinjected although in this case the cells show many of the signs of an advanced prometaphase phenotype. This indicates that the process of spindle formation in somatic interphase cells requires the activation and/or presence of another distinct factor.

Finally, whilst this report has concentrated on the positive and negative roles of protein kinases in mitotic entry, and the effects of PKI we described clearly illustrate the importance of protein dephosphorylation in this process, several lines of evidence imply the crucial role of phosphatase type-1 (PP-1) in the dephosphorylation events involved in metaphase exit. In particular, immunofluorescence analysis reveals an intense staining for PP-1 becoming associated with the chromatin at mitosis, and microinjection experiments indicate a critical requirement for PP-1 in the transition through mitosis after metaphase (Fernandez, Brautigan and Lamb, manuscript in preparation. This latter observation confirms in mammalian cells a notion which was already supported by studies using mutants from lower organisms (39,40,41). It will be of interest to further probe the possibility that PP-1 is responsible for the dephosphorylation of some $p34^{cdc2}$ substrates such as histone H1 or the nuclear lamins and we have preliminary evidences supporting the hypothesis that PP-1 may be a key effecter in processes such as chromosome disjoining, chromatin decondensation or nuclear envelope reformation. What is already clear however, is that the catalytic activity associated with PP-1 is an integral participant in mitotic exit and it now becomes of interest to examine how this is regulated and its interaction with other proteins during mitosis.

Acknowledgements

We would like to thank the many people who have contributed to this work by sharing with us enzymes, expertise and unpublished information. In particular we are grateful to Jean Claude Labbe who has provided us with purified active $p34^{cdc2}$ kinase throughout, Jean Claude Cavadore for synthesizing the many peptides that resulted in active PKI and Richard Maurer (University of Iowa) for providing the plasmid and protein PKI. We would also like to thank Kurt Balmer-Hoffer of FMI in Basel for the plasmids encoding c-src and middle-T. This work was supported throughout by CNRS and INSERM and made possible by a grant from the Association Francaise contre les Myopathies (AFM).

References

1.   Dunphy, W.G., Brizuela, L., Beach. D., and Newport, J. (1988). Cell 54, 423-431.

2.   Gauthier, J., Norbury, C., Lohka, M., Nurse, P. & Maller, J. (1988). Cell 54, 433- 439.

3.   Labbe, J. Lee, M., Nurse, P., Picard, A., and Doree, M. (1988). Nature 335, 251-254.

4. Lohka, M,J. (1989). J. Cell Sci. 92, 131-135.

5. Draetta, G. (1990). T.I.B.S. 15, 378-383.

6. Pines, J and Hunter, T. (1990). New Biol. 2, 389-401.

7. Lamb, N.J.C.., Fernandez, A., Watrin, A., Labbé, J-C., and Cavadore, J-C. (1990). Cell, 60, 151-165.

8. Nurse, P. (1990) Nature 344, 503-508.

9. Draetta, G., and Beach, D. (1988). Cell 54, 17-26.

10. Draetta, G., Luca, F., Westendorf, J., Brizuela, L., Ruderman, J., and Beach, D. (1989). Cell 56, 829-838.

11. Draetta, G., Piwnica-Worms, H., Morrison, D., Druker, B., Roberts, T., and Beach, D. (1988). Nature 336, 738-774.

12. Moreno, S. and Nurse, P. (1990). Cell 61, 549-551

13. Riabowol, K.T., Draetta, G., Brizuela, L., Vandre, D. and Beach, D. 1989. Cell 57, 393-401.

14. Bailly, E., Doree, M., Nurse, P. and Bornens, M. (1989) E.M.B.O. J. 8, 3965-3995.

15. Serrano, L., Hernandez, M.A.; Diaz-Nido, J.; and Avila, J. (1989). Exp. Cell Res. 181, 263-272.

16. Lamb, N.J.C., Fernandez, A., Conti, M-A., Adelstein, R.D., Glass, D.B., Welch W.J. & Feramisco, J.R. (1988). J. Cell. Biol. 106, 1955-1971.

17. Lamb, N.J.C.., Fernandez, A., Feramisco, J.R., and Welch, W.J. (1989). J. Cell Biol. 108, 2409-2423.

18. Chackalaparampil, I., and Shalloway, D. (1988). Cell 52, 801-810.

19. Morgan, D.O., Kaplan, J.M., Bishop, J.M. and Varmus, H.E. (1989). Cell 57, 775-786.

20. Shenoy, S., Chio, J-K., Bagrodia, S., Copeland, T.D., Maller, J.L. and Shalloway, D. (1989). Cell 57, 763-774.

21. Edelman, A. M., Blumenthal, D.K., and Krebs, E.G. (1987). Ann. Rev. Biochem. 56, 567-613.

22. Walsh, D.A., Ashby, C.D., Gonzalez, C., Calkins, D., Fisher, E.D., and Krebs, E.G. (1971). J. Biol. Chem. 246, 1977-1985.

23. Demaille, J.G., Peters, K.A., and Fisher, E.H. (1977). Biochemistry, 16,3080-3086.

24. Cheng, H-C., Kemp, B.E., Pearson, R.B., Smith, A.J., Misconi, L., Van Patten, S.M., and Walsh, D.A. (1986). J. Biol. Chem. 261, 989-992.

25. Maller, J.L. and Krebs, E.G. (1977). J. Biol. Chem. 252, 1712-1718.

26. Meijer, L., Dostman, W., Genieser, H.H., Butt, E. and Jastorff. B. (1989). Dev. Biol. 133, 58-66.

27. Willingham, M.C. and Pastan, I. (1975). J. Cell Biol. 67, 146-159.

28.   Day, R.N., Walder, J.A., and Maurer, R.A. (1989) J. Biol. Chem. 264, 431-436.

29.   Sikorska, M., Whitfield, J.F. and Walker, P.R. (1988). 263. 3005-3011.

30.   Ward, G.E. and Kirschner, M.W. (1990). Cell 61, 561-577.

31.   Peter, M., Nakagawa, J., Doree, M., Labbe, J-C. and Nigg, E.A. (1990). Cell 61, 591-602.

32.   Heald, R. and McKeon, F. (1990). Cell 61, 579-589.

33.   Riabowol, K.T., Fink, J.S., Gilman, M.Z., Walsh, D.A., Goodman, R.H. and Feramisco, J.R. (1988). Nature, 366, 83-86.

34.   Mellon, P.L., Clegg, C.H., Correl, L.A., and McKnight, G.S., (1989). Proc.Natl.Acad.Sci. USA, 86, 4887-4891.

35.   Buchler, W., Meinecke, M., Chakraborty, T., Jahnsen, T., Walter, U. and Lohmann, S.M. (1990). Eur.J.Biochem. 188, 253-259.

36.   Gurley, L.R., D'Anna, J.A., Barham, S.S., Deaven, L.L., and Tobey, R.A. (1978). Eur. J. Biochem. 84, 1-15.

37.   Langan, T.A. (1982). J. Biol. Chem. 257. 14835-14846.

38.   Wu, R.S., Panusz, H.T., Hatch, C.L. and Bonner W.M. (1985). C.R.C Crit. Rev. Biochem. 20, 201-263.

39.   Doonan, J.H. and Morris, N.R. (1989). Cell 57, 987-996

40.   Ohkura, H., Kinoshita, N., Miyatani, S., Toda, T. and Yanagida, M. (1989) Cell 57, 997-1007.

41.   Booher, R. and Beach. D. (1989). Cell 57, 1009-1016.

The maunscript submitted is: Lamb, N.J.C. Labbe, J-C., Maurer R. and Fernandez. A. 1991. Inhibition of cAMP-dependent protein kinase plays a key role in the induction of mitosis and nuclear envelope breakdown in mammalian cells.

REGULATION OF EUKARYOTIC TRANSLATION BY PROTEIN PHOSPHORYLATION

Hans Trachsel
Institut für Biochemie und Molekularbiologie
Universität Bern
Bühlstrasse 28
CH-3012 Bern
Switzerland

1. Introduction

Gene expression in eukaryotes is regulated at multiple levels including transcription, RNA splicing and transport, translation, mRNA stability and protein activity. Regulation at the level of translation is usually characterized by constant concentration of an mRNA species but variable amounts of protein or different forms of protein synthesized from this mRNA. It may be exerted at the level of initiation, elongation or termination. Among other strategies, modification of translational components by protein phosphorylation is used by eukaryotic cells to regulate translation. This type of regulation will be described in more detail in this chapter.

2. Regulation of initiation

Two pathways for translation initiation exist in eukaryotic cells: cap-dependent initiation [Edery et al., 1987; Kozak, 1983; Pain, 1986] and internal initiation [Bienkowska-Szewczyk and Ehrenfeld, 1988; Jang et al., 1989; Pelletier et al., 1988; Pelletier and Sonenberg, 1988; for a review, see Sonenberg and Pelletier, 1989]. In the cap-dependent pathway (presently believed to be the major pathway) ribosomes with

NATO ASI Series, Vol. H 56
Cellular Regulation by Protein Phosphorylation
Edited by L. M. G. Heilmeyer, Jr.
© Springer-Verlag Berlin Heidelberg 1991

their associated initiation factors bind to the mRNA at (or near) the 5' cap structure and scan the leader region until they reach the initiator AUG codon. A model of this pathway is schematically shown in Fig. 1.

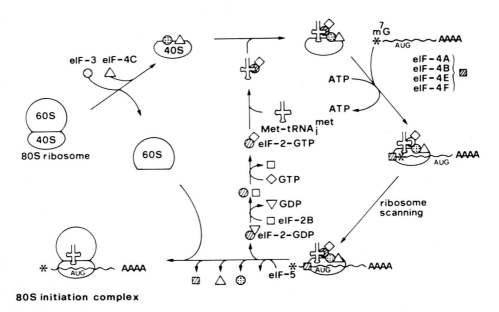

80S initiation complex

Fig. 1: Scheme of eukaryotic translation initiation. The cycle begins with the 80S ribosome and ends with the 80 initiation complex. 40S, small ribosomal subunit; 60S, large ribosomal subunit; eIF, eukaryotic initiation factor; m^{7}G, mRNA cap structure; met-tRNA$_i$^{met}, initiator methionyl transfer RNA.

80S ribosomes dissociate into 40S and 60S subunits and the 40S subunits associate with eukaryotic initiation factor 3 (eIF-3) and eIF-4C. Subsequently, initiation factor eIF-2 carries the initiator methionyl-tRNA (met-tRNA$_i$^{met}) as part of the ternary complex eIF-2-GTP-met-tRNA$_i$^{met} to the 40S subunit. The resulting initiation complex binds at or near the mRNA cap structure in a reaction dependent on ATP, ATP hydrolysis and the factors eIF-4A, B, E, F. This complex then moves on the mRNA in the 5' to 3' direction and positions itself at the AUG initiator codon. At this point, the large ribosomal subunit joins the complex in an eIF-5-catalyzed reaction whereby GTP is hydrolyzed and initiation factors are released.

Internal initiation has so far only been shown to occur on viral mRNAs [Bienkowska-Szewczyk and Ehrenfeld, 1988; Jang et al., 1989; Pelletier et al., 1988; Pelletier and Sonenberg, 1988; Sonenberg and Pelletier, 1989]. It requires unidentified structural features of the mRNA upstream of the initiator AUG and is facilitated by a cellular protein factor [Meerovitch et al., 1989]. Details of the mechanism of internal initiation are not yet known.

Phosphorylation of eIF-2

Regulation of initiation factor activity by protein phosphorylation was most clearly demonstrated for translation initiation factor eIF-2 [for reviews, see Hershey, 1989; Pain, 1986; Safer, 1983]. Phosphorylation of eIF-2 was originally observed in rabbit reticulocyte lysates incubated for translation in the absence of added hemin or in the presence of low concentrations of double-stranded RNA. In these lysates initiation is inhibited after incubation for a few minutes at 30 - 37° C. Activation of a heme-regulated protein kinase or a double-stranded RNA-activated protein kinase [DAI, Kostura and Mathews, 1989] leads to phosphorylation of the α-subunit of eIF-2 at one [or possibly two, Kramer, 1990] serine residues. Phosphorylated factor is released from ribosomes after initiation as an eIF-2-GDP complex (Fig. 1) and sequesters the limiting recycling factor eIF-2B in a stable inactive complex leading to arrest of ternary complex eIF-2-GTP-met-$tRNA_i^{met}$ formation. Phosphorylation of 30 - 40 % of eIF-2 is sufficient to sequester total eIF-2B and to inhibit translation completely. Dephosphorylation of eIF-2 by a type-2A phosphatase [Chen et al., 1989; Crouch and Safer, 1980; Pato et al., 1983] reverses translation inhibition.

To date, the most convincing evidence that phosphorylation of eIF-2 is the cause rather than an effect of translation inhibition comes from experiments in which cell lines were transformed with plasmids encoding wild-type and mutant eIF-2 with serine to alanine substitutions in the phosphorylation sites. Mutation of the phosphorylation sites leads to relief of translation inhibition imposed by activation of DAI [Davies et al., 1989; Kaufman et al., 1989].

Impairment of eIF-2 recycling activity and phosphorylation of eIF-2 occurs in many different cell types and under a variety of conditions including heat shock, limiting nutrient supply and viral infection [Pain, 1986]. Cells treated with interferon synthesize DAI, an eIF-2 kinase which is activated after infection of cells with viruses producing low amounts of double-stranded RNA during their replication cycle. Some viruses like adenovirus [Schneider and Shenk, 1987] and influenza virus [Katze et al., 1986] produce a small RNA in large quantities to inhibit this kinase and to allow virus replication. In uninfected cells, double-stranded RNA segments in the 5' untranslated region of mRNAs may also be able to activate DAI when they are not unwound by translation initiation factors. This may be a way cells use to correlate mRNA unwinding with eIF-2 activity [Edery et al., 1989].

Recent experiments on the regulation of GCN4 mRNA translation in the yeast *Saccharomyces cerevisiae* illustrate possible effects of eIF-2 phosphorylation and/or eIF-2B activity changes on translation of individual mRNAs. The transcription factor GCN4 is encoded by an mRNA with four short upstream open reading frames which repress translation under normal growth conditions. Amino acid starvation of yeast cells leads to enhanced translation of GCN4 mRNA by ribosomes which pass over the inhibitory upstream open reading frames [for a review, see Hinnebusch and Müller, 1987]. Mutations in subunits of eIF-2 [for a review, see Hinnebusch, 1990] and possibly eIF-2B have a similar effect indicating a role of these factors in scanning, reinitiation and AUG recognition by ribosomes.

Phosphorylation of eIF-4E

The initiation factor eIF-4E, a 24 kDa single polypeptide chain binds to the m^7G cap structure of mRNA. It mediates binding of additional initiation factors required for mRNA 5' untranslated region unwinding and mRNA binding to ribosomes during translation initiation [for reviews, see Edery et al., 1987; Hershey, 1989; Rhoads, 1988; Sonenberg, 1988]. In mammalian cells, eIF-4E is phosphorylated at serine 53 [Rychlik et al., 1987] by a so far not well characterized protein kinase. Dephosphorylation of the factor correlates with decreased translation rates during mitosis [Bonneau and Sonenberg, 1987] and in heat shock condition [Duncan et al., 1987]. Experiments with cell-free systems suggest that dephosphorylated eIF-4E is unable to promote mRNA binding to 40S ribosomes [Joshi-Brave et al., 1990], whereas its ability to bind to the mRNA cap structure is not impaired. Interestingly, overexpression of wild-type eIF-4E in mouse 3T3 cells leads to cell transformation while overexpression of eIF-4E with serine 53 mutated to alaninie has no effect on the phenotype of these cells [Lazaris et al., 1990; Smith et al., 1990]. This indicates that phosphorylation of eIF-4E changes its biological activity in vivo.

Phosphorylation of other translational components involved in initiation

A number of other translational components were found to be phosphorylated in mammalian cells [for a review, see Hershey, 1989]. They include the ribosomal protein S6 [reviewed in Kozma et al., 1989], the β-subunit of eIF-2 [Clark et al., 1988], eIF-2B [Dholakia and Wahba, 1988], eIF-3, eIF-4B [Duncan and Hershey, 1987] and the largest subunit of eIF-4F [Morley and Traugh, 1989]. Phosphorylation of ribosomal protein S6 at five serine residues and of eIF-4B at multiple serine residues correlate well with increased translation rates. The roles of these proteins in translation initiation and their modification by phosphorylation are currently not yet understood at the molecular level. For the other components mentioned above a biological function of phosphorylation remains to be established.

3. Regulation of elongation

During the elongation phase of translation the ribosome moves relative to the mRNA in the 5' to 3' direction and incorporates amino acids into the growing polypeptide chain [for a review, see Nygard and Nilsson, 1990]. The elongation factor 1α (EF-1α) carries aminoacyl-tRNA (aa-tRNA) to the ribosome as an EF-1α-aa-tRNA-GTP complex.

416

The ribosome-associated peptidyl-transferase then incorporates the amino acid into the polypeptide chain. Upon GTP hydrolysis, EF-1α-GDP is released from the ribosome and GDP exchanged with GTP through the action of EF-1βγ. Thereafter, an EF-2-GTP compex induces the relative movement of mRNA and ribosome. After GTP hydrolysis EF-2 is released from the ribosome.

The rate of elongation is regulatable by phophorylation of EF-2 by a Ca^{++}/calmodulin-dependent protein kinase [Nairn et al., 1985; Nairn and Palfrey, 1987]. This kinase was found in many cells and tissues. It is specific for EF-2 and phosphorylates this factor at threonine residues. Phosphorylated EF-2 was shown to be inactive in vitro in poly(U)-dependent poly(Phe) synthesis [Nairn and Palfrey, 1987; Ryazanov et al., 1988]. Furthermore, correlation of EF-2 dephosphorylation with increased protein synthesis rates was demonstrated in rabbit reticulocyte lysates [Redpath and Proud, 1989; Sitikov et al., 1988] and in a few in vivo systems such as nerve growth factor-treated rat pheochromocytoma cells [Nairn and Palfrey, 1987].

Since down regulation of the elongation rate is expected to render initiation non-limiting for translation, it should lead to increased translation initiation on mRNAs which are weak competitors under normal conditions. In this way regulation of individual mRNA translation could result.

4. Perspective

Regulation of translation by protein phosphorylation was analyzed, so far, in some detail only for a few translation factors. Even though much remains to be learned about the molecular mechanisms involved and the regulation of other translational components by phosphorylation, it is already clear that protein phosphorylation plays a central role in the regulation of translation in eukaryotes.

References

Bienkowska-Szewczyk K, Ehrenfeld E (1988) An internal 5'-noncoding region required for translation of poliovirus RNA in vitro. J Virol 62:3068-3072

Bonneau AM, Sonenberg N (1987) Involvement of the 24-kDa cap-binding protein in regulation of protein synthesis in mitosis. J Biol Chem 262:11134-11139

Chen S, Kramer G, Hardesty B (1989) Isolation and partial characterization of an Mr 60,000 subunit of a type 2A phosphatase from rabbit reticulocyte. J Biol Chem 264:7267-7275

Clark SJ, Colthurst DR, Proud CG (1988) Structure and phosphorylation of eukaryotic initiation factor 2. Casein kinase 2 and protein kinase C phosphorylate distinct but adjacent sites in the β-subunit. Biochim Biophys Acta 968:211-219

Crouch D, Safer B (1980) Purification and properties of eIF-2 phosphatase. J Biol Chem 255:7918-7924

Davies MV, Furtado M, Hershey JWB, Thimmappaya B, Kaufman RJ (1989) Complementation of adenovirus-associated RNA I gene deletion by expression of a mutant eukaryotic translation initiation factor. Proc Natl Acad Sci USA 86:9163-9167

Dholakia JN, Wahba AJ (1988) Phosphorylation of the guanine nucleotide exchange factor from rabbit reticulocytes regulates its activity in polypeptide chain initiation. Proc Natl Acad Sci USA 85:51-54

Duncan RF, Hershey JWB (1987) Initiation factor protein modifications and inhibition of protein synthesis. Mol Cell Biol 7:1293-1295

Duncan RF, Milburn SC, Hershey JWB (1987) Regulated phosphorylation and low abundance of HeLa cell initiation factor eIF-4F suggest a role in translational control. J Biol Chem 262:380-388

Edery I, Pelletier J, Sonenberg N (1987) Role of eukaryotic messenger RNA cap-binding protein in regulation of translation. Translational Regulation of Gene Expression, ed Ilan J, pp335-366, Plenum Press, New York

Edery I, Petryshyn R, Sonenberg N (1989) Activation of double-stranded RNA-dependent kinase (dsl) by the TAR region of HIV-1 mRNA: a novel translational control mechanism. Cell 56:303-312

Hershey JWB (1989) Protein phosphorylation controls translation rates. J Biol Chem 264:20823-20826

Hinnebusch AG, Müller PP (1987) Translational control of a transcriptional activator in the regulation of amino acid biosynthesis in yeast. Translational Regulation of Gene Expression, ed Ilan J, pp397-412, Plenum Press, New York

Hinnebusch AG (1990) Involvement of an initiation factor and protein phosphorylation in translational control of GCN4 mRNA. Trends Biochem Sci 15:148-152

Jang SK, Davies MV, Kaufman RJ, Wimmer E (1989) Initiation of protein synthesis by internal entry of ribosomes into the 5' nontranslated region of encephalomyocarditis virus RNA in vivo. J Virol 63:1651-1660

Joshi-Brave S, Rychlik W, Rhoads RE (1990) Alteration of the major phosphorylation site of eukaryotic protein synthesis initiation factor 4E prevents its association with the 48 S initiation complex. J Biol Chem 265:2979-2983

Kaufman RJ, Davies MV, Pathak VK, Hershey JWB (1989) The phosphorylation state of eukaryotic initiation factor 2 alters translational efficiency of specific mRNAs. Mol Cell Biol 9:946-958

Katze MG, Detjen BM, Safer B, Krug RM (1986) Translational control by influenza virus: Suppression of the kinase that phosphorylates the alpha subunit of initiation factor eIF-2 and selective translation of influenza viral mRNAs. Mol Cell Biol 6:1741-1750

Kostura M, Mathews MB (1989) Purification and activation of the double-stranded RNA-dependent eIF-2 kinase DAI. Mol Cell Biol 9:1576-1586

Kozak M (1983) Comparison of initiation of protein synthesis in procaryotes, eucaryotes, and organelles. Microbiol Rev 47:1-45

Kozma SC, Ferrair S, Thomas G (1989) Cell Signaling 1:219-225

Kramer G (1990) Two phosphorylation sites on eIF-2α. Febs Lett 267: 181-182

Lazaris A, Montine KS, Sonenberg N (1990) Malignant transformation by a eukaryotic initiation factor subunit that binds to mRNA 5' cap. Nature 345:544-547

Meerovitch K, Pelletier J, Sonenberg N (1989) A cellular protein that binds to the 5'-noncoding region of poliovirus RNA: implications for internal translation initiation. Genes Dev 3:1026-1034

Morley SJ, Traugh JA (1989) Phorbol esters stimulate phosphorylation of eukaryotic initiation factors 3, 4B and 4F. J Biol Chem 264:2401-2404

Nairn AC, Bhagat B, Palfrey HC (1985) Identification of calmodulin-dependent protein kinase III and its major Mr 100,000 substrate in mammalian tissues. Proc Natl Acad Sci USA 82:7939-7943

Nairn AC, Palfrey HC (1987) Identification of the major Mr 100,000 substrate for calmodulin-dependent protein kinase III in mammalian cells as elongation factor-2. J Biol Chem 262:17299-17303

Nygard O, Nilsson L (1990) Translational dynamics. Interactions between the translational factors, tRNA and ribosomes during eukaryotic protein synthesis. Eur J Biochem 191:1-17

Pain VM (1986) Initiation of protein synthesis in mammalian cells. Biochem J 235:625-637

Pato MD, Adelstein RS, Crouch D, Safer B, Ingebritsen TS, Cohen P (1983) The protein phosphates involved in cellular regulation. Eur J Biochem 132:283-287

Pelletier J, Sonenberg N (1988) Internal initiation of translation of eukaryotic mRNA directed by a sequence derived from poliovirus RNA. Nature 334:320-325

Pelletier J, Kaplan G, Racaniello VR, Sonenberg N (1988) Cap-independent translation of poliovirus mRNA is conferred by sequence elements within the 5' noncoding region. Mol Cell Biol 8:1103-1112

Redpath NT, Proud CG (1989) The tumour promoter okadaic acid inhibits reticulocyte-lysate protein synthesis by increasing the net phosphorylation of elongation factor 2. Biochem J 262:69-75

Rhoads RE (1988) Cap recognition and the entry of mRNA into the protein synthesis initiation cycle. Trends Biochem Sci 13:52-56

Ryazanov AG, Shestakova EA, Natapov PG (1988) Phosphorylation of EF-2 kinase affects rate of translation. Nature 334:170-173

Rychlik W, Russ MA, Rhoads RE (1987) Phosphorylation site of eukaryotic initiation factor 4E. J Biol Chem 262:10434-10437

Safer B (1983) 2B or not 2B: Regulation of the catalytic utilization of eIF-2. Cell 33:7-8

Schneider RJ, Shenk T (1987) Translational regulation by adenovirus virus-associated I RNA. Translational Regulation of Gene Expression, ed Ilan J, pp431-445, Plenum Press, New York

Sitikov AS, Simonenko PN, Shestakova EA, Ryazanov AG, Ovchinnikov LP (1988) cAMP-dependent activation of protein synthesis correlates with dephosphorylation of elongation factor 2. Febs Lett 228:327-331

Smith MR, Jaramillo M, Liu Y, Dever TE, Merrick WC, Kung H, Sonenberg N (1990) Translation initiation factors induce DNA synthesis and transform NIH 3T3 cells. The New Biologist 2:648-654

Sonenberg N (1988) Cap-binding proteins of eukaryotic messenger RNA: functions in initiation and control of translation. Prog Nucl Acid Res Mol Biol 35:173-207

Sonenberg N, Pelletier J (1989) Poliovirus translation: A paradigm for a novel initiation mechanism. BioEssays 11:128-132

REGULATION OF MICROFILAMENT ASSEMBLY

Beate Schoepper, Christiane Weigt and Albrecht Wegner
Institute of Physiological Chemistry, Ruhr-University
Bochum, D W-4630 Bochum, F. R. G.

The cytoplasm of eukaryotic cells contains a highly
organized cytoskeleton. Motility and morphological changes
of cells appear to depend on the cytoskeleton. Assembly,
disassembly of the cytoskeleton and translocation of the
constituents of the cytoskeleton underlie motility and shape
changes of cells. The main constituents of the cytoskeleton
are microtubules, microfilaments and intermediate filaments.
These filamentous structures are composed of protein
molecules which associate in a regular and defined manner.
In many types of cells it could be demonstrated that the
turnover of the cytoskeleton is regulated by extracellular
signals. For instance, thrombin stimulation leads to rapid
formation of microfilaments in platelets [1]. During the
last years many proteins of the microfilament system have
been isolated and characterized. The interactions of the
isolated constituents of microfilaments have been
investigated in order to understand regulation of
microfilament assembly. The microfilament proteins have been
classified according to their mode of action on actin
filaments which represent the main constituents of
microfilaments (Fig. 1) [2].Sequestering proteins (e. g.
profilin) bind to actin monomers but not to actin filament
subunits thereby inhibiting actin polymerization. Capping
proteins bind to the "barbed" ends of the polar actin
filaments and prevent binding or dissociation of actin
molecules at these ends. Severing proteins (e. g. gelsolin)
break actin filaments and bind to the newly formed barbed
ends of filaments. Crosslinking proteins (e. g. filamin or
α-actinin) connect actin filaments side by side. Myosin is

the only protein which is known to move actively along actin filaments by consuming ATP. Insertin has been suggested to be an actin monomer-inserting protein that binds to the barbed ends of actin filaments and permits polymerization and depolymerization at these ends [3]. A number of proteins link actin filaments with membranes or are localized near the adhesion sites where actin filaments are linked to membranes (e. g. vinculin).

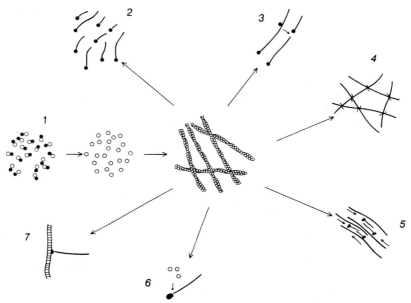

Fig. 1: Actin-binding proteins. 1. sequestring protein (profilin), 2. capping protein, 3. severing protein (gelsolin), 4. crosslinking protein (α-actinin, filamin), 5. myosin, 6. actin monomer-inserting protein (insertin), 7. protein that links actin filaments to membranes.

Isolation of microfilament proteins rendered it possible to investigate regulation of actin polymerization. It turned out that actin polymerization is regulated by actin itself and by actin-binding proteins. The polymerization reaction of actin itself is affected by posttranslational modification of actin and by non-covalent binding of small effector molecules. The assembly of some actin-binding proteins with actin depends on phosphorylation, concentration of free calcium ions and on the presence of phosphatidylinositol-4,5-bisphosphate-containing lipid bilayers. Regulation of actin polymerization by actin-binding proteins is likely to be more important than by actin itself.

A number of clostridial bacteria produce toxins (e. g. botulinum toxin C2 or Clostridium perfringens iota toxin) which cause disruption of microfilaments. These toxins are enzymes which ADP-ribosylate specifically monomeric actin. The ADP-ribosylated monomeric actin is no longer able to polymerize. Furthermore, the ADP-ribosylated actin binds to the barbed ends of actin filaments and inhibits polymerization at these ends in a capping protein-like manner [4,5]. These toxins which ADP-ribosylate actin, bring about disintegration of actin filaments in living cells [6,7]. It has also been reported that monomeric actin is phosphorylated by homogenates of acanthamoeba cells. The phosphorylated actin polymerizes less readily [8]. Recently, it has been shown that phosphate ions and some sugar phosphates bind to actin non-covalently [9,10]. Among these sugars is fructose-1,6-bisphosphate. As there are great differences between the fructose-1,6-bisphosphate concentrations in resting and in activated muscle, it has been proposed that the interaction of actin with other proteins may be regulated by intermediates of glycolytic degradation and, therefore, in an indirect manner by phosphorylation reactions.

Regulation of microfilament assembly by actin-binding proteins is probably physiologically more important than that by small metabolites and posttranslational modification of actin. One of the best understood motile processes in cells is contraction of actin-myosin aggregates by the sliding filament mechanism. Both in smooth muscle cells and in non-muscle cells this contraction is regulated by phosphorylation of the myosin light chain by the calcium calmodulin-dependent myosin light chain kinase. In non-muscle cells phosphorylated myosin tends to form small aggregates consisting of ten or twenty myosin molecules. These myosin minifilaments can interact with actin filaments thereby generating small forces sufficient for contractile processes in non-muscle cells. On dephosphorylation the myosin minifilaments are disintegrated to form single myosin molecules [11].

The assembly of some sequestering and capping proteins with actin filaments is under the control of calcium or phosphatidylinositol phosphates. Profilin binds to monomeric actin thereby inhibiting polymerization of the monomers. The profilin-actin complex can be dissociated on binding to the surface of phosphatidylinositol-4,5-bisphosphate-containing vesicles. This effect might explain how the transition of the profilin-actin complex to actin filaments is controlled in living cells [12]. Gelsolin appears to be regulated both by calcium and by phosphatidylinositol phosphates. In the presence of micromolar Ca^{2+} concentrations this protein severs actin filaments and caps the barbed end of the newly generated filament (Fig. 1). When the gelsolin-capped filaments associate with phosphatidylinositol-4,5-bisphosphate-containing vesicles, gelsolin dissociates from actin filaments and remains bound to the inositol-4,5-bisphosphate head groups of lipid bilayers. The free barbed ends of the short actin filaments associate end to end. When the head groups of phosphatidylinositol-4,5-bisphosphate are hydrolyzed, gelsolin dissociates from membranes. Inositol-1,4,5-trisphosphate which results from the hydrolysis, brings about an increase of the Ca^{2+} concentration. Thus, phosphatidylinositol-4,5-bisphosphate degradation causes disintegration of microfilaments and phosphatidylinositol-4,5-bisphosphate resynthesis gives rise to formation of long actin filaments [13]. There has also some evidence been provided that this mechanism applies to living cells [14].

Several investigations have been reported which suggest that phosphorylation of membrane proteins regulates microfilament assembly. It has been observed that actin polymerization in platelets is inhibited by prostaglandin E_1 that is known to elevate the cytoplasmic concentration of cyclic AMP. Inhibition of microfilament assembly presumably results from the cyclic AMP-stimulated phosphorylation of glycoprotein Ib and other membrane proteins. The existence of Bernard-Soulier syndrome, a hereditary disorder in which platelets lack glycoprotein Ib, rendered it possible to assign inhibition of actin polymerization to phosphorylation of glycoprotein Ib. While prostaglandin E_1 inhibited collagen-induced polymerization of actin in control platelets, it did not inhibit actin polymerization in Bernard-Soulier platelets [15].

Microfilament bundles and their adhesion sites on cell membranes have been found to be rapidly disrupted following tumour virus transformation or exposure of cells to tumour promotors or platelet derived growth factor. Vinculin, a protein localized near the membrane adhesion sites of microfilament bundles, is an in vivo substrate both for protein kinase C and for tyrosine-specific protein kinases encoded by the transforming genes of retroviruses, including pp60^{v-src} [16-18]. Phosphorylation of vinculin attracted considerable interest because it was tempting to speculate that integrity of microfilament adhesion sites could be modulated by phosphorylation of vinculin. However, no simple relationship between phosphorylation of vinculin and disruption of adhesion sites has been found [17,19]. It is possible that phosphorylation of other proteins localized near the adhesion sites (talin, fibronectin receptor) will be more closely related with attachment of microfilaments to membranes [20,21].

Observations on isolated proteins and on living cells
suggest that assembly of microfilaments is regulated by
transmembrane signalling. Interaction of non-muscle myosin
with actin filaments has been relatively well investigated.
However, regulation of assembly of other actin-binding
proteins with actin is not as clearly understood. There are
a number of problems which render investigation of
microfilament assembly difficult. It is still unclear in
which sequence of reactions microfilaments are formed or
disintegrated in living cells. For instance, it is not clear
whether in living cells long microfilaments are formed from
monomers by spontaneous nucleation and subsequent
polymerization, by polymerization of monomers onto existing
actin filaments, by polymerization of monomers onto
nucleating capping proteins or by end to end association of
short actin filaments. As long as even these basic pathways
of microfilament assembly are not known, investigation of
regulation of formation of long actin filaments appears to
be a difficult task. Another problem arises from the fact
that in many cells several actin-binding proteins can serve
the same or similar functions in microfilament assembly; e.
g. both α-actinin and filamin can crosslink actin filaments.
Thus, often it is not known which of the actin-binding
proteins is involved in an event observed in a living cell.
Hopefully, in future it will be possible to make progresses
in elucidating regulation of cytoskeleton assembly by a
skilful combination of different techniques, such as in
vitro investigations, microinjection in living cells,
genetic approaches and other techniques.

References

[1] Carlsson, F., Markey, F., Blikstad, I., Persson, T., &
    Lindberg, U. (1979) Proc. Natl. Acad. Sci. U. S. A. **76**,
    6376-6380.
[2] Weber, K., & Osborn, M. (1985) Sci. Amer. 253 (4),
    92-102.
[3] Ruhnau, K., Gaertner, A., & Wegner, A. (1989) J. Mol.
    Biol. **210**, 141-148
[4] Wegner, A., & Aktories, K. (1988) J. Biol. Chem. **263**,
    13739-13742.

[5] Weigt, C., Just, I., Wegner, A., & Aktories, K. (1989) FEBS Letters **246**, 181–184.

[6] Reuner, K. H., Presek, P., Boschek, C. B., & Aktories, K. (1987) Eur. J. Cell Biol. **43**, 134–140.

[7] Aktories, K., & Wegner, A. (1989) J. Cell Biol. **109**, 1385–1387.

[8] Sonobe, S., Takahashi, S., Hatano, S., & Kuroda, K. (1986) J. Biol. Chem. **261**, 14837–14843.

[9] Rickard, J. E., & Sheterline, P. (1986) J. Mol. Biol. **191**, 273–280.

[10] Gaertner, A., Mayr, G., & Wegner, A. (1990) submitted.

[11] Adelstein, R. S. (1982) Cell **30**, 349–350.

[12] Lassing, I., & Lindberg, U. (1985) Nature **314**, 472–474.

[13] Stossel, T. P. (1989) J. Biol. Chem. **264**, 18261–18264.

[14] Hartwig, J. H., Chambers, K. A., Hopcia, K. L., & Kwiatkowsky, D. J. (1989) J. Cell Biol. **109**, 1571–1579.

[15] Fox, J. E. B., & Berndt, M. C. (1989) J. Biol. Chem. **264**, 9520–9526.

[16] Sefton, B. M., Hunter, T., Ball, E. H., & Singer, S. J. (1981) Cell **24**, 165–174.

[17] Rohrschneider, L., Rosok, M., & Shriver, K. (1982) Cold Spring Harbor Symp. Quant. Biol. **46**, 953–965.

[18] Werth, D. K., & Pastan, I. (1984) J. Biol. Chem. **259**, 5264–5270.

[19] Kellie, S., Patel, B., Mitchell, A., Critchley, D. R., Wigglesworth, N. M., & Wyke, J. A. (1986) J. Cell Sci. **82**, 129–142.

[20] Pasquale, E. B., Maher, P. A., & Singer, S. J. (1986) Proc. Nat. Acad. Sci. U. S. A. **83**, 5507–5511.

[21] Hirst, R., Horwitz, A. F., Buck, C. A., & Rohrschneider L. R. (1986) Proc. Natl. Acad. Sci. U. S. A. **83**, 6470–6474

STIMULATION OF HUMAN DNA TOPOISOMERASE I BY PROTEIN KINASE C

Alsner, J., Kjeldsen, E., Svejstrup, J.Q., Christiansen, K., and Westergaard, O.
Department of Molecular Biology and Plant Physiology.
University of Aarhus.
C.F. Møllers Allé 130.
DK-8000 Århus C.
Denmark.

INTRODUCTION

DNA topoisomerases are ubiquitous enzymes which exert important functions in replication, transcription, and chromatid segregation through regulation of DNA topology (Wang, 1985). Recently, topoisomerase I has also been identified as the primary cellular target for the antitumor drug, camptothecin, and derivatives thereof (Andoh *et al.*, 1987, Zhang *et al.*, 1990). Regulation of topoisomerase I activity might therefore represent one possible mechanism of controlling cellular activity and proliferation as well as directing the cytotoxic effects of clinical relevant antitumor drugs. Topoisomerase I relaxes DNA by concerted single strand cleavage and religation of the DNA backbone (Wang, 1985). The cellular level of mammalian topoisomerase I remains constant throughout the cell cycle and during different proliferative stages, whereas the activity of the enzyme has been found to vary (Hwang *et al.*, 1989), suggesting regulation at the posttranslational level. Thus, the enzyme activity can *in vitro* be modified by poly-(ADP-ribosylation) and phosphorylation (Kasid *et al.*, 1989, Samuels *et al.*, 1989, Pommier *et al.*, 1990). Here, we report that protein kinase C stimulates human topoisomerase I *in vitro* and describe the effect on the enzymatic activity in the absence and presence of camptothecin. Finally, we present a new method for purification of catalytic active forms of topoisomerase I from whole cell extracts.

NATO ASI Series, Vol. H 56
Cellular Regulation by Protein Phosphorylation
Edited by L. M. G. Heilmeyer, Jr.
© Springer-Verlag Berlin Heidelberg 1991

RESULTS AND DISCUSSION

Protein kinase C stimulates human topoisomerase I in vitro. Human topoisomerase I was purified from exponentially growing Daudi cells as described by Thomsen *et al.* (1987) and phosphorylated by incubation with protein kinase C (PKC) and ATP. Figure 1-A demonstrates that the DNA relaxation activity of the PKC treated topoisomerase I is stimulated approximately three-fold. This observation is in agreement with previous findings which demonstrate that purified topoisomerase I is a phosphoprotein and that the relaxation activity can be modulated by either phosphorylation or dephosphorylation (Kaiserman *et al.*, 1988, Samuels *et al.*, 1989, Pommier *et al.*, 1990). To further examine the effect of phosphorylation of topoisomerase I, the interaction of the enzyme with DNA was investigated by trapping the covalent topoisomerase I-DNA intermediate with SDS (Bonven *et al.*, 1985). It has from previous studies been shown that topoisomerase I interacts preferentially with a high affinity DNA binding sequence (Thomsen *et al.*, 1987, Busk *et al.*, 1987), while, in the presence of the antitumor drug, camptothecin, a number of additional DNA cleavage sites are seen (Kjeldsen *et al.*, 1988). Both the high sequence preference and the alteration of the cleavage pattern in the presence of camptothecin is conserved for the phosphorylated enzyme (lanes 5-12). The specific activity of the phosphorylated enzyme, however, is approximately five-fold increased relative to the untreated topoisomerase I (lanes 2-4 and 13-14). Thus, phosphorylation of topoisomerase I by PKC enhances the enzymatic activity *in vitro*, suggesting a possible regulatory role of phosphorylation on topoisomerase I action and its sensitivity to camptothecin.

Isolation of catalytic active topoisomerase I from whole cell extracts. Traditionally, topoisomerase I is purified by ion exchange chromatography resulting in a mixture of catalytic active and inactive forms of the enzyme. To investigate the *in vivo* effect of phosphorylation, a new purification method was developed for isolating catalytic active forms of topoisomerase I (fig. 2-A). The method takes advantage of a previously described specific DNA substrate by which it is possible to catch topoisomerase I in an active conformation covalently bound to DNA (Svejstrup *et al.*, 1990a and 1990b). When the DNA substrate, coupled to magnetic particles (Dynabeads™), are incubated with whole cell extract, catalytic active topoisomerase I in the extract will form covalent complexes with the DNA substrate. By magnetic detention of the particles, these can be washed extensively. At the final step,

A

1 0' 2' 4' 8' 16' **2** 0' 2' 4' 8' 16'

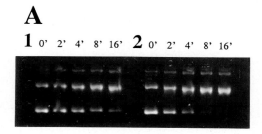

B

1 3 5 7 9 11 13
 2 4 6 8 10 12 14

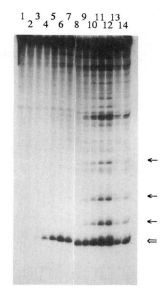

Figure 1 **A.** PKC stimulates the relaxation activity of human topoisomerase I. 300 U human topoisomerase I are incubated for 30 min at 30°C in 20 mM Tris·HCl pH 7.5, 10% glycerol, 10 mM MgCl$_2$, 1 mM CaCl$_2$, 0.5 mM EDTA, 2 mM DTT, 0.01% Triton X-100, and 100 µg/ml phosphatidylserine in a total volume of 40 µl in the absence or presence, panel 1 and 2 respectively, of 1 mM ATP and 2 U PKC, an α/β/γ mixture kindly provided by Dr. Peter J. Parker. 2 µl of the reactions are then incubated with 2 µg supercoiled pBR322 in 10 mM tris·HCl pH 7.5, 150 mM NaCl, 0.5 mM EDTA, 1 mM spermidine, 50 µg/ml BSA, and 10% glycerol at 30°C. Aliquots are taken at the indicated time points, terminated by addition of SDS to 1%, and loaded on a 1% agarose gel. **B.** PKC stimulates the cleavage activity of human topoisomerase I in the absence and presence of camptothecin. 3-5 fmol end-labeled *Hind* III - *Pvu* II DNA fragment from pNC1 (Thomsen *et al.*, 1987) are incubated with various amounts of the untreated and PKC treated topoisomerase I described above in Tris·HCl pH 7.5, 60 mM NaCl, 5 mM MgCl$_2$, 5 mM CaCl$_2$ for 10 min at 30°C. SDS is added to 1%, the samples are treated with proteinase K, and loaded on a 6% denaturing polyacrylamide gel. Lane 1: DNA fragment; lanes 2-4: 1, 2, and 4 µl, respectively, of the untreated topoisomerase I; lanes 5-7: same as lanes 2-4 but with PKC treated topoisomerase I; lanes 8-12: 2 µl of the PKC treated topoisomerase I in the presence of 1% DMSO and 0, 0.01, 0.1, 1, and 10 µM camptothecin, respectively; lanes 13-14: 4 and 1 µl of untreated and PKC treated topoisomerase I, respectively, in the presence of 1 µM camptothecin. ⇐ high affinity DNA binding site; ← additional DNA cleavage sites in the presence of camptothecin.

the covalently bound topoisomerase I is released from the particles by addition of a dinucleotide thereby completing the catalytic cycle (Svejstrup *et al.*, 1990b). The activity of the enzyme in the different fractions was monitored by DNA relaxation (fig. 2-B). The figure demonstrates that all of the relaxation activity can be recovered from the extract. To examine

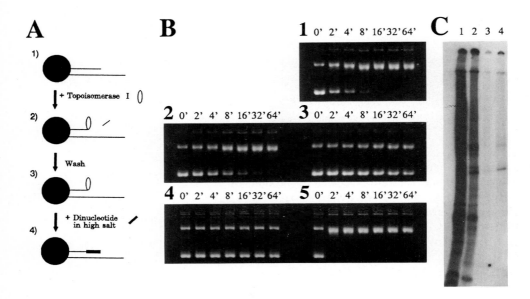

Figure 2 A. Scheme for isolation of catalytic active topoisomerase I. 1) DNA substrate (see text for details) bound to a magnetic particle (●); 2) formation of a covalent complex between DNA and catalytic active topoisomerase I; 3) washing of particles; 4) release of topoisomerase I by addition of a high concentration of dinucleotides in high salt. **B.** Relaxation assay of samples from different steps in the isolation of catalytic active topoisomerase I. Whole cell extracts are prepared from an exponentially grown human lymphoblastoid cell line, RPMI 8402, by: 1) lysis of cell-membranes with Triton X-100, 2) a high salt nuclear extraction (1 M NaCl), and 3) PEG precipitation of DNA. After each purification step, the final volume is adjusted to 300 μl before the relaxation assay (same as in fig. 1-A). Panel 1: 10 U human topoisomerase I; panel 2: 1 μl of the cell extract incubated with the particles; panel 3: 3 μl of the material not bound to particles; panel 4: 12 μl of the washing buffer; panel 5: 30 μl of the purified enzyme. **C.** SDS-PAGE of samples from purification of topoisomerase I from nuclear extract labeled with ^{32}P-γ-ATP by endogenous kinases. Lane 1: labeled nuclear extract; lane 2: material not bound to the particles; lane 3: washing step; lane 4: purified topoisomerase I.

the efficiency of the purification method, nuclear extract was incubated with ^{32}P-γ-ATP, resulting in phosphorylation of multiple protein substrates in the extract by endogenous kinases. As shown in fig. 2-B, a single protein, identified as topoisomerase I, was purified by the described technique. Thus, it is possible, in a single-step procedure, to purify the catalytic active form of topoisomerase I to homogeneity.

Based on the existing data (Kaiserman *et al.*, 1988, Samuels *et al.*, 1989, Pommier *et al.*, 1990), phosphorylation seems to play an important role in the physiological regulation of topoisomerase I activity. The observation that PKC sensitizes topoisomerase I to camptothecin indicates that phosphorylation might also be involved in modulating the action of topoisomerase I-targeting antitumor drugs.

REFERENCES

Andoh T, Ishii K, Suzuki Y, Ikegami Y, Kusunoki Y, Takemoto Y, Okada K (1987) Characterization of a mammalian mutant with a camptothecin-resistant DNA topoisomerase I. Proc Natl Acad Sci USA 84:5565-5569

Bonven BJ, Gocke E, Westergaard, O (1985) A high affinity topoisomerase I binding sequence is clustered at DNAase I hypersensitive sites in *Tetrahymena* R-chromatin. Cell 41:541-551

Busk H, Thomsen B, Bonven BJ, Kjeldsen E, Nielsen OF, Westergaard O (1987) Preferential relaxation of supercoiled DNA containing a hexedecameric recognition sequence for topoisomerase I. Nature 327:638-640

Hwang J, Shyy S, Chen AY, Juan C-C, Whang-Peng J (1989) Studies of topoisomerase-specific antitumor drugs in human lymphocytes using rabbit antisera against recombinant human topoisomerase II polypeptide. Cancer Res 49:958-962

Kasid VN, Halligan B, Liu LF, Dritschilo A, Smulson M (1989) Poly(ADP-ribose)-mediated post-translational modification of chromatin-associated human topoisomerase I. J Biol Chem 264(31):18687-18692

Kjeldsen E, Mollerup S, Thomsen B, Bonven BJ, Bolund L, Westergaard O (1988) Sequence-dependent effect of camptothecin on human topoisomerase I DNA cleavage. J Mol Biol 202:333-342

Pommier Y, Kerrigan D, Hartman KD, Glazer RI (1990) Phosphorylation of mammalian DNA topoisomerase I and activation by protein kinase C. J Biol Chem 265(16):9418-9422

Samuels DS, Shimizu Y, Shimizu N (1989) Protein kinase C phosphorylates DNA topoisomerase I. FEBS Lett 259(1):57-60

Svejstrup JQ, Christiansen K, Andersen AH, Lund K, Westergaard O (1990a) Minnimal DNA duplex requirements for topoisomerase I-mediated cleavage *in vitro*. J Biol Chem 265(21):12529-12535

Svejstrup JQ, Christiansen K, Gromova II, Andersen AH, Westergaard O (1990b) Topoisomerase I action characterized by the use of a new technique to separate the cleavage and religation reactions of the enzyme. Submitted to Biochemistry

Thomsen B, Mollerup S, Bonven BJ, Frank R, Blöcker H, Nielsen OF, Westergaard O (1987) Sequence specificity of DNA topoisomerase I in the presence and absence of camptothecin. EMBO 6:1817-1823

Wang JC (1985) DNA topoisomerases. Annu Rev Biochem 54:665-697

Zhang H, D'Arpa P, Liu LF (1990) A model for tumor cell killing by topoisomerase poisons. Cancer Cells 2(1):23-27

PROTEIN PHOSPHORYLATION IN PARTIALLY SYNCHRONIZED CELL SUSPENSION CULTURE OF ALFALFA

László Bakó, László Bögre and Dénes Dudits
Institute of Plant Physiology
Biological Research Center of the Hungarian Academy of Sciences
6701 Szeged, P.O.B. 521
Hungary

INTRODUCTION

Protein phosphorylation plays an essential part in the regulation of cell cycle of eukaryotes (Norbury and Nurse, 1989; Nurse, 1990). Several components of this regulatory mechanism are well characterized in fission and budding yeasts (Moreno et al., 1989; Wittenberg and Reed, 1988), Xenopus oocytes (Cyert and Kirschner, 1988) and mammalian cells (Morla et al., 1989; Draetta and Beach, 1988). Concerning our knowledge about the cell cycle regulation of higher plants, the presence of a $p34^{cdc2}$ homolog protein kinase has been shown (John et al., 1989; Feiler and Jacobs, 1990; Hirt et al., 1990). In order to obtain more information about the control mechanism of cell cycle in plants we studied the changes of in vitro phosphorylation pattern in both suspension cultured alfalfa cells partially synchronized by phosphate starvation and in alfalfa leaf protoplast culture.

RESULTS

To synchronize alfalfa cells, we applied the phosphate starvation method, which reversibly stops cells in G1 phase (Amino et al., 1983). After phosphate readdition we extracted proteins from cells collected at 3 hour intervals and assayed for endogenous phosphorylation and histone H1 kinase activity. In crude extracts we

NATO ASI Series, Vol. H 56
Cellular Regulation by Protein Phosphorylation
Edited by L. M. G. Heilmeyer, Jr.
© Springer-Verlag Berlin Heidelberg 1991

observed a phosphoprotein with an apparent molecular mass of 65 kD of which phosphorylation has been progressively increased throughout the first twelve hours and then remained at a constant level. In neither of extracts could we detect histone H1 kinase activity (Figure 1.).

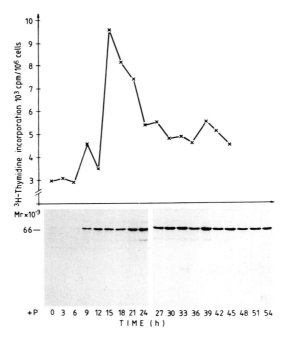

Figure 1. Synchronization of alfalfa cells.
The upper panel shows the ^{3}H-thymidine incorporation curve, the lower panel is an autoradiograph of a 10 % SDS-PAGE showing phosphorylating activity towards endogenous proteins.

In a parallel experiment we prepared protoplasts from leaves, that consists of non-dividing cells considered to be stopped in early G1 during differentiation. By enzymatic treatment we isolated a relatively homogeneous population of mesophyll leaf cells. Hormone treatment (auxin and cytokinin) of these cells resulted in the initiation of cell cycle. The study of the phosphorylation pattern showed several characteristic changes. In leaf mesophyll cells phosphorylation of a 62 and a 65 kD protein is equally prominent. However, when the cell cycle is initiated the phosphorylation of the 62 kD protein is gradually diminished. After 6 days, when the leaf protoplast culture consists of dividing cells, mainly the 65 kD protein phosphorylation can be observed (Figure 2.)

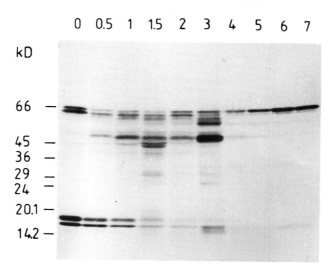

Figure 2. In vitro phosphorylation in protoplast culture.
Autoradiograph of a 5-15 % gradient polyacrylamide gel showing endogenous phosphorylation in samples taken at the indicated times (days).

To purify the p62 and p65 we chromatographed proteins extracted from cycling cells of a suspension culture on Q-Sepharose column and fractions containing the p62-p65 phosphorylation activity were pooled and applied to a Superose 12 FPLC gel filtration column. Testing the phosphorylation pattern of the fractions we found that a 35 kD protein simultaneously with histone H1 was phosphorylated in addition to p65. Immunoblotting with the PSTAIR antibody showed recognition of a protein at 34 kD and a slower migrating band at 35 kD (Figure 3.).
Since we could not resolve p62 and p65 by this purification scheme we included chromatography on Phenyl Superose column after the first ion-exchange step. By this means we could separate the two proteins and with subsequent gel filtration chromatography we determined their molecular weights. p65 eluted from the column at 60 kD while p62 at around 100 kD. Figure 4 shows that anti cdc13 antibody reacted with a protein of 62 kD on immunoblot of Superose 12 fractions containing p62.

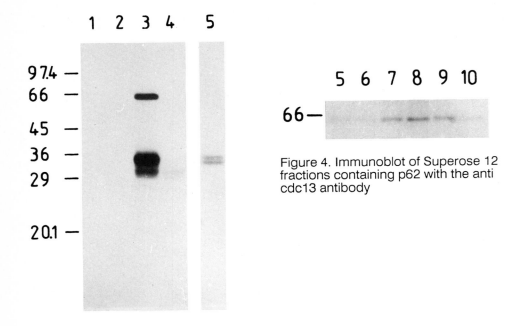

Figure 4. Immunoblot of Superose 12 fractions containing p62 with the anti cdc13 antibody

Figure 3. Superose 12 chromatography of p62-p65
Lane 1-4: Autoradiography of the fractions after in vitro phosphorylation assay in the presence of calf thymus histone H1.
Lane 5: Western blot of the third fraction with the PSTAIR antibody.

Discussion

In the course of our work we applied two synchronized cell culture to observe changes in the in vitro phosphorylation pattern during the initiation of cell cycle in resting plant cells. At the phosphate starvation experiment a 65 kD protein with increasing phosphorylation from the onset of phosphate feeding could be noticed. In addition to this p65, a 62 kD protein showed altering phsophorylation in the leaf protoplast culture. Contrary to p65, the phosphorylation of p62 was at the highest level in differentiated cells blocked in G1 then decreased as they entered the cycle. In cycling alfalfa cell suspension culture the 62 kD protein forms complex with the plant p34^{cdc2} homolog kinase. p13^{suc1}-Sepharose binding experiments confirmed also this fact, since p62 bound to the matrix while p65 remained completely unbound (data not shown). Immunoblotting of p62 with the anti cdc13 antibody showed recognition of this protein. We have no direct evidence that the p62 phosphoprotein we purified from cell suspension culture is identical to the 62 kD phosphoprotein detected in mesophyll leaf protoplast lysates. Should the case occur, this 62 kD plant cyclin together with the plant p34^{cdc2} protein kinase may be required for the initiation of cell cycle in resting cells.

Acknowledgement

We thank P.Nurse and C. Norbury (Oxford University) for providing the PSTAIR antibody and H. Hirt (University of Vienna) and J. Hayles (Oxford University) for providing the cyclin B antibody and the bacterial strain overexpressing SUC1.

REFERENCES

Amino,S., Fujimura,T. and Komamine,A. (1983) Physiol.Plant. **59**: 393-396
Cyert,M.S. and Kirschner,M.W. (1988) Cell **53**: 185-195
Draetta,G. and Beach,D. (1988) Cell **54**: 17-26
Feiler,H.S. and Jacobs,T.W. (1990) Proc.Natl.Acad.Sci. USA **87**: 5397-5401
Hirt,H. et al. (1990) PNAS in press
John,P.C.L., Sek,F.J. and Lee,M.G. (1989) The Plant Cell **1**: 1185-1193
Moreno,S., Hayles,J. and Nurse,P. (1989) Cell 58, 361-372
Morla,A.O., Draetta,G., Beach,D. and Wang,J.Y.J. (1989) Cell **58**: 193-203
Norbury,,C.J. and Nurse,P. (1989) Biochem,Biophys.Acta **989**: 85-95
Nurse,P. (1990) Nature **344**: 503-508
Wittenberg,C. and Reed,S.I. (1988) Cell **54**: 1061-1072

PHOSPHORYLATION OF ELONGATION FACTOR TU *IN VITRO* AND *IN VIVO*

C.Lippmann[1], C.Lindschau[2], K.Buchner[1], V.A.Erdmann[1]

[1]Institut für Biochemie
Fachbereich Chemie
Freie Universität Berlin
Thielallee 63
D-1000 Berlin 33
FRG

[2]Abt. für allgemeine Innere Medizin und Nephrologie
Klinikum Steglitz
Freie Universität Berlin
Hindenburgdamm 30
D-1000 Berlin 45
FRG

INTRODUCTION

Phosphate transfer reactions followed by covalent modifications of proteins constitute a major mechanism of regulation in higher eukaryotes. Generally the second messengers (Ca2+, diacylglycerol, cAMP) submit their signal to the cell by activating serine and threonine specific protein kinases, whereas growth factors induce phosphorylation of tyrosines directly through their receptors. In eukaryotic protein biosynthesis regulation by phosphorylation has been shown for the initiation factor 2 (IF-2), elongation factor 2 (EF-2) and several ribosomal proteins. The phosphorylation of elongation factor 1α (EF-1α) and 1β (EF-1β) was also reported, but has been demonstrated only *in vitro*. Limited knowledge exists about the phosphorylation and dephosphorylation of prokaryotic proteins by protein kinases and phosphatases (e.g. osmoregulation, nitrogen regulation and chemotaxis). The known phosphate acceptor residues in prokaryotes are histidine and aspartate, but the existence of phosphotyrosine, phosphothreonine and phosphoserine has been demonstrated, too.

The role of EF-Tu in protein biosynthesis seems to be well understood. Here we report on the phosphorylation *in vivo* and *in vitro* of prokaryotic elongation factor Tu (EF-Tu). Our results suggest that there are additional events in the action of this multifunctional protein.

IN VITRO PHOSPHORYLATION

Different protein kinases were examined for their ability to phosphorylate EF-Tu.GDP. Until now the only enzyme found to be suitable for in vitro studies is the eukaryotic protein kinase C (PKC), whereas the cAMP-dependent kinase is unable to use EF-Tu as a substrate. Protein kinase C plays a central role in transmembrane signalling. It has been demonstrated recently that G-proteins can be phosphorylated through C-kinase action and EF-Tu is known to be a model for structure-function relationships of G-proteins.

The in vitro phosphorylation of purified EF-Tu from *E.coli* MRE 600 (Fig. 1) revealed an interesting result. Although one discrete band or spot of the purified EF-Tu could be detected on the overloaded stained SDS-PAGE (Fig. 1a, lane 4), a second band with an apparent Mr of 45 kD was visualized by autoradiography after phosphorylation (Fig. 1b, lane 4).

NATO ASI Series, Vol. H 56
Cellular Regulation by Protein Phosphorylation
Edited by L. M. G. Heilmeyer, Jr.
© Springer-Verlag Berlin Heidelberg 1991

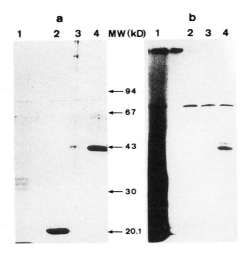

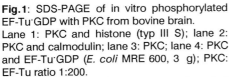

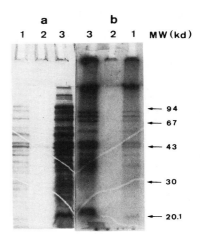

Fig.1: SDS-PAGE of in vitro phosphorylated EF-Tu·GDP with PKC from bovine brain.
Lane 1: PKC and histone (typ III S); lane 2: PKC and calmodulin; lane 3: PKC; lane 4: PKC and EF-Tu·GDP (E. coli MRE 600, 3 g); PKC: EF-Tu ratio 1:200.
Proteins were visualized by staining (panel a) and autoradiography (panel b).

Fig.2: Analysis of total cell extract from *Escherichia coli* MRE 600 by SDS-PAGE
Lane 1, 3: Cell lysate, 0.2 and 0.4 A_{600}; lane 2: purified EF-Tu.
Panel a: Commassie blue-staining; panel b: Autoradiography.
Cells were grown in minimal medium containing [^{32}P] ortho-phosphate (20 MBq/ml) and lysed by sonication. After electrophoresis the gel was incubated for 30 min. at 90 °C in 5 % TCA to remove nucleic acids.

IN VIVO PHOSPHORYLATION

Our first studies on in vivo phosphorylation of EF-Tu gave only poor results. *E.coli* cells grown in ^{32}P-containing medium were analyzed on SDS-gels. Only a faint band of the molecular weight of EF-Tu was visible. Several buffer conditions were tested for their ability to stabilize the *in vivo* phosphorylation (Fig. 2). To verify, that the radioactive band corresponds with EF-Tu, the complete cell lysate was applied on to DEAE-Sepharose CL-6B and separated by FPLC (Fig. 3). Gradient-eluted fractions were precipitated with ethanol and subjected to SDS-PAGE. Immunoreactivity was checked by western-blot analysis. Antibody-binding was visualized by alkaline phosphatase reaction (Fig. 4a) and phosphorylation was confirmed by autoradiography (Fig. 4b). The antibody recognized two bands with apparent Mrs of 43 and 45 kD. The "45 kD" band was eluted in front of the main EF-Tu peak, which coincide with a large radioactive peak, maybe bound [^{32}P]-GDP (Fig. 3).

To confirm that the in vivo phosphorylation of EF-Tu is not restricted to *E.coli*, the thermophilic eubacterium *Thermus aquaticus* EP00276, was analyzed and two phosphorylated proteins with apparent

Mrs of 45 and 47 kD were found, corresponding to the 43 and 45 kD EF-Tu from *E.coli* (data not shown).

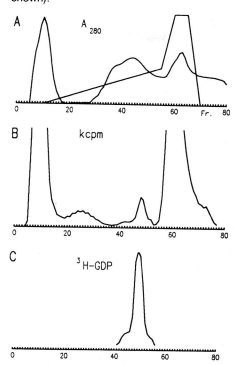

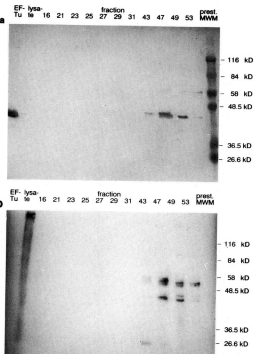

Fig. 3: Purification of *in vivo* phosphorylated EF-Tu. Total cell lysate was loaded onto a DEAE Sepharose CL-6B column (0.5 x 5 cm) and fractionated by using a HPLC-System. The column was eluted with a linear NaCl-Gradient of 0-400 mM. 0.2 ml fractions were collected (A). The radioactivity profile (B) was determined by Cerenkov counting and GDP-binding activity was estimated by a standard nitrocellulose filter assay (C).

Fig. 4: Western immunoblot analysis of DEAE eluted fractions from Figure 2. Proteins were precipitated with ethanol and subjected to SDS-PAGE (10 %). After electro-transfer to the PVDF-membrane anti-EF-Tu reactive bands were visualized by alkaline phosphatase staining (panel a). The phosphorylation of EF-Tu was confirmed by autoradiography (panel b).

IS PHOSPHORYLATION OF EF-TU A REGULATORY OR A FUNCTIONAL EVENT?

To get more information about the nature of the phosphorylation we analyzed the overexpression of the *tufA* gene. If phosphorylation is a regulatory effect, may be inactivation of "not used" EF-Tu, one would expect a greater extend of phosphorylation. In case of a functional phosphorylation the opposite should be true, because the active fraction is decreasing. Cells from *E.coli* MRE 600 transformed with pQECT1 (coding for wild type *tufA* under control of the lac-repressor) were grown under identical conditions as described above. Analysis was done as described elsewhere; in brief: cells were grown to 0.5 A_{600}, pelleted and resuspended in the same volume of medium containing [^{32}P] orthophosphate. After 0 and 2.5 h, proteins from induced and uninduced cultures were analyzed by SDS-PAGE (Fig. 6). At 0 h, immediately after *in vivo* labeling with [^{32}P] orthophosphate, the first radioactive band appears, indicating a fast phosphorylation of EF-Tu. After 2.5 h the IPTG-induced culture shows a slight decrease in phosphorylation, indicating a functional role of phosphorylation.

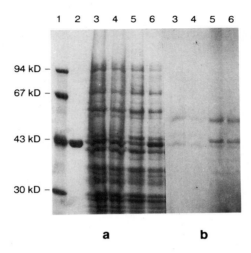

Fig.5: Overexpression and phosphorylation of EF-Tu
Lane 1: molecular weight marker proteins (94 kD, 67 kD, 43 kD, 30 kD); lane 2: EF-Tu; lane 3 and 5: proteins from uninduced cultures (0 h and 2.5 h); lane 4 and 6: proteins from induced cultures (0 h and 2.5 h).
Left panel: Commassie staining; right panel: Autoradiography

DISCUSSION

The 45 kD form of EF-Tu observed in all experiments leads to the the question of the nature of this shift in molcular weight. There are at least two possible explanations:

1. The 45 kD protein presents the *tuf*B gene product. The aminoacid exchange from glycine to serine at position 393 leads to an additional phosphorylation site. It has been demonstrated that different degrees of phosphorylation contributes to differences in mobility in SDS-PAGE. The higher radio-activity/protein ratio compared to the 43 kD band substanciate this hypothesis.

2. EF-Tu is modified through methylation at position 56 (lysine) to an increasing extent during growth phase. The methylation of lysine 56 increases during growth. These results agree with our observation of increasing amounts of the 45 kD protein (Fig. 6), leading to a nearly complete loss of the 43 kD protein. The 45 kD protein may represent the methylated form of EF-Tu, which coincides with the elution profile of the DEAE column, where the 45 kD band elutes in front of the 43 kD EF-Tu (see Fig. 3).

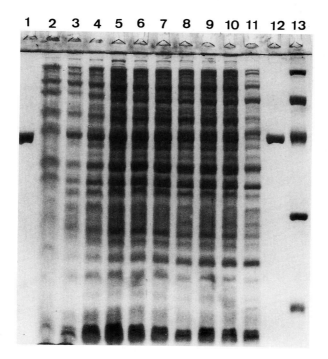

Fig.6: Proteins from *E. coli* MRE 600 harvested at different growth stages
Lane 1, 12: Purified EF-Tu; lane 2-11: Cells harvested after 0, 0.75, 1.0, 1.5, 2.0, 2.5, 3.0, 3.5, 4.0 and 20 hours; lane 13: molecular weight marker proteins (94 kD, 67 kD, 43 kD, 30 kD, 20.1 kD).
Each lane contained the same amount of A_{650} of cells.

CONCLUSIONS

The ability of the eukaryotic protein kinase C to modify prokaryotic EF-Tu, establishes the high conservation among the family of G-proteins. We have started to investigate the influence of *in vitro* phosphorylation on the functions of EF-Tu. The analysis of the phosphorylation site(s) is in progress.The complete set of data will be published elsewhere (Lippmann et al., submitted).

The first experiment carried out with overexpression of EF-Tu impliclates that the *in vivo* phosphorylation is of functional importance. At time we only can speculate about the role of phosphorylation in prokaryotic protein biosynthesis.

ACKNOWLEDGEMENTS
We thank Dr.E.Vijgenboom for his kind gift of the anti-EF-Tu antibody and the plasmid pQECT1. Many thanks are due to Prof.L.Bosch and Dr.E.Vijgenboom (University of Leiden, Gorlaeus Laboratories) for helpful discussions. The in vitro phosphorylations were carried out in the laboratory from Prof.F.Hucho. We thank H.Glowa and A.Schreiber for photography. This work was supported in part by grants to V.A.E. from the Deutsche Forschungsgemeinschaft (DFG), Fonds der chemischen Industrie e.V. and the Bundesministerium für Forschung und Technologie (BMFT).

ABOUT A CONTROVERSY CONCERNING THE EXISTENCE OF A MITOGEN-RESPONSIVE S6-KINASE (LATE-ELUTING FROM DEAE-SEPHACEL)

O.H.W. Martini*, A. Lawen, M. Burger
* Biochemisches Institut
Zülpicher Straße 47
Universität zu Köln
D-5000 Köln
Germany

Phosphorylase kinase and the first phosphorylation cascade were discovered by tackling intracellular signal flux from the distal end of a signalling pathway, i.e. by asking for cause, not effect. During the last decade a comparable approach has been successfully applied in studies of anabolic signal flux towards the ribosomal protein S6. The approach utilizes the fact that serum (Lastick et al., 1977), epidermal growth factor (EGF), insulin and other anabolic agents (Thomas et al., 1982) - and also certain transforming proteins, notably of the tyrosine kinase family, - elicit phosphorylation of several S6 serines.

We used serum-treated S49 kin⁻ cells (murine T-cell lymphoma, defective in cA-PrK expression) to seek S6-phosphorylating activities. The reason: S6 is an in vitro substrate for the cA-PrK, and cA-PrK could, theoretically, modulate the activities of cAMP-independent pathways to S6. Indicating the existence of such pathways, cells of the cyc⁻ mutant line had been observed to contain 4fold phosphorylated S6 despite their subnormal cAMP levels (O.M., unpublished). - Examining post-DEAE fractions from serum-treated log phase S49 kin⁻ cells, we could detect an S6-phosphorylating activity which seemed novel. It required EDTA to show up strongly, was serine-specific, independent of effectors such as cNMP, Ca^{2+} etc, and was not activatable with trypsin.

Our second model were HeLa cells to which we applied the shift protocol with which serum-induced S6 phosphorylation had been discovered (Lastick et al., 1977). Utilizing some technical improvements, 2 well-separated activities were seen in DEAE-

NATO ASI Series, Vol. H 56
Cellular Regulation by Protein Phosphorylation
Edited by L. M. G. Heilmeyer, Jr.
© Springer-Verlag Berlin Heidelberg 1991

Sephacel eluates (Fig. 1). The 'late-eluting' one (200 mM KCl, 3 mM $MgCl_2$, 25 mM potassium phosphate buffer, pH 7.4) was strongly increased when cells synchronized by medium exhaustion were transferred to fresh medium and serum (stimulated cells). The enzyme's properties closely resembled those of the S49 cell kinase. The M_r (gel filtration) was 56 000. In vitro phosphorylation was faster than the apparent in vivo rate. Insulin activated an enzyme with indistinguishable chromatographic behaviour. We termed the enzyme 'mitogen-responsive S6 kinase' (presented at the 1984 Ste. Odile Meeting on Hormones and Cell Regulation (Martini and Lawen, 1985). Publication in detail (Lawen et al.) succeeded in 1989 (cf. discussion).

We could detect late-eluting S6 kinase activity in further systems: Mouse liver, chick embryo fibroblasts both normal (Lawen and Martini, 1985) and transformed, rat hepatocytes.

Discussion. Our working hypothesis which introduced the S6 kinase as a novel enzyme proved controversial, since there existed a different interpretation of a similar profile, observed in insulin-treated 3T3-L1 cells (Perisic and Traugh, 1983). Insulin, through its receptor, is speculated to cause the activation of a protease, which in turn cleaves a protein kinase (protease-activated kinase II, PAK II) the catalytic fragment of which then phosphorylates S6. The speculation has been adopted by Donahue and Masaracchia (1984) for H4P kinase which resembles PAK II (critical discussion in Burger et al., 1989). This 'fragment hypothesis' assigns biological significance to the fact that PAK II and histone-4 kinase, like protein kinase C (PKC), can be activated by trypsin to multiply phosphorylate S6. Judging from reported data, and confirmed for PKC, the intact progenitors of these fragments would, in chromatographic behaviour, resemble our early-eluting activity (Fig. 1). This seems to apply also to an enzyme denoted S6-kinase II (S6-KII) (Sweet et al., 1990). If the fragment hypothesis were correct, the late-eluting activity would have to represent a fragment of an early-eluting enzyme.

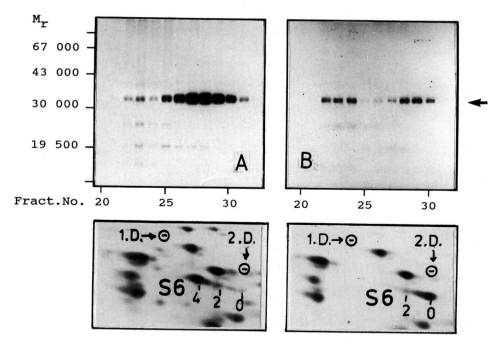

Fig. 1. Top panels: S6-phosphorylating kinase activities from HeLa cells. - A suspension of HeLa cells (a variant of the S3 line) which had exhausted their medium was devided in two portions. (A) Cells from one portion were transferred to fresh medium (JMEM) supplemented with 13 % horse serum of selected quality (stimulated cells). (B) Unstimulated control cells. - 30 min after serum stimulation, cell extracts were prepared (Dounce homogenizer; phosphate buffer; 10 000 x g centrifugation; for details see (Lawen et al. 1989)), and 1 ml portions (from 0.3 g cells) were immediately loaded on 2 ml DEAE-Sephacel columns. The columns were washed. The catalytic subunit of cA-PrK was removed with 50 µM cAMP. Further protein kinases were eluted with a 0-580 mM KCl gradient (the phosphate buffer adds 60 mM K^+). 1.1 ml fractions were collected, and 5 µl of each were examined for protein kinase activity using 40S subunits (0.3 A_{260} units/ml assay; from reticulocyte polysomes) and $[\gamma-^{32}P]$ATP (0.5 mM; 0.3 Ci/mmol; 60 µl assays); details in (Lawen et al., 1989). After 30 min incubation at 35 °C, reaction products were precipitated with $MgCl_2$/ethanol, sedimented and treated with sample buffer for SDS-PAGE (15% acrylamide). Autoradiograms of dried gels are shown. The arrow indicates S6. (Reprinted, with permission, from Lawen et al. 1989)
Bottom panels: S6-phosphorylation in stimulated (left) and control (right) cells. - Stained two-dimensional electropherograms of ribosomal proteins are shown (phosphoryl groups reduce the mobility of S6). The numbers refer to degrees of phosphorylation. A 2 h stimulus was applied; but the serum effect on S6 phosphorylation was essentially complete within 45 min.

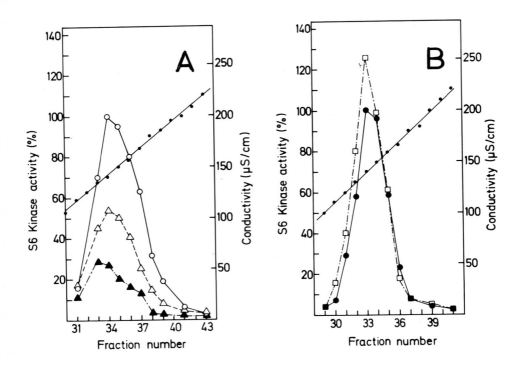

Fig. 2. <u>Effect of heat shock on insulin-responsive S6-kinase</u>
<u>activity.</u> - Four equal portions of HeLa S3 cells suspended in
JMEM containing 0.2 % BSA received insulin (5 x 10^{-8} M); a
further portion (A;Δ--Δ; control cells; 30 min incubation at 36
°C) stayed insulin-free. Cells treated with insulin were
incubated as follows: (A;O—O) 30 min at 36 °C;
(A;▲-·▲) 20 min at 36 °C, then 20 min at 42 °C;
(B;●—●) 70 min at 36 °C;
(B;□-··-□) 20 min at 36 °C, then 20 min at 42 °C; finally 20 min
at 36 °C. - Temperature adjustment between incubation periods
required 5 min. - Cells were lysed with Triton X-100; extracts
were prepared (cf. Fig. 1), fractionated by DEAE-Sephacel
chromatography and assayed (cf. Fig.1) for late-eluting S6-
phosphorylating activity. Following SDS-PAGE, S6-bands were
identified and their ^{32}P contents determined. The highest
insulin-induced activity measured after continuous incubation
at 36 °C for either 30 min (panel A) or 70 min (panel B) was
set 100 %. The conductivity data refer to 250fold deluted
eluate. (Reprinted, with permission, from Burger et al., 1989).

Support for viewing the S6-kinase as enzyme sui generis came
from heat shock experiments: As shown in Fig. 2, insulin-
induced S6-kinase activation is rapidly reversed upon exposure
of cells to hyperthermic stress, and this reverse is itself
reversible (Burger et al, 1989). Thus the mitogen-responsive S6

kinase could be an interconvertible enzyme, possibly the terminal enzyme (like phosphorylase kinase) in a phosphoryl-ation cascade. Several other groups, using further cell types and the same as well as different growth-stimulatory agents, e.g. EGF (Novak-Hofer and Thomas, 1985), have described similar S6-kinases. There is evidence that they are regulated by phosphorylation (cf. G. Thomas, this volume).

Acknowledgements. Fig. 1 was obtained in 1983 at the Institut für Virologie und Immunbiologie, Würzburg, FRG, on a 1981-83 DFG-grant. Fig. 2 was obtained in 1986 at the Physiologisch-Chemisches Institut, Würzburg, on a 1985-87 project grant from the Hermann-und-Lilly-Schilling-Stiftung. Hospitality of the Institut f. Biochemie, Würzburg, and of the Abt. f. Molekular-biologie, Hamburg, to A.L. (supported by Stiftung Cusanuswerk) and O.M., respectively, also helped us to secure this work.

REFERENCES:
Burger M, Lawen A, Martini OHW (1989) Insulin-induced S6 kinase ativation and its reversal by hyperthermic stress. Eur J Biochem 183:255-262
Donahue MJ, Masaracchia RA (1984) Phosphorylation of ribosomal protein S6 at multiple sites by a cyclic AMP-independent protein kinase from lymphoid cells. J Biol Chem 259:435-440
Lastick SM, Nielsen PJ, McConkey EH (1977) Phosphorylation of ribosomal protein S6 in suspension cultured HeLa cells. Mol Gen Genet 152:223-230
Lawen A, Burger M, Martini OHW (1989) Mitogen-responsive S6 kinase. Eur J Biochem 183:245-253
Lawen A, Martini, O (1985) A chick embryo fibroblast protein kinase recognizing ribosomal protein S6. Activity increase after serum stimulation. FEBS Lett 185:272-276
Martini OHW, Lawen A (1985) Mitogen-responsive protein kinase activity of the 10 000 x g supernatant of HeLa cells. In: Hormones and cell regulation (Dumont JE, Hamprecht B, Nunez J eds) 9:411-412
Novak-Hofer I, Thomas G (1985) Epidermal growth factor-mediated activation of an S6 kinase in Swiss 3T3 cells. J Biol Chem 260:
Perisic O, Traugh JA (1983) Protease-activated kinase II as the potential mediator of insulin-stimulated phosphorylation of ribosomal protein S6. J Biol Chem 259:9589-9592
Sweet LJ, Alcorta DA, Erikson RL (1990) Two distinct enzymes contribute to biphasic S6 phosphorylation in serum-stimu-lated chick embryo fibroblasts. Mol Cell Biol 10:2787-2792
Thomas G, Martin-Pérez J, Siegmann M, Otto AM (1982) The effect of serum, EGF, $PGF_{2\alpha}$ and insulin on S6 phosphorylation and the initiation of protein synthesis. Cell 30: 235-242

EXPRESSION OF FERRITIN MESSENGER RNA IN AN in vitro MODEL OF HUMAN CD3-MEDIATED T-LYMPHOCYTE ACTIVATION

Leandro de Oliveira and Maria de Sousa

Abel Salazar Institute of Biomedical Sciences
University of Oporto
Lg. Prof. Abel Salazar 2
4000 PORTO
Portugal.

Differentiation of mammalian leukocytes is associated with the expression of surface antigens which have been catalogued as "clusters of differentiation" (CD; Pallesen and Plesner, 1987). This designation indicates that, using appropriate antibodies, or combinations among them, it is possible to identify cell types bearing distinct antigens. CD3 antibodies bind to polypeptides forming the T-cell antigen-receptor complex; under suitable conditions in vitro, this binding will direct T-lymphocytes to a transient programme of "lymphoblastoid" maturation, including mitosis and the expression of some immunological effector functions (Imrie and Mueller, 1968; Biberfeld, 1971; Meuer et al., 1984).

Human peripheral blood mononuclear cells, largely containing T-lymphocytes (and 20-30% "accessory cells", such as monocytes and B-lymphocytes), can be obtained easily for the study of short-term cellular responses in primary cultures. We have used this experimental model to monitor the steady-state mRNA expression for ferritin during the process of activation triggered by DAKO-T3, a CD3 monoclonal antibody. Ferritin is a virtually ubiquitous iron-storage protein which cooperates in the maintenance of non-toxic levels of soluble iron within the cell (Mattia and van Renswoude, 1988). Its 24 subunits form a tight shell where a micellar iron core is enclosed, and it is known that the entry of iron through this shell occurs by ferrous iron oxidation reversibly catalysed by ferritin proper (Levi et al., 1988). For different types of cells, ferritin displays a broad heterogeneity, depending on the varying proportions of the two types (H and L) of its subunits, yet the significance of this heterogeneity is not quite understood (Worwood, 1989).

We report here preliminary observations of changes in the relative expression of the H and L subunits, at the mRNA level, that follow CD3-mediated activation of peripheral blood mononuclear cells.

Reagents. DAKO-T3 was purchased from DAKO-PATTS, Denmark. Phorbol dibutyrate (PdB) is from Sigma (USA) and is kept in a stock at 1mM in dimethylsulfoxide (Merck, Germany) stored at -70 degrees C. Hybridization probes were prepared from pHF16 and pLF108 cDNA clones (Boyd et al., 1985) using alpha-32P-dCTP and a multiprime-labelling kit (Amersham, UK), with specific activities at 0.5-1 million dpm/nanog. DNA.

Blood samples. Buffy-coats, prepared from peripheral blood donated by patients with hemochromatosis or normal blood donors, were obtained in the Hemotherapy Sector at Sto. Antonio General Hospital, Oporto. For each buffy-coat an accompanying blood sample was collected in an acid-treated glass tube (Brock, 1989) to prepare autologous serum. A total of 24 samples were analysed.

Cells and cultures. Mononuclear cells were separated by density centrifugation on Lymphoprep (Nycomed, Norway) according to the method of Boyum (1968) and resuspended in RPMI 1640 culture medium (Moore et al., 1967) supplemented with 25mM HEPES and antibiotics. Cultures were set in 24-well trays (Nunc, Denmark), with 1mL medium containing 1.5 million mononuclear cells and 10% heat-inactivated (56 degrees C, 30 min.) autologous serum in each well. Further additions (0.1% (v/v) DAKO-T3, 10nM PdB) were made at the beginning of the incubations (37 degrees C, 5% CO_2 water-saturated atmosphere). All cultures were microscopically checked, using a WILD (Switzerland) inverted model.

RNA extraction and dot-blot analysis. At 48 hours of incubation, cells were collected, and cytoplasmic RNA prepared therefrom according to the method of Pearse and Wu (1988). Formaldehyde-denatured RNA samples were transferred to nylon membranes (Amersham) using a dot-blot apparatus (Minifold I, Schleicher and Schuell, USA) and hybridized according to an Amersham protocol ("Blotting and hybridization protocols for Hybond^{TM} membranes"). Autoradiography was carried out at -70 degrees C using intensifying screens on Hyperfilm-MP films (Amersham) preflashed essentially as described by Laskey (1984), and measured by densitometry.

RESULTS

In previous experiments with this model of activation we have studied tritiated thymidine uptake by the cells as an indicator of DNA synthesis rates. While in control (unstimulated) cultures these rates remain at very low levels, close to background, in cultures set in the presence of DAKO-T3 a sharp increase in thymidine uptake is observed after 24 hours of culture, reaching a plateau by

2 to 3 days and declining thereafter. Morphological changes, namely cell aggregation and blast cell formation, are seen only in the activated cultures.

In the control cultures there is already a marked predominance (about one order of magnitude higher) of messenger RNA for the H-subunit of ferritin, by comparison to the L-subunit ferritin message. These "H/L" values displayed a broad variation amongst individuals, either with hemochromatosis or normal, and did not seem to be significantly different between the two groups.

In the corresponding CD3-activated cultures the H/L values were raised several times (figure 1). With three subjects that did not respond to DAKO-T3, H/L remained at the level of control cultures (without the antibody). In two of these cases, however, addition of 10nM PdB in a parallel culture (with DAKO-T3) showed the potential of the cells to become activated, with a concommitant raise of H/L to values consistent with those of other activated cultures. Cultures with 10nM PdB alone (made from three subjects, including two that typically responded to DAKO-T3) had a H/L value only slightly above the control cultures. The presence of PdB at this concentration did not alter the full effect of DAKO-T3 (results not shown).

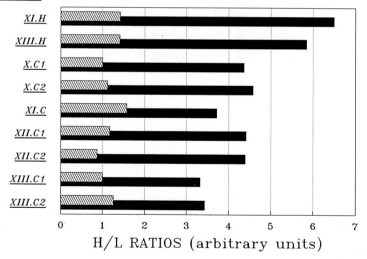

Figure 1 – Ratios between densitometer measurements of autoradiographic signals resulting from hybridizations with pHF16 (H) and pLF108 (L) of the same RNA sample, obtained from 9 separate individuals. Experiments X-XIII. RNA samples from patients with hemochromatosis (.H) or controls (.C). Filled bars represent CD3-activated samples, hatched bars control cultures. The results shown were obtained from a single hybridization and are representative of a total of 21 samples tested.

DISCUSSION

The present results suggest that, during the process of activation of human T-lymphocytes in mononuclear cell populations, the isoferritin distribution is shifted towards a greater relative content of H-type subunits of this iron-storage protein. Several studies (Levi et al., 1988; Worwood, 1989) indicate that this shift would mean a greater rate of iron uptake and/or release by the ferritin compartment. In view of the importance of iron in heme-containing proteins like cytochromes, and some nonheme enzymes (namely ribonucleotide reductase), such an effect on the isoferritin expression might be closely linked to metabolic changes associated with the new steady-state resulting from activation.

Nevertheless, the fact that ferritin is subject to a refined translational control (Aziz and Munro, 1987; Hentze et al., 1987) does not allow us, based on the present results, to infer changes in the characteristics of the protein. Data from similar experiments, examining the protein levels are, however, in line with the interpretation suggested here (Dorner et al., 1983; Pattanapanyasat et al., 1988).

Acknowledgements. This work was supported by the JNICT grant 86/26 and JNICT fellowship BIC 294. We thank Dr. G. Porto for her helpful support for obtaining the buffy-coats, and Cmdr. Contreras de Passos for the useful advices in his field which contibuted to make this work possible.

REFERENCES

1. Aziz N, Munro HN (1987) Iron regulates ferritin mRNA translation through a segment of its 5' untranslated region. Proc Natl Acad Sci USA 84: 8478-8482.

2. Biberfeld P (1971) Morphogenesis in blood lymphocytes stimulated with phytohemagglutinin (PHA). A light and electron microscopy study. Acta path Scand suppl A: 223-229.

3. Boyd D, Vecoli C, Belcher DM, Jain SK, Drysdale JW (1985) Structural and Functional Relationships of Human Ferritin H and L Chains Deduced from cDNA Clones. J Biol Chem 260: 11755-11761.

4. Boyum A (1968) Isolation of Mononuclear Cells and Granulocytes from Human Blood. Scand J Clin Lab Invest Suppl 97: 77-87.

5. Brock JL (1989) Problems with Iron and Iron-Binding Proteins in Tissue Culture. In: de Sousa M, Brock JL (eds) Iron in Immunity, Cancer and Inflammation. Wiley, Chichester New York Brisbane Toronto Singapore, p 399-408.

6. Dorner MH, Silverstone AE, de Sostoa A, Munn G, de Sousa M (1983) Relative Subunit Composition of the Ferritin Synthesized by Selected Human Lymphomyeloid

Cell Populations. Exp Hematol 11: 866-872.

7. Hentze MW, Rouault TA, Caughman SW, Dancis A, Harford JB, Klausner RD [1987] A cis-acting element is necessary and sufficient for transcriptional regulation of human ferritin expression in response to iron. Proc Natl Acad Sci USA 84: 6730–6734.

8. Imrie RC, Mueller GC [1968] Release of a Lymphocyte Growth Promoter in Leukocyte Cultures. Nature 219: 1277-1279.

9. Laskey RA [1984] Radioisotope detection by fluorography and intensifying screens. Amersham Review 23.

10. Levi S, Luzzago A, Cesarini G, Cozzi A, Franceschinelli F, Albertini A, Arosio P [1988] Mechanism of Ferritin Iron Uptake: Activity of the H-Chain and Deletion Mapping of the Ferro-oxidase Site. J Biol Chem 263: 18086-18092.

11. Mattia E, van Renswoude J [1988] The Pivotal Role of Ferritin in Cellular Iron Homeostasis. Bioessays 8: 107-111.

12. Meuer SC, Hassey RE, Cantrell DA, Hodgdon JC, Sclossman SF, Smith KA, Reinherz EL [1984] Triggering of the T3-Ti antigen-receptor complex results in clonal T-cell proliferation through an interleukin-dependent autocrine pathway. Proc Natl Acad Sci USA 81: 1509-1513.

13. Moore GE, Gerner RE, Franklin HA [1967] Culture of normal human leukocytes. J Am Med Assoc 199: 519-524.

14. Pallesen G, Plesner T [1987] The Third International Workshop and Conference on Human Leukocyte Differentiation Antigens with an Up-to-date Overview of the CD Nomenclature. Leukemia 1: 231-234.

15. Pattanapanyasat K, Hoy TG, Jacobs A [1988] Effect of phytohaemagglutinin on the synthesis and secretion of ferritin in peripheral blood lymphocytes. Br J Haematol 69: 565-570.

16. Pearse MJ, Wu L [1988] Preparation of both DNA and RNA for hybridization analysis from limiting quantities of lymphoid cells. Immunol Letts 18: 219-224.

17. Worwood M [1989] An overview of iron metabolism at a molecular level. J Internal Med 226: 381-391.

RELAXATION OF SMOOTH MUSCLE AT HIGH LEVELS OF MYOSIN LIGHT CHAIN (MLC) PHOSPHORYLATION

G. Pfitzer and S. Katoch

Gabriele Pfitzer
II. Physiologisches Institut
Universität Heidelberg
Im Neuenheimer Feld 326
D-6900 Heidelberg

INTRODUCTION

It is generally believed that smooth muscle contraction is initiated by phosphorylation of the 20 kDa light chains of myosin (MLC). One of the important discoveries in smooth muscle research was the finding of the variable phosphorylation contraction coupling (Kamm and Stull, 1985). The state where maintained force is associated with low levels of MLC and low shortening velocity has been termed the `latch` state (reviewed in Hai and Murphy, 1989). Current hypothesis of the regulation of this state may be divided into two groups (Martson, 1989): 1) regulation of the latch state requires a second Ca^{2+}-dependent regulatory mechanism and 2) MLC phosphorylation is sufficient to regulate both contractions at high and low levels of MLC phosphorylation (Hai and Murphy, 1989). In the latter model, so called latch bridges are generated by dephosphorylation of attached phosphorylated crossbridges from which they differ only by their lower rate of detachment from actin. This requires that the rate of dephosphorylation is higher than the rate of detachment of a phosphorylated crossbridge from actin. We show here, that inhibition of the MLC phosphatase by okadaic acid (Takai et al. 1987) indeed prevents the formation of the latch state under certain conditions. However, we also found conditions where smooth muscle relaxes at high levels of MLC

NATO ASI Series, Vol. H 56
Cellular Regulation by Protein Phosphorylation
Edited by L. M. G. Heilmeyer, Jr.
© Springer-Verlag Berlin Heidelberg 1991

low levels of phosphorylation obtained in the absence of okadaic acid, v_{us} is low (fig 2), although tension is nearly maximal (cf. fig.1). In the presence of okadaic acid, v_{us} increased
in parallel with the increase in MLC phosphophorylation.

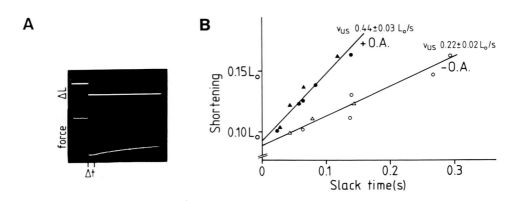

Figure. 2. Determination of unloaded shorten velocity, v_{us}, with
the slack test. A: Fibre bundles, activated as in fig. 1, were
subjected to quick releases, so that they were slack causing
force to drop to zero. They shortened now under no load until
they attained the new length and then redeveloped tension. The
time of unloaded shortening, the slack time Δt, depends on the
amplitudes of the release as well as on the velocity whith
which the preparations shorten. B: The slope of the plot of the
length step versus the slack time is a measure of v_{us}. The values given are the mean +/- S.E.M., n=11, while the plotted
data are examples from two different fibres (Δ,O).

Thus, the development of a latch-like state, characterized by
high force while MLC phosphorylation and v_{us} are low, seems to
require a high MLC phosphatase activity. The experiments
further show, that v_{us} may be affected by the level of
phosphorylation independent of changes in the activators Ca^{2+}
and CaM, suggesting that no additional Ca-dependent regulator
has to be involved as proposed Hai and Murphy (1989).
However, in intact preparations of chicken gizzard, the rate of
relaxation is regulated at apparently basal levels of MLC phosphorylation (Fischer and Pfitzer, 1989) suggesting the necessity of an additional regulatory mechanism. This is supported by the fact that relaxation may also occur without apparent

dephosphorylation of MLC (fig.3). In intact smooth muscle strips from lamb trachea, endothelin elicited a slowly developing contraction with a steady state force comparable to the force elicited by K^+-depolarization. Tension development was preceded by a rise in MLC phosphorylation. The K-channel agonist EMD 52 692 (E. Merck, Darmstadt, F.R.G.) induces a rapid relaxation which is associated with dephosphorylation of MLC. In contrast, the Ca-channel antagonist nitrendipine induces a slow relaxation which is *not* associated with dephosphorylation.

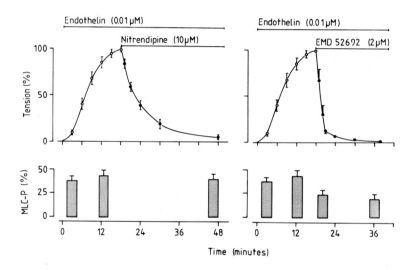

Figure 3: Relaxation of endothelin induced contractions in intact smooth muscle strips from lamb trachea by Nitrendipine and the K-channel opener EMD 52 692. MLC-P: phosphorylated MLC.

In conclusion, evidence has been presented that the phosphorylation theory may account for the the regulation of the latch state. However, relaxation may occur without apparent changes in MLC phosphorylation (fig. 3 and Fischer and Pfitzer, 1989) suggesting that the net detachment of phosphorylated and dephosphorylated crossbridges may be regulated requiring additional regulatory mechanisms.

REFERENCES

Fabiato A, Fabiato F (1979) Calculator programs for computing the composition of solution containing multiple metals and ligands used for experiments in skinned muscle cells. J Physiol (Paris) 75:463-505.

Fischer W, Pfitzer G (1989) Rapid myosin phosphorylation transients in phasic contractions in chicken gizzard smooth muscle. FEBS Lett 258:59-62.

Hai C-M, Murphy RA (1989) Ca^{2+}, crossbridge phosphorylation, and contraction. Annu Rev Physiol 51:285-298.

Kamm KE, Stull JT (1985) The function of myosin and myosin light chain kinase phosphorylation in smooth muscle. Annu Rev Pharmacol Toxicol 25:593-620

Marston SB (1989) What is latch? New ideas about tonic contraction in smooth muscle. J Muscle Res Cell Mot 10:97-100

Paul RJ, Doerman G, Zeugner C, Rüegg JC (1983) The dependence of unloaded shortening velocity on Ca^{++}, calmodulin, and duration of contraction in "chemically skinned" smooth muscle. Circ Res 53:342-351.

Takai A, Bialojan C, Troschka M, Rüegg JC (1987) Smooth muscle myosin phosphatase inhibition and force enhancement by block sponge toxin. FEBS Lett 217:81-84.

EVIDENCE FOR SITE- AND DOMAIN-SPECIFIC PHOSPHORYLATION OF THE 145-kDa NEUROFILAMENT SUBUNIT IN VIVO

Ram K. Sihag and Ralph A. Nixon
Ralph Lowell Laboratories
McLean Hospital
Harvard Medical School
Belmont, MA 02178

A comparison of two-dimensional phosphopeptide maps of NF-M phosphorylated in vivo and in vitro in combination with protein sequence data showed that phosphate groups on the amino terminal head domain are added by protein kinase A and protein kinase C. By contrast, the phosphate groups on the carboxy-terminal domain, which account for about 90% of the phosphate on NF-M, are added by heparin-sensitive second-messenger-independent protein kinase(s).

INTRODUCTION

The three neurofilament protein subunits, particularly NF-M and NF-H, are extensively phosphorylated. Available evidence has suggested that the phosphate groups on NF-M and NF-H are localized only on their extended C-terminal tail domain (Julien and Mushynski, 1983). Recent studies have shown that the NF-L subunit of neurofilaments as well as vimentin and desmin (Sihag et al., 1988; Geisler and Weber, 1988; Inagaki et al., 1987; Sihag and Nixon, 1989) are phosphorylated by protein kinase A and protein kinase C exclusively on the NH_2-terminal domain and suggested a possible conservation of function of this region among subunits that compose the core of intermediate filaments. In this communication we show that protein kinase A and protein kinase C phosphorylate NF-M on the N-terminal head domain sites whereas the neurofilament-associated second-messenger- independent protein kinase(s) phosphorylate the sites on the acidic C-terminal tail domain.

NATO ASI Series, Vol. H 56
Cellular Regulation by Protein Phosphorylation
Edited by L. M. G. Heilmeyer, Jr.
© Springer-Verlag Berlin Heidelberg 1991

MATERIALS AND METHODS

The neurofilament proteins from retinal ganglion cell neurons were phosphorylated in vivo or in vitro (Sihag and Nixon, 1989). Proteins were separated on 320-mm 5-15% linear gradient polyacrylamide gels and the bands identified as NF-M were analyzed by two-dimensional phosphopeptide mapping). For protein sequencing, α-chymotryptic peptides of NF-M, after phosphorylation by protein kinase A, were separated on a reverse-phase C_8 column (Sihag and Nixon, 1990).

RESULTS AND DISCUSSION

Neurofilament proteins are major acceptors of ^{32}P-phosphate in vivo and in vitro (Sihag and Nixon, 1989). Two-dimensional phosphopeptide map analysis of NF-M phosphorylated in vivo revealed at least 15 phosphopeptides after digestion of the isolated polypeptide with TLCK-α-chymotrypsin and TPCK-trypsin (Fig. 1). Most ^{32}P-phosphate groups were located on seven phosphopeptides (M_8-M_{14}). Phosphopeptide M_1-M_7 contained only 5-15% of the total ^{32}P-radioactivity incorporated into NF-M, depending on the post-injection interval (Sihag and Nixon, 1990).

To identify kinases that may mediate the addition of phosphates to these polypeptide domains, we compared two-dimensional phosphopeptide maps of NF-M after axonal neurofilaments were radiolabeled in vitro in the presence of either the endogenous kinases associated with the cytoskeleton or individual purified protein kinases (Fig. 1B-D). In the absence of calcium and cyclic nucleotides, endogenous protein kinases associated with the cytoskeleton preferentially added phosphates to phosphopeptides M_9-M_{13} (Fig. 1B). Heparin, an inhibitor of the calcium- and cyclic-nucleotide-independent kinase associated with the axonal cytoskeleton (Sihag and Nixon, 1989), inhibited the phosphorylation of the major phosphopeptides M_9-M_{13}. Protein kinase A preferentially phosphorylated sites located on phosphopeptides M_1-M_6 and M_8 but not on M_9-M_{13}. Protein kinase C phosphorylated peptides M_1-M_7.

467

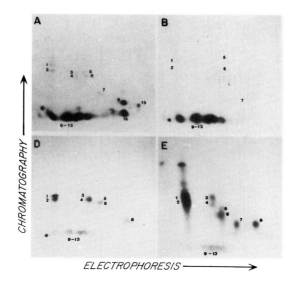

Fig. 1. Two-dimensional phosphopeptide map analyses of NF-M phosphorylated in vivo and in vitro. The phosphopeptides labeled in vivo are designated 1 through 15 (panel A). The in vitro phosphorylation conditions were as follows: no additions (B); 20 units of catalytic subunit of cAMP-dependent protein kinase and 100 μg/ml heparin (C); 0.5 μg of protein kinase C from mouse brain, 100 μM Ca²⁺, 100 μg/ml phosphatidylserine and 100 μg/ml heparin (D).

For isolation and sequencing of protein kinase A phosphorylated domains, neurofilaments were phosphorylated by exogenous protein kinase A in the presence of 100 μg/ml heparin or by cytoskeleton-associated protein kinase(s). Following separation on SDS-PAGE, NF-M was digested with TLCK-α-chymotrypsin and the peptides were separated by HPLC on a reverse-phase C_8 column. The α-chymotryptic peptides phosphorylated by the two kinases eluted as distinct radioactive peaks on a C_8 column (Fig. 2A). On further purification by HPLC, the radioactive peak containing most of the ^{32}P-phosphates added by protein kinase A was resolved into two peptides, C_1 and C_2. Peptide C_2 was identified as a breakdown product of peptide C_1. The amino acid sequence analysis showed that the N-termini of peptides C_1 (S R V S G P . . .) and C_2 (S R G S P S . . .) localized at residues 25 and 41 on NF-M, respectively. Digestion of ^{32}P-phosphopeptide C_1 with TLCK-α-chymotrypsin and TPCK-trypsin and analysis by two-dimensional phosphopeptide mapping on TLC plates showed that peptides M_1-M_6 were located on this peptide (Fig. 2B). Peptides (M_1-M_6) phosphorylated by protein kinase C were also located on the N-terminus. Also, available evidence supports the view that the phosphopeptides M_9-M_{13} are located on the C-terminal domain (Geisler et al., 1987; Lee et al., 1988; Sihag and Nixon, 1990). These data indicate that the NF-M subunit of neurofilaments is phosphorylated by multiple kinases in situ in a domain- and site-specific manner.

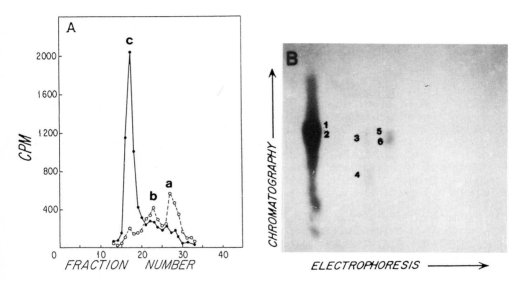

Fig. 2A. Purification of α-chymotryptic phosphopeptides by HPLC on C_8 reverse phase column. The phosphopeptides were separated at a flow rate of 0.2 ml/min with a linear gradient of acetonitrile in 0.08% trifluoroacetic acid. The radio-activity of each fraction was determined by measuring Cerenkov radiation and is shown here. B. Two-dimensional phosphopeptide map of peptide C_1.

The results show that NF-M resembles NF-L and other intermediate filament classes in being phosphorylated on the N-terminal head-domain by protein kinase A and protein kinase C (Sihag and Nixon, 1989). The apparent conserved regulation of phosphate groups on the N-terminal domain of subunits from different intermediate filament classes suggests a general role for these posttranslational modifications in inter-mediate filament dynamics, possibly in regulating aspects of subunit assembly and disassembly. By contrast, the observed addition of phosphate groups on the C-terminal domain by a neurofilament-associated second-messenger-independent kinase(s) may imply neurofilament-specific roles critical for meeting the unique structural needs of neurons (Lewis and Nixon, 1988).

ACKNOWLEDGMENTS

We wish to thank the American Society for Biological Chemists and Molecular Biology for permission to reproduce Figures 1 and 2, and Johanne H. Khan for help in preparing the manuscript. Supported by AG02126 and AG05604.

REFERENCES

Geisler N, Vanderkerckhove J, Weber K (1987) Location and sequence characterization of the major phosphorylation sites of the high molecular mass neurofilament proteins M and H. FEBS Lett 221:403-407

Geisler N, Weber K (1988) Phosphorylation of desmin in vitro inhibits formation of intermediate filaments; Identification of three kinase A sites in the amino-terminal head domain. EMBO J 7:15-20

Inagaki M, Nishi Y, Nishizawa K, Matsuyama M, Sato C (1987) Site-specific phosphorylation induces disassembly of vimentin filaments in vitro. Nature 328:649-652

Julien J-P, Mushynski WE (1983) The distribution of phosphorylation sites among identified proteolytic fragments of mammalian neurofilaments. J Biol Chem 258: 4019-4025

Lee V M-Y Otvos L Carden MJ Hollosi M Dietzschold B Lazzarini RA (1888) Identification of the major multiphosphorylation site in mammalian neurofilaments. Proc Natl Acad Sci USA 85:1998-2002

Lewis SE, Nixon RA (1988) Multiple phosphorylated variants of the high molecular mass subunit of neurofilaments in axons of retinal cell neurons: Characterization and evidence for their differential association with stationary and moving neurofilaments. J Cell Biol 107: 2689-2701

Sihag RK, Nixon RA (1988) Phosphorylation of neurofilament proteins by protein kinase C . FEBS Lett 233:181-185

Sihag RK, Nixon RA (1989) In vivo phosphorylation of distinct domains of the 70 kilodalton neurofilament subunit involves different protein kinases. J Biol Chem 264: 457-464

Sihag RK, Nixon RA (1990) Phosphorylation of the amino-terminal head domain of the middle molecular mass 145 kDa subunit of neurofilaments: Evidence for regulation by second messenger-dependent protein kinases. J Biol Chem 265:4166-4171

TUMOR NECROSIS FACTOR-INDUCED GENE EXPRESSION AND CYTOTOXICITY SHARE A SIGNAL TRANSDUCTION PATHWAY

G. Haegeman, V. Vandevoorde & W. Fiers
Laboratory of Molecular Biology
State University
K.L. Ledeganckstraat 35
9000 Gent
Belgium

Tumor Necrosis Factor (TNF) is a multifunctional cytokine which mediates a variety of biological effects both in vivo and in vitro (Fiers et al., 1987 & to be published). More specifically, TNF displays a cytostatic/cytotoxic effect on several transformed cells in vitro, while leaving most normal cells unaffected (Carswell et al., 1975; Fransen et al., 1986). This cellular toxicity is mainly the result of reactions taking place at the cell membrane/cytosol level, in which (a) GTP-binding protein(s) may be involved (Hepburn et al., 1987) and a phospholipase A_2 becomes activated (Hepburn et al., 1987; Suffys et al., 1987 & to be published). Besides this cytotoxic action, which is transcription-independent, TNF also affects the nuclear programming and induces a number of cellular genes. One of these is the gene coding for the secretory protein interleukin-6 (IL6), of which the complete coding information as well as the preceding promoter region were characterized (Haegeman et al., 1986; Hirano et al., 1986; Yasukawa et al., 1987).

The IL6 promoter contains a variety of well-characterized sequence elements (Ray et al., 1988; Haegeman et al., 1989) and accordingly can be induced by a diversity of chemical and biological agents (Billiau, 1987; Defilippi et al., 1987). We and others (Libermann & Baltimore, 1990; Zhang et al., 1990) obtained evidence that induction of the IL6 gene by treatment of mammalian cells with TNF takes place mainly via activation of a factor called NFκB. This factor, which is normally present in the cytoplasm of unstimulated cells as part of an inactive complex with a corresponding inhibitor molecule IκB (Baeuerle & Baltimore, 1988), is released upon TNF induction from this inhibitory complex and migrates towards the nucleus (Lenardo & Baltimore, 1989), where it recognizes and binds to the corresponding NFκB-responsive sequence in the IL6 promoter in order to stimulate IL6 gene expression. At present, there is some preliminary evidence that phosphorylation of the inhibitory subunit IκB of the cytoplasmic complex is the driving force for release of the active compound NFκB. Protein kinase C, for example, is supposed to be the acting enzyme in case of NFκB activation by phorbol ester treatment (Ghosh & Baltimore, 1990). However, the actual activation step in response to TNF induction and the corresponding signal transduction pathway from the cell membrane (where TNF contacts its receptor) into the cytoplasm are not known so far.

Therefore, the question can be raised whether there is any relationship between the signal for TNF-mediated gene induction and the cellular reactions, going on within the cytoplasm and leading to cellular toxicity, and finally to cell death. This problem was studied with the murine fibrosarcoma cell line L929, which is very sensitive to TNF cytotoxicity. Moreover, two particular TNF-resistant L929 subclones (Vanhaesebroeck et al., submitted) could serve as a control and as a model for analogous cells, where the cytotoxic reactions are absent or counteracted.

Upon studying and comparing the TNF effects on the cell lines mentioned above, we observed a clear TNF-mediated IL6 gene induction in the sensitive L929 cell line, while little or no IL6 expression could be detected by TNF treatment of the TNF-resistant lines. Furthermore, inhibitors of the cellular toxicity, such as pertussis toxin (acting on some G-proteins) or dexamethasone (an indirect inhibitor of phospholipase A_2), correspondingly decreased the TNF-directed IL6 induction in the sensitive L929 cell line. Inversely, activation of the cytotoxic mechanism by combined action of TNF and LiCl (Beyaert et al., 1989) resulted in a corresponding dramatic increase in

TNF-mediated IL6 production. As a matter of fact, this increased cytotoxicity and the corresponding increase in IL6 stimulation by TNF and LiCl could be counteracted again by pretreatment of the cells with dexamethasone. In all these experiments, IL6 gene induction was determined at the mRNA level by dot-blot hybridization as well as by measuring the biological activity of the secreted IL6 in the growth medium.

The results so far obtained permit us to conclude that, at least in the murine L929 cell line, there is a partial sharing of the signal transduction pathway leading to cytotoxicity and to TNF-induced gene activation. As far as our present evidence goes, and although the molecular mechanism for the cytotoxic effect is not yet understood, it seems that the activation of the phospholipase A_2 itself, or the appearance of one or more of its reaction products might be (a) crucial intermediate(s) involved in IL6 gene induction. Whether these observations reflect a general phenomenon of signal transduction, is not known at present. Anyhow, it may be mentioned that in a few human cell lines studied so far, an inverse correlation between TNF sensitivity and TNF-induced IL6 gene expression was observed (Defilippi et al., 1987).

GH is a Research Director with the NFWO. Research was supported by the Belgian FGWO, ASLK, OOA and Lotto.

References

Baeuerle PA, Baltimore D (1988) IκB: A specific inhibitor of the NF-κB transcription factor. Science 242:540-546
Beyaert R, Vanhaesebroeck B, Suffys P, Van Roy F, Fiers W (1989) Lithium chloride potentiates tumor necrosis factor-mediated cytotoxicity in vitro and in vivo. Proc Natl Acad Sci USA 86:9494-9498
Billiau A (1987) Interferon ß₂ as a promoter of growth and differentiation of B cells. Immunol Today 8:84-87
Carswell EA, Old LJ, Kassel RL, Green S, Fiore N, Williamson B (1975) An endotoxin-induced serum factor that causes necrosis of tumors. Proc Natl Acad Sci USA 72:3666-3670
Defilippi P, Poupart P, Haegeman G, Tavernier J, Fiers W, Content J (1987) Induction of 26 kDa protein mRNA in human cells treated with recombinant human tumour necrosis factor (rTNF-α). In: Cantell K, Schellekens H (eds) The biology of the interferon system, Martinus Nijhoff Publishers, Dordrecht Boston Lancaster, p 217
Fiers W, Brouckaert P, Devos R, Fransen L, Haegeman G, Leroux-

Roels G, Marmenout A, Remaut E, Suffys P, Tavernier J, Van der Heyden J, Van Roy F (1987) Molecular and cellular biology of tumor necrosis factor, interferon-γ, and their synergism. In: Cantell K, Schellekens H (eds) The biology of the interferon system, Martinus Nijhoff Publishers, Dordrecht Boston Lancaster, p 205

Fiers W, Beyaert R, Brouckaert P, Everaerdt B, Libert C, Suffys P, Takahashi N, Vanhaesebroeck B, Van Roy F (to be published) Mechanism of action of tumour necrosis factor and its implications for synergizing and antagonizing drugs. Médecine-Sciences

Fransen L, Van der Heyden J, Ruysschaert R, Fiers W (1986) Recombinant tumor necrosis factor: Its effect and its synergism with interferon-γ on a variety of normal and transformed human cell lines. Eur J Cancer Clin Oncol 22:419-426

Ghosh S, Baltimore D (1990) Activation in vitro of NF-κB by phosphorylation of its inhibitor IκB. Nature 344:678-682

Haegeman G, Content J, Volckaert G, Derynck R, Tavernier J, Fiers W (1986) Structural analysis of the sequence coding for an inducible 26-kDa protein in human fibroblasts. Eur J Biochem 159:625-632

Haegeman G, Lesage A, Fiers W (1989) Characterization of the regulatory elements involved in induction and cell-specific expression of interleukin-6. In: Abstracts, Sardinia Symposium on Advances in Biotechnology. Control of Gene Expression. 18-23 May 1989. p 198

Hepburn A, Boeynaems JM, Fiers W, Dumont JE (1987) Modulation of tumor necrosis factor-α cytotoxicity in L929 cells by bacterial toxins, hydrocortisone and inhibitors of arachidonic acid metabolism. Biochem Biophys Res Commun 149:815-822

Hirano T, Yasukawa K, Harada H, Taga T, Watanabe Y, Matsuda T, Kashiwamura S, Nakajima K, Koyama K, Iwamatsu A, Tsunasawa S, Sakiyama F, Matsui H, Takahara Y, Taniguchi T, Kishimoto T (1986) Complementary DNA for a novel human interleukin (BSF-2) that induces B lymphocytes to produce immunoglobulin. Nature 324:73-76

Lenardo MJ, Baltimore D (1989) NF-κB: A pleiotropic mediator of inducible and tissue-specific gene control. Cell 58:227-229

Libermann TA, Baltimore D (1990) Activation of interleukin-6 gene expression through the NF-κB transcription factor. Mol Cell Biol 10:2327-2334

Ray A, Tatter SB, May LT, Sehgal PB (1988) Activation of the human "ß₂-interferon/hepatocyte-stimulating factor/interleukin 6" promoter by cytokines, viruses, and second messenger agonists. Proc Natl Acad Sci USA 85:6701-6705

Suffys P, Beyaert R, Van Roy F, Fiers W (1987) Reduced tumour necrosis factor-induced cytotoxicity by inhibitors of the arachidonic acid metabolism. Biochem Biophys Res Commun 149:735-743

Suffys P, De Valck D, Beyaert R, Vanhaesebroeck B, Van Roy F, Fiers W (to be published) Tumour-necrosis-factor-mediated cytotoxicity is correlated with phospholipase-A2 activity, but not with arachidonic acid release per se. Eur J Biochem

Vanhaesebroeck B, Van Bladel S, Lenaerts A, Suffys P, Beyaert R, Lucas R, Van Roy F, Fiers W (submitted) Two discrete types of tumor necrosis factor-resistant cells derived from the same cell line

Yasukawa K, Hirano T, Watanabe Y, Muratani K, Matsuda T, Nakai S, Kishimoto T (1987) Structure and expression of human B cell stimulatory factor-2 (BSF-2/IL-6) gene. EMBO J. 6:2939-2945

Zhang Y, Lin JX, Vilček J (1990) Interleukin-6 induction by tumor necrosis factor and interleukin-1 in human fibroblasts involves activation of a nuclear factor binding to a κB-like sequence. Mol Cell Biol 10:3818-3823

PROTEIN PHOSPHORYLATION IS INVOLVED IN THE RECOGNITION OF PATHOGEN-DERIVED SIGNALS BY PLANT CELLS

D.G. Grosskopf, G. Felix and T. Boller
Friedrich Miescher-Institut
Postfach 2543
4002 BASEL
Switzerland

Plants react to invasion by fungal pathogens with a range of biochemical responses. They recognize the presence of pathogens using a sensitive perception system for chemical signals derived from fungi, so-called elicitors. These elicitors, which in most cases have not been chemically well defined, comprise carbohydrates, peptides and glycopeptides (Dixon and Lamb, 1990).

Protein phosphorylation plays a central role for signal transduction in animals (Hunter, 1987). Plant cells, when treated with elicitors or related compounds, were observed to exhibit changes in the phosphorylation status of specific proteins (Dietrich et al., 1990) and in protein kinase activities (Farmer et al., 1989; Feller, 1989). The functional significance of these changes remains, however, unknown. We recently found that K-252a, a potent inhibitor of several animal kinases (Burgess and Rüegg, 1989), in submicromolar concentrations was able to block early responses of tomato cells to elicitors (Grosskopf et al., 1990). We have studied, in particular, the rapid increase in production of the plant stress hormone, ethylene, in response to elicitors. Treatment of suspension-cultured tomato cells for 4 h with an elicitor derived from yeast extract resulted in a 8-fold increase in ethylene biosynthesis, and we found that this induction was blocked by K-252a. A dose-response curve showed 50% inhibition at a concentration of about 100 nM K-252a and complete inhibition at concentrations of 300 nM and higher (Fig. 1).

K-252a at these concentrations showed no general toxicity, judged by its effect on protein synthesis and the growth of the cells (Grosskopf et al., 1990).

NATO ASI Series, Vol. H 56
Cellular Regulation by Protein Phosphorylation
Edited by L. M. G. Heilmeyer, Jr.
© Springer-Verlag Berlin Heidelberg 1991

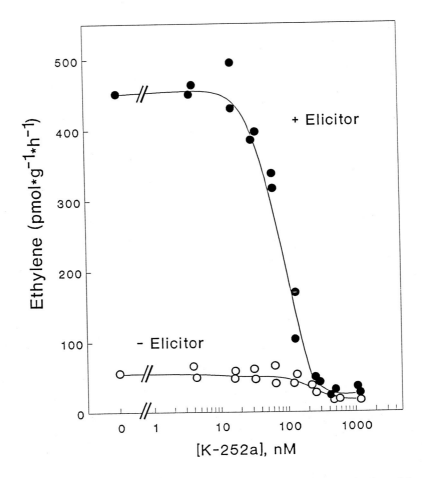

Fig.1. Dose-response curve for the inhibition of elicitor-stimulated ethylene production by K-252a. Cells were treated with different concentrations of K-252a and with 10 μg/ml elicitor derived from yeast extract (closed circles) or with water (open circles), and incubated for 4 h [reproduced from Grosskopf et al. (1990) with permission].

To study if K-252a inhibits plant protein kinases, we examined protein kinase activity in microsomal preparations of tomato cells using histone III-S as exogenous substrate. A V_{max} of 25 pkat/mg protein was measured in microsomes from both elicitor-treated and untreated cells. The K_m for ATP was 100 μM in both instances. The inhibition by K-252a was competitive with respect to ATP and revealed a K_i of about 15 nM (Fig.2).

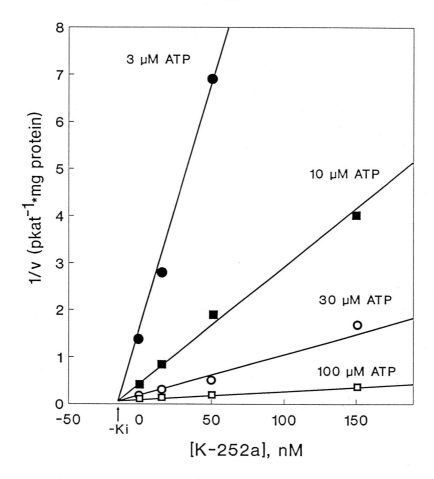

Fig.2. Inhibition of protein kinase in tomato microsomes by K-252a, presented in a Dixon plot. Histone III-S served as a protein kinase substrate [reproduced from Grosskopf et al. (1990) with permission].

Thus, the protein kinase of plant microsomes is as sensitive to K-252a as the most sensitive mammalian kinases (Burgess and Rüegg, 1989).

In order to characterize the K-252a-sensitive protein kinase(s) we separated the microsomal proteins by electrophoresis on a histone III-S-containing polyacrylamide gel according to the method described by Geahlen et al. (1986).

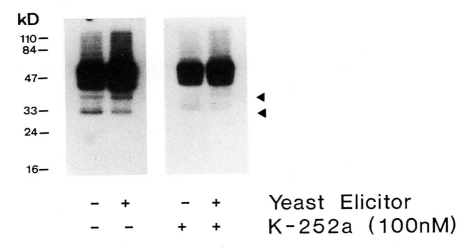

kD
110—
84—

47—

33—

24—

16—

| - | + | - | + | Yeast Elicitor |
| - | - | + | + | K-252a (100nM) |

Fig.3. Detection of protein kinase activity following SDS-PAGE. Microsomes were isolated from untreated tomato cells and from cells stimulated with yeast elicitor (10 µg/ml) for 15 min. Microsomal proteins (20 µg per lane) were separated by SDS-PAGE on gels containing histone III-S polymerized into the matrix. After renaturation the gels were incubated with and without K-252a (100 nM) in the presence of 10µCi [γ-^{32}P]ATP.

Replicates of the renatured gels were incubated with [γ-^{32}P]ATP in either the presence or absence of 100 nM K-252a, respectively (Fig.3). The phosphorylation pattern of microsomes from elicitor-treated or control cells was similar. The bands that gave the strongest signals were little affected by K-252a. However, several bands (apparent M_r of 33 and 38 kD) showed considerably reduced histone phosphorylation activity in the K-252a treatments. We are currently investigating these K-252a sensitive protein kinases in order to study their possible involvement in the perception of elicitor signals.

Acknowledgement: We thank HA Lane and H Macdonald for critically reading the manuscript.

REFERENCES

Burgess GM, Rüegg UT (1989) Staurosporine, K-252a and UCN-01: Potent non-specific inhibitors of protein kinases.Trends Pharm Sci 10:218-220

Chappell J, Hahlbrock K, Boller T (1984) Rapid induction of ethylene biosynthesis in cultured parsley cells by fungal elicitor and its relationship to the induction of phenylalanine ammonia-lyase. Planta 161:457-480

Dixon RA, Lamb CJ (1990) Molecular communication in interactions between plants and microbial pathogens. Annu Rev Plant Physiol Plant Mol Biol 41:339-367

Dietrich A, Mayer JE, Hahlbrock K (1990) Fungal elicitor triggers rapid, transient, and specific protein phosphorylation in parsley cell suspension cultures. J Biol Chem 265:6360-6368

Farmer EE, Pearce G, Ryan CA (1989) *In vitro* phosphorylation of plasma membrane proteins in response to the proteinase inhibitor inducing factor. Proc Natl Acad Sci USA 86:1539-1542

Feller K (1989) Characterization of a protein kinase from soybean seedlings. Study of the variation of calcium-regulated kinase activity during infection with the incompatible race 1 and the compatible race 3 of *Phytophtora megasperma* f. sp. *glycinea* by *in vitro* phosphorylation of calf thymus histone H1. Plant Sci 60:67-75

Geahlen RL, Anostario M, Jr, Low PS, Harrison ML (1986) Detection of protein kinase activity in sodium dodecyl sulfate-polyacrylamide gels. Anal Biochem 153:151-158

Grosskopf DG, Felix J, Boller T (1990) K-252a inhibits the response of tomato cells to fungal elicitors *in vivo* and their microsomal protein kinase *in vitro*. FEBS Lett, in press

Hunter TA (1987) Thousand and one protein kinases. Cell 50:823-829

Effect of membrane modifiers on polyphosphoinositide synthesis in rat heart sarcolemma

Nasrin Mesaeli and Vincenzo Panagia
Division of Cardiovascular Sciences,
St. Boniface General Hospital Research Centre,
351 Tache Avenue, Winnipeg, Manitoba,
Canada, R2H 2A6

INTRODUCTION

The positive inotropic response associated with α_1-adrenoceptor stimulation has been related to Ca^{2+} mobilization from intracellular stores and/or influx of extracellular Ca^{2+} coupled to changes in phosphoinositide turnover (Brückner et al., 1985). Phosphoinositides constitute less than 10 % of the cell membrane lipids with phosphatidylinositol (PtdIns) being predominant. In cardiac membrane PtdIns is transformed in stepwise reactions utilizing two ATP molecules and yielding phosphatidylinositol 4-phosphate (PtdIns(4)P) and phosphatidylinositol 4,5-bisphosphate (PtdIns(4,5)P_2) subsequentially. The presence of the three phosphoinositides (PtdIns, PtdIns(4)P and PtdIns(4,5)P_2) in the heart was first shown by Gaut and Huggins (1966). Evidence for the existence of the phosphorylation pathways was first provided in rabbit cardiac sarcoplasmic reticulum (Enyedi et al., 1984) and in rabbit cardiac sarcolemma (Varsanyi et al., 1986), with the highest enzyme activity in the sarcolemma (Quist et al., 1989).

The phosphoinositides are not evenly distributed in the phospholipid bilayer but are mainly located in the inner, cytoplasmic leaflet of the sarcolemmal membrane (Post et al., 1988). Nevertheless, in previous studies the detergent Triton X-100 was found to have an inhibitory effect on the phosphoinositide kinases especially on the PtdIns(4)P kinase (Quist et al., 1989). Therefore, the purpose of the present investigation was to study the effect of membrane modifiers on the phosphorylation of endogenous phosphoinositides by the membrane associated phosphoinositide kinases. The effect of the antibiotic alamethicin, which was previously shown to be superior to detergents for

NATO ASI Series, Vol. H 56
Cellular Regulation by Protein Phosphorylation
Edited by L. M. G. Heilmeyer, Jr.
© Springer-Verlag Berlin Heidelberg 1991

unmasking the enzyme activities in heart sarcolemma (Jones *et al.*, 1980), was compared with that of the detergent Triton X-100.

MATERIALS AND METHODS

Preparation of cardiac sarcolemmal membranes

Male Sprague Dawley rats weighing 350-400 g were sacrificed, hearts were excised and atria were removed. Ventricular tissue was used to isolate highly purified sarcolemmal membrane according to the method described by Pitts (1979). The protein concentrations of the sarcolemmal preparations were measured by the method of Lowry *et al.* (1951), using serum albumin as standard.

PtdIns and PtdIns4P phosphorylation assay.

The effect of membrane modifiers was studied by preincubating 30 μg of sarcolemmal protein for 30 min in a final volume of 100 μl of 40 mM Hepes-Tris (pH 7.4), 2.5 mM $MgCl_2$, 2 mM EGTA and different concentration of either alamethicin or Triton X-100. The phosphorylation was started by addition of 0.25 mM [^{32}P]ATP (0.4 mCi/ml). The reaction was terminated after 1 min by adding 2 ml of ice cold methanol: 13 N HCl (100:1), and vortexing for 2 min. For extraction of polyphosphoinositides 2 ml chloroform and 1 ml of 2.5 N HCl were added to each tube. After vortexing and centrifugation at 1000^*g_{av} the aqueous methanol was discarded and the chloroform phase was washed with 2 ml chloroform: methanol: 0.6 N HCl (3:48:47). After a second vortexing and centrifugation the chloroform phase was removed, an aliquot of it was dried under nitrogen. The residue was then dissolved in chloroform: methanol: water (75:25:2) and applied to HPTLC plates impregnated with 1 % potassium oxalate in methanol: water (2:3) and activated at 110°C for at least 1 hour. The plates were developed in chloroform: acetone: methanol: glacial acetic acid: water (40:15:13:12:8). The ^{32}P labelled spots were visualized by autoradiography. PtdIns(4)P and PtdIns(4,5)P_2, as identified from the autoradiograph and co-chromatographed standards, were scraped from the plates and counted in 5 ml Ecolume by liquid scintillation.

RESULTS AND DISCUSSION

The effect of two membrane modifiers (alamethicin and Triton X-100) on the phosphorylation of endogenous phosphoinositides was investigated. Table 1 shows the effect of the antibiotic alamethicin. The antibiotic alamethicin acts as an ionophore and permeabilizes the sarcolemmal vesicles. By doing so alamethicin can unmask the activity of membrane enzymes which are latent due to the right-side out formation of the sarcolemmal vesicles (Jones et al. 1980). Alamethicin can probably also act as a detergent and modify the interaction between the enzyme and its intramembranal cofactors. Increasing the ratio between alamethicin and protein from 0 to 2 increased both PtdIns(4)P and PtdIns(4,5)P_2 formation with an optimal effect at a ratio of 1:1. This finding is in accordance with the earlier report (Pitts et al., 1980) that the sarcolemmal vesicles prepared according to Pitts method (1979) are mainly right-side out. As can be seen in Table 1, at a higher alamethicin/protein ratio both enzyme activities decline, with a more drastic effect on PtdIns(4)P kinase. Apparently, alamethicin perturbs the interaction between PtdIns(4)P kinase and the sarcolemmal membrane at this ratio.

Table 2 shows PtdIns and PtdIns(4)P phosphorylation in the presence of Triton X-100. Addition of Triton X-100 at concentrations up to 0.1 % increased ^{32}P incorporation in both PtdIns and PtdIns(4)P. Above this concentration (0.25 %) the ^{32}P incorporation declined drastically but was still higher than control (Table 2). This is in contrast to other reports on the effect of Triton X-100 on cardiac sarcolemmal PtdIns and PtdIns(4)P kinases. Quist and coworkers (1989) found that 0.25 % Triton X-100 abolished PtdIns(4)P kinase activity and reduced PtdIns kinase activity by 80 %. The presence of exogenous substrate enhanced PtdIns kinase activity but failed to restore PtdIns(4)P kinase activity. Kasinathan and coworkers also were unable to show any PtdIns(4)P kinase activity in the presence of 0.1 % Triton X-100, however this concentration of Triton X-100 increased [^{32}P]PtdIns(4)P formation. The discrepancy between our results and those of the above-mentioned studies might be due to differences in phosphorylation assay conditions and animal species (rat vs. canine). Furthermore, Quist et al. (1989) studied the effect of Triton X-100 in the presence of alamethicin. The concerted action of these substances probably is responsible for the lack of [^{32}P]PtdIns(4)P and [^{32}P]PtdIns(4,5)P_2 formation.

Comparison of the effects of the two membrane modifiers reveals that the optimal

Table 1. Effect of alamethicin on phosphorylation of PtdIns and PtdIns(4)P in rat heart sarcolemma.

Alamethicin/Protein Ratio	PtdIns(4)P % of Control	PtdIns(4,5)P_2 % of Control
Control	100	100
0.5	643	587
1.0	1801	1159
2.0	1617	757

Membranes were preincubated at 30°C for 30 min. with or without alamethicin. Reactions were terminated after 1 min. The control values for PtdIns(4)P and PtdIns(4,5)P_2 were 61.05 and 10.28 pmol ^{32}P/mg/min respectively. Results are average of two separate experiments with triplicate determination in which the variation was less than 10 %.

Table 2. Effect of Triton X-100 on phosphorylation of PtdIns and PtdIns(4)P in rat heart sarcolemma.

Triton X-100 %	PtdIns(4)P % of Control	PtdIns(4,5)P_2 % of Control
Control	100	100
0.05	1180	396
0.10	1232	500
0.25	659	402

Membranes were preincubated at 30°C for 30 min. with or without Triton X-100. Reactions were terminated after 1 min. The control values for PtdIns(4)P and PtdIns(4,5)P_2 were 61.05 and 10.28 pmol ^{32}P/mg/min, respectively. Results are average of two separate experiments with triplicate determination in which the variation was less than 10 %.

concentration for PtdIns kinase stimulation corresponds with that for PtdIns(4)P kinase. The explanation for this could be that the product of PtdIns kinase, PtdIns(4)P, is the limiting factor for the rate of [^{32}P]PtdIns(4,5)P_2 formation. However, this is contradicted by the fact that a supraoptimal concentration of Triton X-100, in contrast to alamethicin, induced a steeper decline in ^{32}P incorporation in PtdIns(4)P than in PtdIns(4,5)P_2.

In general, the presented experiments showed that alamethicin is more effective in unmasking the latent enzyme activities than Triton X-100.

Both PtdIns and PtdIns(4)P kinases are membrane bound enzymes (Downes *et al.* 1989). The subsequent action of these two enzymes results in the formation of PtdIns(4,5)P_2. The sarcolemmal PtdIns(4,5)P_2 is essential in the signal transduction pathway. Receptor mediated activation of phospholipase C results in the hydrolysis of PtdIns(4,5)P_2 forming two second messengers, diacylglycerol and inositoltrisphosphate. Therefore the activity of these kinases is an important regulatory point in the signal transduction pathway. Little is known about these enzymes in the heart tissue. In future more attention should be directed to the kinetics and the physiological regulation of these enzymes.

ACKNOWLEDGMENTS

This work was supported by a grant to Dr. V. Panagia from the Medical Research Council of Canada.

REFERENCES

Brückner R, Scholz H (1984) Effects of α-adrenoceptor stimulation with phenylephrine in the presence of propranolol on force of contraction, slow inward current and cyclic AMP content in the bovine heart. Br J Pharmacol 82: 223-232.

Downes CP, Hawkins PT, Stephens L (1989) Identification of the stimulated reaction in intact cells, its substrate supply and the metabolism of inositol phosphates. In: Michell RH, Drummond AH, Downes CP (eds) Inositol lipids in cell signalling. Academic press, London, p 3.

Enyedi A, Farago A, Sarkadi B, Gardos G (1984) Cyclic AMP-dependent protein kinase and Ca^{2+}-calmodulin stimulate the formation of phosphoinositides in sarcoplasmic reticulum preparation of rabbit heart. FEBS Lett 176: 235-238.

Gaut ZN, Huggins CG (1966) Effect of epinephrine on the metabolism of the inositol phosphatides in rat heart *in vivo*. Nature 212: 612-613.

Jones LR, Maddock SW, Besch HR (1980) Unmasking effect of alamethicin on the (Na^+, K^+)-ATPase, β-adrenergic receptor-coupled adenylate cyclase, and cAMP-dependent protein kinase activities of cardiac sarcolemmal vesicles. J Biol Chem 255: 9971-9980.

Kasinathan C, Xu ZC, Kirchberger MA (1989) Polyphosphoinositide formation in isolated cardiac plasma membranes. Lipids 24: 818-823.

Lowry OH, Rosenbrough NJ, Fare AL, Ransall RJ (1951) Protein measurement with the folin phenol reagent. J Biol Chem 193: 325-334.

Pitts BJR (1979) Stoichiometry of sodium-calcium exchange in cardiac sarcolemmal vesicles. J Biol Chem 254: 6232-6235.

Pitts BJR, Okhuysen CH (1980) Sodium-calcium exchange and sodium pump activities in cardiac sarcolemmal vesicles: Estimation of sidedness. Ann NY Acad Sci 358: 357-358.

Post JA, Langer GA, Op Den Kamp JAF, Verkleij AJ (1988) Phospholipid asymmetry in cardiac sarcolemma- Analysis of intact- Cells and gas-dissected membranes. Biochim Biophys Acta 943: 256-266.

Quist E, Satumtira N, Powell P (1989) Regulation of polyphosphoinositide synthesis in cardiac membranes. Arch Biochem Biophys 271: 21-32.

Varsanyi M, Messer M, Brandt NR, Heilmeyer LMG (1986) Phosphatidylinositol 4,5-bisphosphate formation in rabbit skeletal and heart muscle membranes. Biochem Biophys Res Comm 138: 1395-1404.

PHOSPHORYLATION AND ACYLATION OF THE GROWTH-RELATED MURINE SMALL STRESS PROTEIN P25

Steffi Oesterreich[1], Rainer Benndorf[1], Gunther Reichmann[2,3], Heinz Bielka[1]

[1]Institute of Molecular Biology, Department of Cell Physiology, Robert-Rössle-Str. 10, 1115 Berlin-Buch , FRG

Alterations of cell proliferation are correlated with changes in the rate of synthesis or in the abundance of proteins (Benndorf et al., 1988a). After an early exponential phase, the in vivo growth rate of the Ehrlich ascites tumor (EAT) declines continuously leading finally to the stationary phase. This transition is accompanied by the accumulation of 3 isoforms of a 25kDa protein (p25) (Fig. 1). Two of the isoforms (p25/1 and /2) were isolated and partially sequenced (Benndorf et al., 1988b). Furthermore p25cDNA was cloned and sequenced (Gaestel et al., 1989). At the amino acid level the sequencing data revealed that p25 has an about 80% homology to the human small heat shock protein hsp27 (Gaestel et al., 1989) and 94% to the hamster small heat shock protein hsp27 (Lavoie et al., 1990). In further experiments (Oesterreich et al., 1990) it could be shown that the synthesis of p25 is induced following heat shock treatment.

---

[2]Institute of Pathological Biochemistry, Department of Medicine (Charité), Humboldt University Berlin, Hermann-Matern-Str., 1040 Berlin, FRG
[3]deceased

NATO ASI Series, Vol. H 56
Cellular Regulation by Protein Phosphorylation
Edited by L. M. G. Heilmeyer, Jr.
© Springer-Verlag Berlin Heidelberg 1991

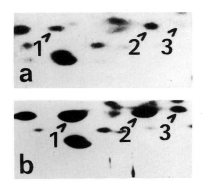

Fig.1: Occurrence of p25 isoforms in exponential (a) and stationary (b) phase EAT cells. Only the relevant sections of silver stained gels are shown. Numbers refer to p25 isoforms. (From Bielka et al., 1988)

HEAT SHOCK DEPENDENT P25 PHOSPHORYLATION

Similar to hsp27 in human (Arrigo et al., 1987) and rat (Kim et al., 1984) cell lines, p25 occurs in EAT cells in one non-phosphorylated (p25/1) and two phosphorylated (p25/2 and /3) isoforms (Benndorf et al., 1988b; Bielka et al., 1988). The phosphorylated isoforms can be discerned by incorporation of radioactive phosphorous (Fig. 2a). Exposure of EAT cells to heat shock results in various characteristic patterns of p25 isoforms as demonstrated by pulse incorporation of [3H]leucine (Oesterreich et al., 1990). A single heat treatment (2h 41.5°C) leads to a labeling of p25/2 and /3 (Fig. 3b), whereas after a twofold heat-shock with an intermittent recovery period (1h 41.5°C, 2h 37°C, 2h 43.5°C) p25/1 is predominantly labeled (Fig. 3c).

The analysis of the phosphoamino acids revealed that both isoforms, p25/2 and p25/3, are phosphorylated exclusively at serine residues (Benndorf et al., 1988b; Bielka et al., 1988). However, a large portion of radioactive phosphorous was released even after gentle acidic hydrolysis as inorganic phosphate the origin of which is not yet clear. Similar results were obtained in the case of rat small stress proteins (Kim et al., 1984).

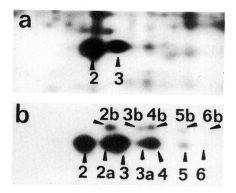

Fig. 2: Comparison of phosphorylated p25 isoforms labeled in EAT cells (a) and in the ammonium sulfate precipitate fraction (b). Numbers refer to p25 isoforms.

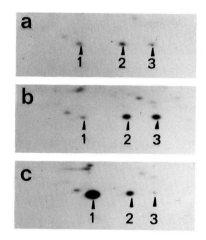

Fig. 3: Synthesis of p25 isoforms after exposure of EAT cells to different heat shock conditions. a, control; b, 2h 41.5°C; c, 1h 41.5°C, 2h 37.0°C, 2h 43.5°C. (From Oesterreich et al., 1990)

CELL-FREE PHOSPHORYLATION OF P25 BY ENDOGENOUS KINASES

From stationary phase EAT cells an ammonium sulfate precipitate (45% saturation) was prepared containing mainly non-phosphory-lated p25 and kinases using p25 as substrate (Benndorf et al., unpublished). Incubation of this fraction with 2 mM γ-[^{32}P]ATP results in a complex pattern of p25 isoforms (Fig. 2b). The most prominent isoforms are, according to 2D data identical with p25/2 and /3 occurring naturally in EAT cells. By immunological methods evidence was obtained that all labeled proteins designated in Fig. 2 are isoforms of p25 (not shown). Interestingly, in this assay p25 isoforms with elevated molecular weights (p25/2b-6b) are detectable, the nature and origin of which are not yet known. In a further cell-free approach recombinant p25 was used as substrate for protein kinases A and C (Gaestel M, Schröder W, to be published). Both kinases were found to phosphorylate the recombinant p25. However, according to 2D data the resulting isoforms are different from those óccurring in tumor cells.

PHOSPHORYLATION AND ACYLATION WITH FATTY ACIDS

When EAT cells are incubated with [³H]stearic acid (Fig. 4a, lane 1) or oleic acid (Fig. 4a, lane 2) p25 becomes labeled. In the case of stearic acid it was shown that preferentially the phosphorylated isoforms p25/2 and /3 become acylated (Fig. 4b). In a further approach the fatty acid modification of p25 was directly determined. After extraction of non-covalently associated fatty acids, p25 (mixture of p25/1 and /2) was hydrolyzed using 6N HCl/ Methanol and the released fatty acids were analyzed as their methyl esters by gas-liquid-chromatography. Four different fatty acids (palmitic acid: 43.1%; stearic acid: 19.7%; oleic acid: 24.0%; linolic acid: 13.2%) were shown to be covalently bound to p25. Regarding the stochiometry about one third of the isolated p25 appears to be acylated.

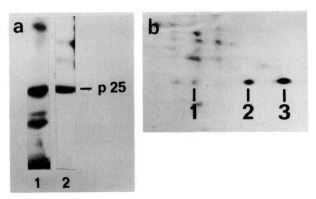

Fig. 4: Incorporation of tritium-labeled fatty acids into p25. a, stearic acid (lane 1), oleic acid (lane 2); b, stearic acid. In b, numbers refer to p25 isoforms.

CONCLUSIONS

Two of the three isoforms of the growth-related stress protein p25 are phosphorylated. The degree of phosphorylation is related to the physiological state of the cells. In a cell-free reaction using endogenous kinases, a complex pattern of phosphorylated p25 isoforms is obtained. The main isoforms are identical with those occurring naturally in the tumor. Besides phosphorylation, p25 is acylated by fatty acids; predominantly the phosphorylated isoforms are acylated.

LITERATURE

Arrigo A-P, Welch JW (1987) Characterization and purification of the small 28,000-dalton mammalian heat shock protein. J Biol Chem 262:15359-15369

Benndorf R, Nürnberg P, Bielka H (1988a) Growth phase-dependent proteins of the Ehrlich ascites tumor analyzed by one- and two-dimensional electrophoresis. Exp Cell Res 174:130-138

Benndorf R, Kraft R, Otto A, Stahl J, Böhm H, Bielka H (1988b) Purification of the growth-related protein p25 of the Ehrlich ascites tumor and analysis of its isoforms. Biochem Int 17:225-234

Bielka H, Benndorf R, Junghahn I (1988) Growth related changes in protein synthesis and in a 25 kDa protein of Ehrlich ascites tumor cells. Biomed Biochim Acta 47:557-563

Gaestel M, Gross B, Benndorf R, Strauss M, Schunk W-H, Kraft R, Otto A, Böhm H, Stahl J, Drabsch H, Bielka H (1989) Molecular cloning, seqencing and expression in Escherichia coli of the 25-kDa growth-related protein of Ehrlich ascites tumor and its homology to mammalian stress proteins. Eur J Biochem 179:209-213

Kim Y-J, Shuman J, Sette , Pryzbyla A (1984) Nuclear localization and phosphorylation of three 25-kilodalton rat stress proteins. Mol Cell Biol 4:468-474

Lavoie J, Chrétien P,Landry J (1990) Sequence of the Chinese hamster small heat shock protein HSP 27. Nucl Acids Res 18:1637

Oesterreich St, Benndorf R, Bielka H (1990) The expression of the growth-related 25kDa protein (p25) of Ehrlich ascites tumor cells is increased by hyperthermic treatment (heat shock). Biomed Biochim Acta 49:219-226

INDEX

NATO ASI Series H

NATO ASI Series H

NATO ASI Series H

Printing: Druckhaus Beltz, Hemsbach
Binding: Buchbinderei Schäffer, Grünstadt

DATE DUE